Trace Metals Selenium, Chromium and Vanadium

Trace elements selenium, vanadium and chromium play an essential role in the nutrition of both animals and humans. Their accumulation in various environments and the subsequent transition of these elements into the food chain significantly affect human and environmental health. This volume emphasizes the integrative aspects of selenium, vanadium and chromium with chapters dedicated to their fundamental chemistry, biochemistry, clinical science and environmental effects. Each chapter focuses on the advancement of scientific knowledge about these trace elements. Important studies on these elements will be described through interdisciplinary approaches.

Trace Metals Selenium, Chromium and Vanadium

Chemistry, Biology and Human Health

Bimal Krishna Banik, Sangeeta Bajpai,
Monika Kamboj, and Garima Pandey

CRC Press
Taylor & Francis Group
Boca Raton London New York

CRC Press is an imprint of the
Taylor & Francis Group, an **informa** business

First edition published 2026
by CRC Press
2385 Executive Center Drive, Suite 320, Boca Raton, FL 33431

and by CRC Press
4 Park Square, Milton Park, Abingdon, Oxon, OX14 4RN

CRC Press is an imprint of Taylor & Francis Group, LLC

ISBN: 9781032613710 (hbk)
ISBN: 9781032615103 (pbk)
ISBN: 9781032615110 (ebk)

DOI: 10.1201/9781032615110

Typeset in Times
by Newgen Publishing UK

Contents

Section III Role of Trace Metals (Se, V and Cr) in Human Health

Preface

Trace metals play an important role in the proper functioning of lipid and carbohydrate metabolism in humans. This book focuses on the roles of three trace metals vanadium, chromium, and selenium. These elements play an essential role in the nutrition of animals and humans. Their accumulation in various environments and the subsequent transition of these elements into the food and food chain severely affect human and environmental health. They are also essential for the proper functioning of living organisms. However, the disproportionate consumption of these metals can lead to health problems.

This book incorporates the essential aspects of selenium, vanadium, and chromium, with chapters dedicated to their fundamental chemistry, biochemistry, clinical science, and environmental effects. It is divided into three sections, each with three chapters describing chemistry, biology, and the impact of these trace metals on human health.

This book has been divided into nine chapters focusing on the diversity of research on trace metals. The authors expect this book will help students, scientists, and academic and business professionals in their education, research, and professional careers. Information provided in this book may enhance people's curiosity about the benefits of these compounds.

Chapter 1 discusses the chemistry of a naturally occurring rare mineral, selenium, which exists in several allotropic forms. Selenium and its reagents catalyze various organic reactions, showing their wide application in organic chemistry, which are discussed in this chapter. Also, various organoselenium compounds are covered due to their biological significance.

Chapter 2 focuses on the history, occurrence, atomic, and chemical and physical properties of vanadium. A study on the coordination chemistry of vanadium and its complexes is also discussed.

Chapter 3 focuses on the application of chromium (Cr). This chapter also pays attention to its two major oxidation states: trivalent chromium Cr(III), which is vital for glucose and lipid metabolism, and hexavalent chromium Cr(VI), a hazardous form associated with industrial pollution.

Chapter 4 deals with the biological properties of selenium. Selenoprotein, which contains amino acids, selenocysteine, plays a role in maintaining the various physiological functions of human beings. Selenoprotein regulates the metabolism of the thyroid gland, immune system, defense, and reproductive system. This chapter also discusses the metabolism of selenium, the biosynthesis of selenoprotein, and the different functions of various selenoproteins.

Chapter 5 summarizes the current understanding of vanadium's biological properties, highlighting its effects on human health and biochemical processes.

Chapter 6 focuses on the significance of chromium on human health and its importance in daily nutrition. Trivalent chromium is integral to insulin signalling, assisting in glucose metabolism and blood sugar regulation. It also contributes to lipid metabolism, protein synthesis, and cardiovascular health.

Chapter 7 highlights the effect of selenium and its compounds on human health. Exposure to selenium or its compounds through breathing or by consuming selenium-contaminated food and water may have adverse effects on the body and result in cancer and other diseases. This chapter highlights the benefits and effects of selenium and its compounds on human health.

Chapter 8 aims to provide a comprehensive overview of the dual roles of vanadium compounds on human health, highlighting both their beneficial and adverse effects and emphasizing the importance of considering individual factors such as dose, duration of exposure, and individual susceptibility when evaluating the role of vanadium compounds in disease prevention and treatment.

Chapter 9 aims to highlight the crucial role of chromium in human health, industry, and environmental systems. While Cr(III) is essential for metabolism, Cr(VI) is highly toxic, causing respiratory issues, skin irritation, and an increased cancer risk. This chapter also summarizes the effective safety measures, stringent regulations, and remediation efforts necessary to mitigate the harmful effects on human health and ecosystems.

This book is published by CRC, Taylor & Francis. The authors are tremendously grateful to Ms. Hilary Lafoe, Ms. Varalika Kathuria, and Ms. Sukriti Singh for their science knowledge, helping attitude, participation in the process, and encouragement.

About the Authors

Bimal Krishna Banik has had a remarkable career, filled with significant achievements in teaching, research, mentoring, and professional service. He completed his Ph.D. in chemistry from the Indian Association for the Cultivation of Science, Calcutta. Then, he pursued post-doctoral research at Case Western Reserve University, Ohio, and Stevens Institute of Technology, New Jersey. Prof. Banik was a tenured full professor in chemistry and First President's Endowed Professor in Science and Engineering at the University of Texas – Pan American. He was the Vice President of Research and Education Development at the Community Health Systems of Texas. Currently, Prof. Banik is a full professor at the Prince Mohammad Bin Fahd University, Kingdom of Saudi Arabia. Notably, he has been selected as a fellow of five international scientific societies and has been designated as a distinguished researcher/distinguished scientist.

Prof. Banik is engaged in studies on synthetic chemistry and medicinal research on cancers, antibiotics, natural products, hormones, and catalysis. His extensive and impactful contributions are reflected through the $7.25 million grant from US National Institutes of Health (NIH), US National Cancer Institute (NCI), and US private foundations as the principal investigator. Prof. Banik is the author of more than 740 publications including books, book chapters, name reactions, perspectives, reviews, and papers with more than 10,000 citations. In addition, he has 520 presentation abstracts to his credit. Many of his international presentations at various conferences have been designated as keynote and plenary lectures. Notably, he has authored and edited 23 books, which have been published by Elsevier, Springer, Springer Nature, CRC Taylor & Francis, De Gruyter, ICS, PMU, and Nova.

Prof. Banik's leadership roles in organizing international and national conferences highlight his influence on scientific communities. He chaired 20 symposiums at the American Chemical Society national meetings and over two dozen conferences at the international level, including one at the Nobel Prize celebration meeting in Germany. In his capacity as the chair, he was able to introduce approximately 300 speakers. He has been deeply involved in the publication process of other authors. For example, Prof. Banik has been a reviewer of 93 journals, editorial board member of 28 journals, editor-in-chief of 12 journals, founder of eight journals, and guest editor of 10 research journals. Due to his leading role in journals administration, Prof. Banik has recruited approximately 200 associate editors and editorial board members for these journals. Importantly, he has served as a reviewer of National Science Foundation (NSF), National Institutes of Health-National Cancer Institute (NIH-NCI), National Research Council of the National Academies (NRC), American Chemical Society (ACS), US research corporations and international grant applications. Prof. Banik has evaluated approximately 2,500 research grants and manuscripts. He was the chair of the M.D. Anderson Cancer Center's drug discovery symposiums and director of their analytical chemistry laboratory. He served as a chair or member of more than 100 committees; reviewer of promotion and tenure of faculties in US and international universities; chair of the Faculty Research Council of the University of Texas – Pan American (UTPA), and an examiner of doctoral thesis.

Prof. Banik has been involved in teaching science to medical and engineering students on a regular basis for the past many years. He has taught more than 3,000 students in the United States and Saudi Arabia. His brilliant teaching skills were reflected and proven in several thousand students', peers', and administrators' outstanding evaluations as well as by independent comments posted on the "Rate My Professor" website by the US students. Numerous distinguished leaders in US universities constantly rated Prof. Banik as a "truly exceptional teacher" and "excellent teacher" by rating him as *excellent* and *good* for the past many years. He has established two science societies that have 1,400 students at the University of Texas. Many research students from his group have received Ph.D., M.D., D.D.S., D.Pharm, M.D. Ph.D., M.S., and P.A. degrees from the US universities. Some of his students are working as university/hospital faculty members, clinicians, directors, and principal scientists in pharmaceutical/chemical companies around the world. Prof. Banik has mentored 20 postdoctoral fellows, eight research scientists, and approximately 300 students in research. Importantly, he has mentored 28 faculty members.

Prof. Banik was honored with several prestigious awards. He is consistently ranked among the top 2% researchers in the world. He received the Indian Chemical Society's Lifetime Achievement Award; Mahatma Gandhi Pravasi Honor Medal from the UK Parliament; US National Society of Collegiate Scholars' Best Advisor Award for students; Professor P.K. Bose Endowment Medal; Dr. M.N. Ghosh Gold Medal; University of Texas Board of Regents' Outstanding Teaching Award; ACS Member Service Award; and Best Researcher, Teacher and Mentor Awards by the UTPA. He is recognized by Burdwan University and Itachuna College as a distinguished eminent alumni and also by ACS News, Elsevier, Royal Society of Chemistry (RSC), US NCI, US Research Foundation, University of Texas Board of Regents, *Times of India*, Stevens Institute of Technology, ACS SEED, NCI, NIH and by Indian and US newspapers, and *Down to the Earth* magazine. Prof. Banik has received 200 invitations to deliver international lectures. He is also being invited to write books by publishers of repute in the world.

Sangeeta Bajpai is a faculty of Chemistry, working with Amity University, Lucknow, India for the last 17 years. She has 30 years of academic research experience. She was awarded a doctorate degree in chemistry from Lucknow University in 1998. Her academic journey includes more than 50 research publications in various journals of National and international repute, various authored and edited books, book chapters in books published by leading publishers (Elsevier, Bentham and DeGruyter). She has participated in various international and national conferences as an invited speaker. As the principal investigator, she completed two major research projects (Fast

Track Young Scientist and Women Scientist). They were duly funded by Department of Science and Technology, Government of India. She has been a lifetime member of reputed organizations, Indian Science Congress, Association of Chemistry Teachers, and editorial board member of reputed scientific journals. Under her supervision, two students have been awarded their Ph.D. degrees. Many more students are pursuing their research work under her guidance. Her research areas of interest include organometallic chemistry, coordination chemistry, supramolecular chemistry, green chemistry and macrocyclic chemistry. She has been engaged in the studies of the chemistry of organotellurium compounds, mainly to explore newer bonding perspectives in its complexes via spectral and X-ray diffraction studies.

Monika Kamboj, a faculty of chemistry, working with Amity University, Lucknow, India for the last 15 years. She was awarded a doctorate in chemistry from the National Institute of Technology Kurukshetra in 2012. Her academic journey includes more than 40 research publications in various journals of national and international repute and authored book chapters in books published by leading publishers. She has participated in various international and national conferences as an oral paper presenter, poster presenter, and as a part of the faculty development program. She was awarded the Dr. R.K. Aggarwal Award 2016 for best poster presentation by the Indian Council of Chemists, India. She also received third prize in the oral presentation at the International Conference on Chemical Sciences: National and Global Prospective 2018, held in Lucknow. She has to her credit the prestigious Indian Academics Summer Research Fellowship 2024. She has lifetime memberships with reputed organizations, the Indian Science Congress, the Association of Chemistry Teachers and Indian Society for Photochemical and Radiation Sciences. She is also an editorial board member and a reviewer in reputed scientific journals. Under her supervision, one student has been awarded his Ph.D. degree. Her area of interest is coordination chemistry, green chemistry, and macrocyclic chemistry. Her research interest includes the chemistry of macrocyclic metal complexes, their characterization, and biological aspects.

Garima Pandey is an Associate Professor in the Department of Chemistry at the SRM Institute of Science and Technology, Modinagar, Ghaziabad. She received her doctorate in coordination chemistry in 2008. Dr. Garima has a rich teaching experience of more than 15 years with core areas being coordination chemistry, nanotechnology, environmental chemistry, and education. Dr. Pandey has more than 30 publications in various national and international journals of repute and has attended more than 20 national and international conferences, workshops, symposiums, and faculty development programs. All the way through her career, she has taken up a number of challenging assignments serving in the capacities of a teacher, counselor, mentor, anti-ragging committee member, and girls hostel warden. As a core member of Women Grievance Cell, she is always working toward promoting a healthy working environment for the female staff, students, and faculty, building a gender-sensitized environment at the institute, organizing workshops and awareness programs toward building a gender-neutral workplace. She has always earned appreciation for her active involvement in student welfare, community development, and women empowerment programs.

Section I

Chemistry (Elemental Properties, Coordination Chemistry and Heterocyclic Chemistry) of Se, V and Cr

1 Chemistry of Selenium

Monika Kamboj, Sangeeta Bajpai, Garima Pandey and Bimal Krishan Banik

1.1 INTRODUCTION

Selenium is an element that belongs to the chalcogen family (group 16) and shows similarities with sulfur and oxygen. In 1817, Jons Jacob Berzelius discovered a new element at Stockholm that was like sulfur and had properties intermediate between sulfur and tellurium. This new element was discovered during the preparation of sulfuric acid as red sludge at the bottom of the acid chamber. He named it selenium, derived from the Greek word *selene* (the name for the moon). This element has a characteristic smell of garlic and can be absorbed through the skin and cause bad breath. It is a non-metal and is placed in the 4th period along with the 16th group chalcogen but got its identification as "selenium metal". It is found in nature as a trace element. It is an essential micronutrient for human, plant, and animal health. It can be toxic and problematic if taken in excess. The deficiency of selenium affects the function of our body. The atomic and physical properties of selenium are represented in Table 1.1.

1.1.1 Sources

It is a naturally occurring element found in rare minerals like crookesite, berzelianite, tiemannite, antimonselite, clausthalite, and naumannite [1,2] as inorganic forms, selenide, selenate, and selenite. In nature, it exists as an impurity in sulfide ores of metals [3,4]. Soils contain selenium in the form of soluble selenate. The selenium can enter the rivers through leaching from soil [3,4]. Ocean water also contains selenium in significant amounts [5,6]. Soil becomes selenium-rich by the process of bioconcentration by some plants. The bioavailability and toxicity of selenium depend on its chemical forms. Selenium also finds a special place in living systems. In all living organisms, selenium is present in the form of inorganic [selenide (Se^{2-}), selenite ($SeO3^{2-}$), and selenate ($SeO4^{-}$)] and organic compound (selenomethionine, selenocysteine, methylselenocysteine, selenocystathionine and as amino acids in proteins).

Selenium concentration varies with nature of the soil and has been found in the range of 0.125–0.3 mg/kg in world soil [7]. The selenium level in the food chain is also affected by selenium concentration in soil. The selenium level in the atmosphere

DOI: 10.1201/9781032615110-2

TABLE 1.1
Atomic and physical properties of selenium

Symbol	Se
Atomic number	34
Atomic mass	78.96 g mol^{-1}
Density	4.79 g cm^{-3} at 20°C
Electronic configuration	$[Ar]3d^{10}4s^2\ 4p^4$
Electronegativity according to Pauling Scale	2.55
Van der Waals radius	190 pm
Atomic radius	103 pm
Covalent radius	116 pm
Common oxidation states	−2,0,+4,+6
Electron affinity	−4.21 eV
Ionization potential (first)	941.0 kJ/mol
Ionization potential (second)	2045.0 kJ/mol
Ionization potential (third)	2973.7 kJ/mol
Melting point	221°C
Boiling point	685°C

should not exceed 1 ng/m^3 and in water by 0.5 μg/L [8]. Burning of fossil fuel, coal, smelting of metal, and mining of ores are some of the human activities that accumulate selenium in soil [9,10]. Good sources of selenium are whole grains, vegetables, oil, fruits, and dairy products. The main sources of atmospheric selenium are activities like volcanic eruptions and fossil fuel burning [11]. The main producers of selenium are Finland, Belgium, Germany, and England. Elemental selenium is readily available, cheap, bench-stable, and easy to handle and has been used as a selenium source in organic synthesis [12]. Selenium occurs along with sulfur minerals as selenides (H_2Se) or selenites (SeO_3^{-2}) and is produced as a by-product during sulfur refining [13]. Selenate and selenite are inorganic forms of selenium, found in soil and water with varying contents of Se. Selenium has great utility in electronics, showing its ability to convert light into electrical energy. So it has been used in photocells and solar cells. Compounds of selenium such as sodium selenate are used in the glass industry for decolorizing glass [14].

1.1.2 Isolation

It is obtained during the electrolytic refining of copper and nickel [3]. Selenium was recovered from copper refinery anode slimes for industrial production [15–17]. Metallurgical-grade selenium of 90–99.5 percent yield is obtained but preference is for higher purity [18]. The oxidation–reduction of solar cell waste material, thermal decomposition, and leeching–precipitation chemical methods were utilized to refine high-purity selenium [19–23]. Impurities in selenium are removed by the zone refining process [24–27].

1.1.3 Allotropes

Different allotropes of selenium exist in gray, black, and red color [28]. These allotropes are interconvertible with temperature. The most stable of these allotropes is crystalline, gray selenium, having a hexagonal crystal lattice that consists of the helical polymeric chain. It is formed by the slow heating of black selenium. Amorphous red selenium is the most reactive allotrope of this element and appears as a powdery substance. It is converted to vitreous black selenium beads on rapid heating, which are brittle and lustrous. On heating, it softens and is converted to gray selenium at 180°C. The red and black Se behaves as insulators while gray selenium has semiconducting and photoconduction abilities. This feature has been utilized in electronics for photoelectronic and solar cells (Figure 1.1).

Selenium shows resistance to oxidation in air and other oxidizing acids. It forms poly-selenides with strong reducing agents.

1.1.4 Isotopes

There are six isotopes of natural selenium and only five are very stable: ^{74}Se, ^{76}Se, ^{77}Se, ^{78}Se, and ^{80}Se [29]. Table 1.2 shows the isotopic composition of Se.

Audi et al. described 24 unstable radioactive isotopes of selenium with lifetimes ranging from even shorter than seconds to years [30]. Kondev et al. described the isotopes of Se with half-lives longer than 100 ns [31]. The radioactive isotopes of

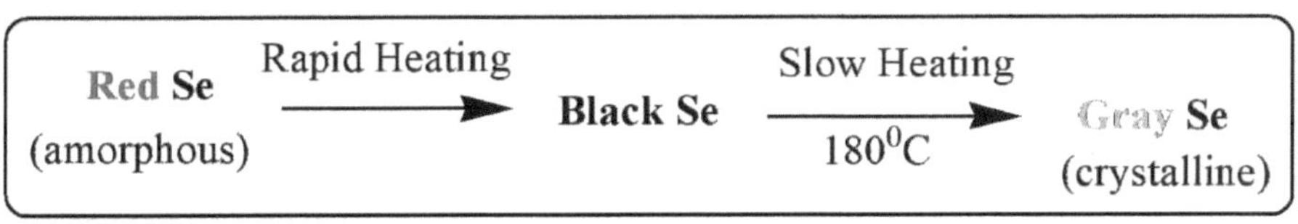

FIGURE 1.1 Effect of temperature on the allotropes of selenium.

TABLE 1.2
Naturally occurring isotopic composition of selenium

Isotope	Half-life ($t_{1/2}$)	Representative Isotopic Composition (atom%)	Nuclear Spin
^{74}Se	Stable	0.89(4)	0
^{76}Se	Stable	9.37(29)	0
^{77}Se	Stable	7.63(16)	½
^{78}Se	Stable	23.77(28)	0
^{80}Se	Stable	49.61(41)	0
^{82}Se	10^{20} years (practically stable) undergoes β-β-decay to yield ^{82}Kr	8.73(22)	0

selenium find application in the manufacture of bromine radioisotopes for medical and industrial use.

1.2 CHEMICAL REACTIONS OF SELENIUM

Selenium in its elemental form is low cost, easily available, stable, and easy to handle. It is utilized for the synthesis of a variety of chemical compounds including organoselenium compounds [12, 32]. Figure 1.2 depicts the chemical reaction of selenium.

1. *Reaction with hydrogen*: Selenium reacts with hydrogen to form a colorless gaseous inflammable hydrogen selenide (Figure 1.3).
2. *Reaction with oxygen*: Elemental Se undergoes combustion in the air to form two oxides (SeO_2 and SeO_3). It burns with a blue flame to form monomer selenium dioxide in the gas phase and polymerizes in the solid phase (Figure 1.4).

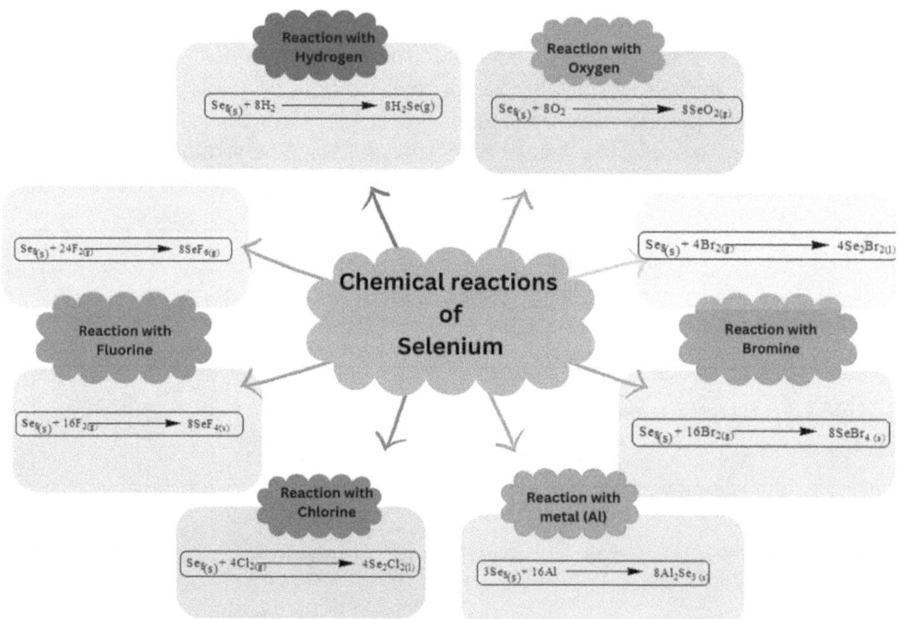

FIGURE 1.2 Chemical reactions of selenium.

$$Se_{8(s)} + 8H_2 \longrightarrow 8H_2Se_{(g)}$$

FIGURE 1.3 Reaction of Se with hydrogen.

$$Se_{8(s)} + 8O_2 \longrightarrow 8SeO_{2(g)}$$

FIGURE 1.4 Reaction of Se with oxygen.

$$2SeO_3 \longrightarrow 2SeO_2 + O_2 \quad \Delta H = -54 kJ/mol$$

FIGURE 1.5 Decomposition of SeO_3.

$$H_2SeO_3 + 2\,H_2S \longrightarrow SeS_2 + 3H_2O$$

FIGURE 1.6 Formation of selenium sulfide.

$$H_2SeO_3 + 2\,H_2S \longrightarrow SeS_2 + 3H_2O$$

FIGURE 1.7 Formation of selenium hexafluoride.

$$Se_{8(s)} + 16F_{2(g)} \longrightarrow 8SeF_{4(s)}$$

FIGURE 1.8 Formation of selenium tetrafluoride.

It is also produced by reaction of SO_3 with anhydrous K_2SeO_4. Selenous acid, H_2SeO_3 is formed by dissolving SeO_2 in water. SeO_2 is also formed from SeO_3, which is unstable and decomposes above 185 ^{0}C to SeO_2 (Figure 1.5).

3. *Reaction with sulfur compounds*: Selenium sulfide is produced by reacting hydrogen disulfide with an aqueous solution of selenous acid (Figure 1.6).

Selenium sulfide forms an eight-membered ring having Se and S in variable numbers (Se_nS_{8-n}). Selenium sulfide has numerous applications and has been utilized as an antidandruff agent in shampoos and for dying glass, reductant in fireworks, and as a polymer chain inhibitor.

4. *Reaction with halogen* (F_2, Cl_2, and Br_2): Selenium forms selenium hexafluoride on reacting with fluorine (Figure 1.7). It also forms selenium tetrafluorides (Figure 1.8). SeF_4 is solid while SeF_6 is liquid.

$$Se_{8(s)} + 4Cl_{2(g)} \longrightarrow 4Se_2Cl_{2(l)}$$

FIGURE 1.9 Formation of diselenium dichloride.

$$Se_{8(s)} + 4Br_{2(g)} \longrightarrow 4Se_2Br_{2(l)}$$

FIGURE 1.10 Formation of diselenium dibromide.

$$Se_{8(s)} + 16Br_{2(g)} \longrightarrow 8SeBr_{4\,(s)}$$

FIGURE 1.11 Formation of selenium tetrabromide.

$$3Se_{8(s)} + 16Al \longrightarrow 8Al_2Se_{3\,(s)}$$

FIGURE 1.12 Formation of aluminum selenide.

Se reacts with chlorine gas to form diselenium dichloride, selenium dichloride, and selenium tetrachloride (Figure 1.9).

With bromine, it forms diselenium dibromide (Figure 1.10) and selenium tetrabromide (Figure 1.11).

$SeBr_4$ decomposes to form diselenium dibromide and bromine. Similarly, selenium reacts with Iodine to form solid selenium tetraiodide.

5. *Reaction with metals*: Se reacts with metals like Al, Ga, Zn, Cd, Cu, etc. to form selenides (MSe) in which the oxidation number of Se is –2. Example: Aluminum selenide is formed from reacting Al with Se metal (Figure 1.12).
6. *Reaction with salts*: On reacting with salts, Se forms selenites. Selenites are the salts of selenous acid (anion SeO^{2-}), for example, silver selenite (Ag SeO) and sodium selenite (Na_2SeO_3).

1.3 ORGANOSELENIUM COMPOUNDS

Organoselenium compounds in which selenium forms a bond with carbon are present in water, soil, and sediments in trace amounts [33,34] and are considered essential nutrients. These organoselenium compounds also have sulfur, nitrogen, oxygen, and halogen as additional elements. Since the nineteenth century, organoselenium compounds have been known with a wide range of functional groups and remarkable biological properties [35–39]. Lowig and Wohler reported the first organoselenium compounds. Organoselenium derivatives have gained the interest of several research groups due to their role in the synthesis, catalysis, and biological activities [40]. Dimethyl selenide, a selenium ether, is another organoselenium compound found in nature [41,42].

Some of the organoselenium compounds are as follows:

3a. Selenols (RSeH) are analogs of alcohols and thiols. They are strong nucleophiles and reducing agents and undergo a variety of reactions. They are strong acids and, in alkaline solutions, readily donate protons to form the selenolate (RSe^-) conjugate bases.

Preparation of selenols

(a) *From Grignard reagent/alkylithium/aryllithium*: Elemental selenium reacts with Grignard reagent/alkylithium/aryllithium to form selenols (Figure 1.13).
(b) *From diselenides*: Diselenides are reduced to selenols and selenolates on reacting with reducing agents such as alkali metals, NaH (Figure 1.14).
(c) *From sodium hydrogen selenide*: The reaction of alkyl halides with sodium hydrogen selenide results in the formation of selenols (Figure 1.15).
(d) *From selenocyanates (RSeCN)*: Selenols and selenolates are formed by reducing or hydrolyzing selenocyanates (RSeCN) (Figure 1.16).

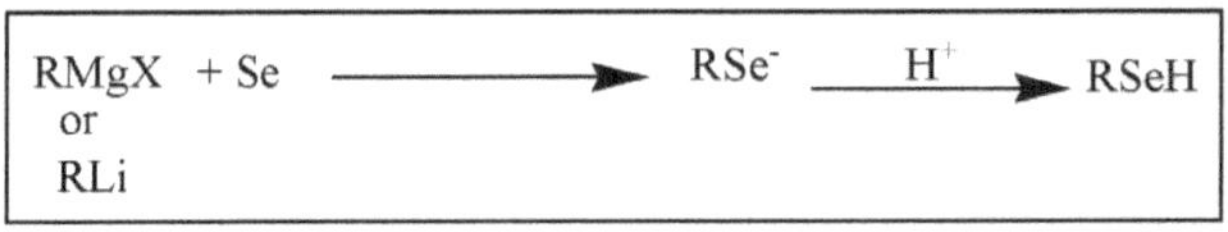

FIGURE 1.13 Preparation of selenols from RMgX/RLi.

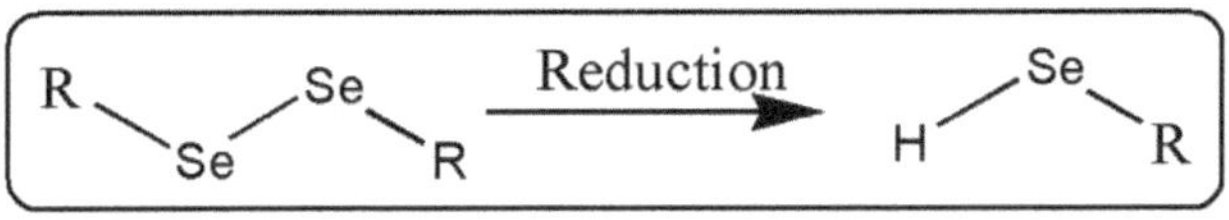

FIGURE 1.14 Preparation of selenols from diselenides.

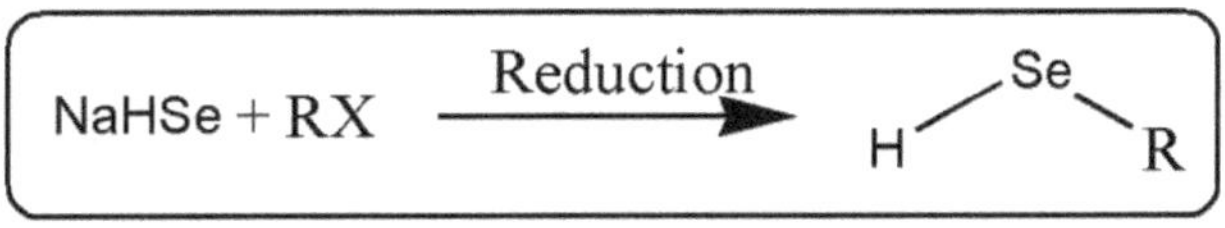

FIGURE 1.15 Preparation of selenols from alkyl halide.

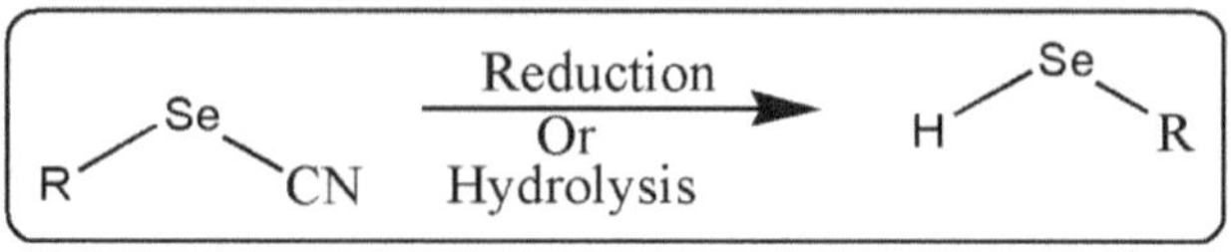

FIGURE 1.16 Preparation of selenols from selenocyanate.

Many chemists have reported synthetic approaches for forming selenol [43].

Chemical properties of selenols and selenolates

The presence of Se–H moiety in selenols makes them acidic and nucleophilic. Under mild reaction conditions, react with electrophiles and result in the formation of new Se–C and Se–X bonds. Selenols are stored under an inert atmosphere as they have a strong affinity for oxygen, so they undergo oxidation to form diselenides. As compared to thiols, they are more dynamic reducing agents [43] so convert N–O and N–N compounds into amines, and dihalides into alkenes. N-containing aromatic compounds can be reduced by selenols via the SET mechanism. Also, they can convert disulfides into thiols and azobenzene into hydrazobenzene [44]. This shows selenols are less stable and more reactive. All the reactions related to selenols must be carried in a ventilated fume hood.

The most important selenol is selenocysteine, an analog of cysteine. It is the only amino acid with a Se atom and is the 21st amino acid [45]. Mammals contain many enzymes (glutathione peroxidase, iodothyronine deiodinase, and thioredoxin) that contain selenols at their active site [46]. They are strong nucleophiles and yield selenides on reaction with alkyl halides and epoxides [47].

Application of selenols

They act as nucleophiles, react with a variety of electrophiles, and bring about important organic transformations. They act as catalysts for the reaction resulting in the formation of metal complexes, polymer, and selenium-containing small molecules and metal nanoclusters. The electronic feature of Se is utilized for applications such as semiconductors, photovoltaic and photocell devices, and photographic toner [48]. They act as a protecting group in organic synthesis [49].

3b. Selenophenol (C_6H_5SeH), also known as benzeneselenol, is an analog of phenols and thiophenols. It is a colorless, foul-smelling compound, toxic, shortlived, and used as an in-situ reagent in organic synthesis. It is seven times more acidic than thiophenols.

Preparation of selenophenol: The reaction of phenylmagnesium bromide and selenium results in the formation of benzeneselenol. Being reactive, it is generated in situ (Figure 1.17).

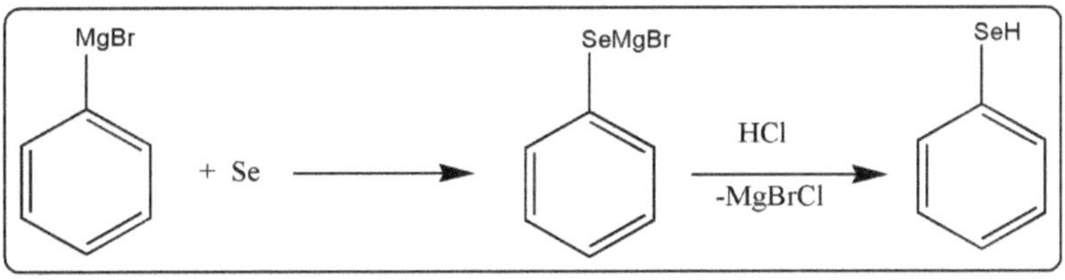

FIGURE 1.17 Preparation of selenophenol.

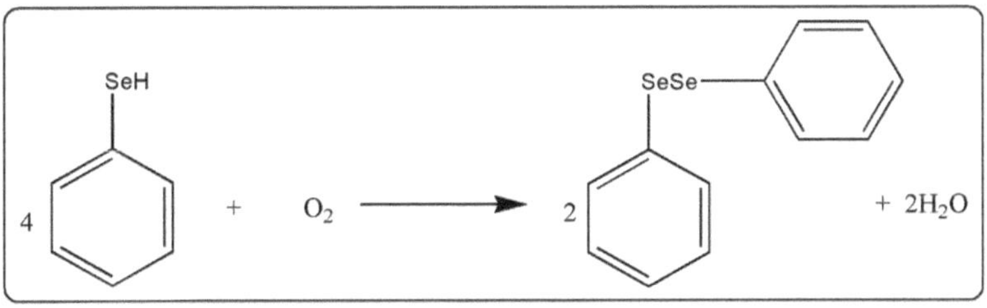

FIGURE 1.18 Oxidation of benzeneselenol.

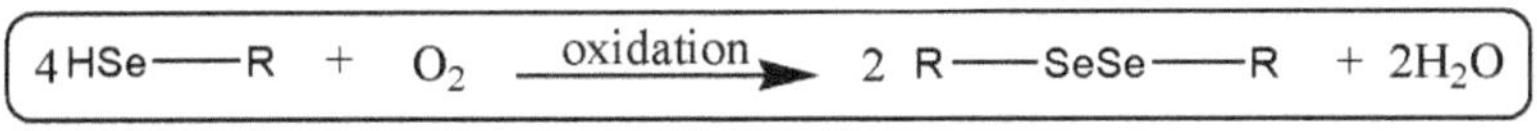

FIGURE 1.19 Preparation of diselenides.

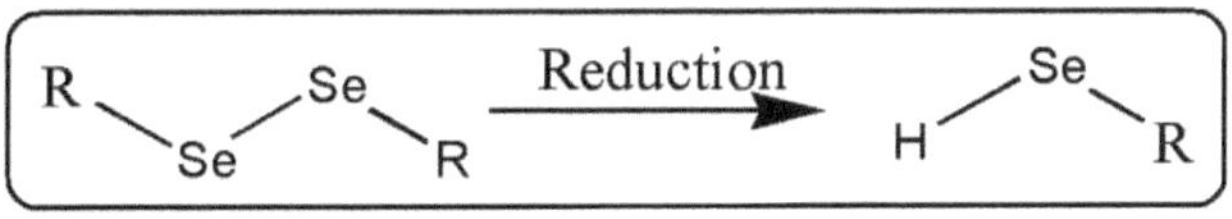

FIGURE 1.20 Reduction of diselenides.

Benzeneselenol is more prone to oxidation than thiophenol due to less bond dissociation energy of Se–H bond as compared to S–H bond. The oxidation of benzeneselenol results in yellow color diphenyl diselenide, which can be further reduced to selenol (Figure 1.18).

3c. Diselenides (RSeSeR) are analogs of peroxides and disulfides. They have a Se–Se bond. Diselenides are stable under atmospheric conditions and are considerably easier to handle [50, 51].

Preparation of diselenides

1. *From selenols*: The oxidation of selenols gives diselenides. Selenols are readily oxidized to diselenides (RSeSeR) by exposure to oxygen (Figure 1.19).

Diselenelides on reduction form selenols (Figure 1.20).

2. *From disodium diselanediide*: Disodium diselanediide reacts with alkyl or aryl halides to form diselenides in the presence of DMF at high temperatures. (Figure 1.21). Aryl diselenides are formed by the reaction of the diazonium salt with disodium diselanediide (Figure 1.22).

   Chemical properties: In-situ selenium nucleophilic reagents are formed by the reductive cleavage of diselenides. Such reagents help to introduce selenium moiety into organic substrate [50]. Selenates are prepared from diselenides using various reducing agents [52,53]. Diselenides are utilized for synthesizing other organoselenium compounds (selenols, selenenic acid, selenonic acids, and selenenyl halides).

Applications of diselenides: They are useful stereoselective reagents for asymmetric synthesis, act as catalysts, and also show therapeutic and hormetic effect [50,54,55].

3d. *Selanyl halides* (R–Se–Cl, R–Se–Br): Selanyl halides are formed when diselenides are halogenated with sulfuryl chloride, chlorine, or bromine (Figure 1.23).

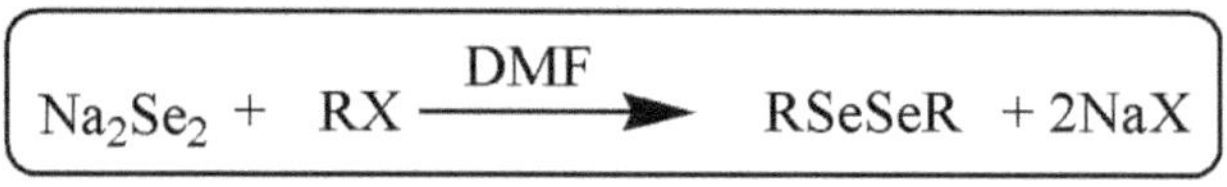

FIGURE 1.21 Preparation of diselenides from disodium diselanediide.

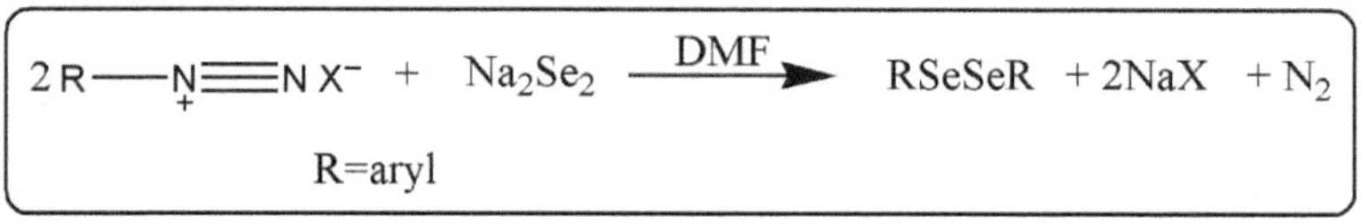

FIGURE 1.22 Preparation of diselenides from diazonium salts.

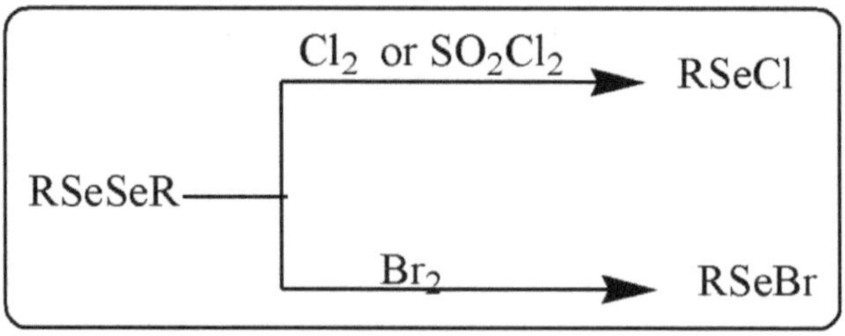

FIGURE 1.23 Preparation of selanyl halides.

Chemical properties: Selanyl halides are electrophilic, relatively stable, and sensitive to moisture [56–58]. It is a very useful and versatile compound and is converted into products such as selenenamides, (phenylseleno)phthalimide, selenocyanates, selenosulfonates, benzeneselenenyl acetate, trifluoroacetate, and tosylate [59–63] (Figure 1.24).

3e. *Selenides or selenoethers* (RSeR): They are analogs of ether. Among organoselenium compounds, they are the most prevalent stable compounds in organic chemistry with nonlinear structures. They are both cyclic and acyclic and have two C–Se bonds.

Preparation:

The dialkylation of sodium selenide yields symmetrical selenides while the alkylation of selenoates yields unsymmetrical selenides (Figure 1.25).

Aryl derivatives of selenides are prepared by reacting selenolate with a diazonium salt or aryl iodide. They are also prepared by reacting aryl selenocyanates with aryllithium. Other synthetic routes have also been reported for the selenoethers by using tri-*n*-butylphosphine [64] (Figure 1.26).

Chemical properties: Selenides are typically nucleophilic in nature and in some conditions act as electrophiles. Stable selenonium salts are formed on reacting alkyl halides with nucleophilic selenides. They are oxidized easily

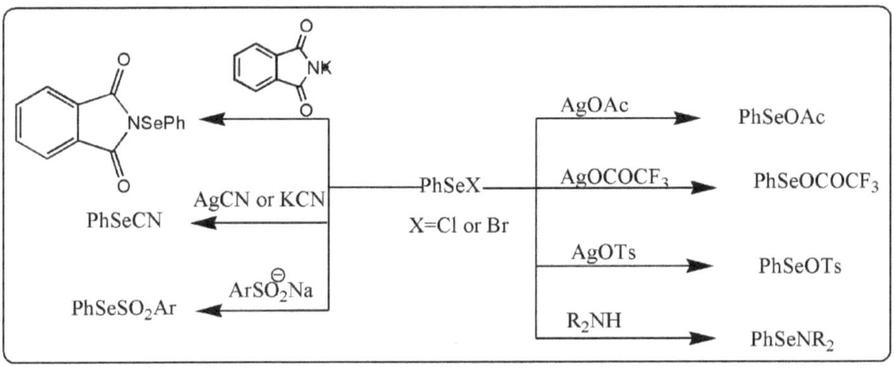

FIGURE 1.24 Chemical reaction of phenylselanyl halide.

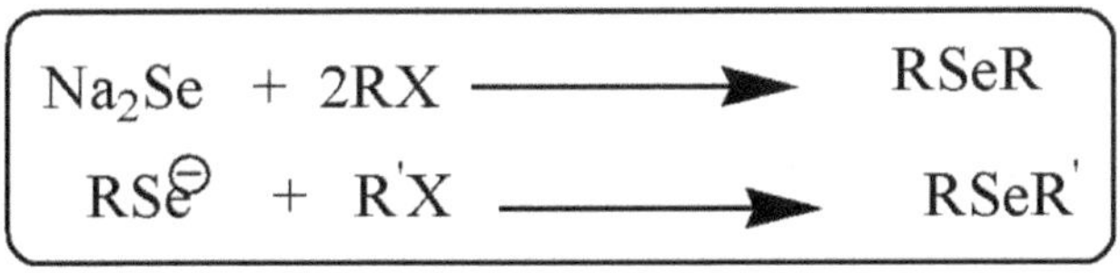

FIGURE 1.25 Preparation of selenides.

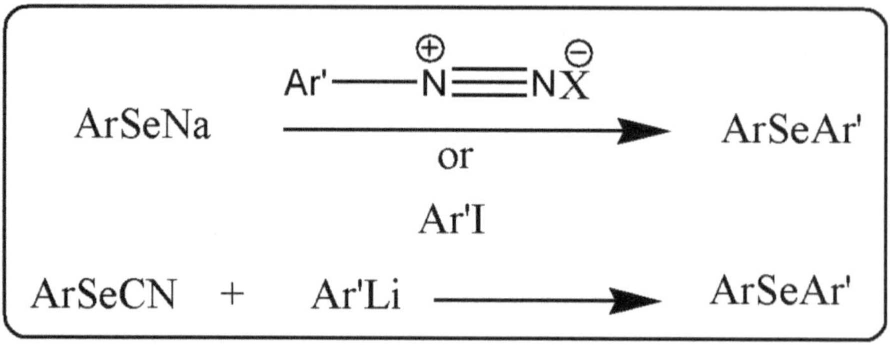

FIGURE 1.26 Preparation of aryl selenides.

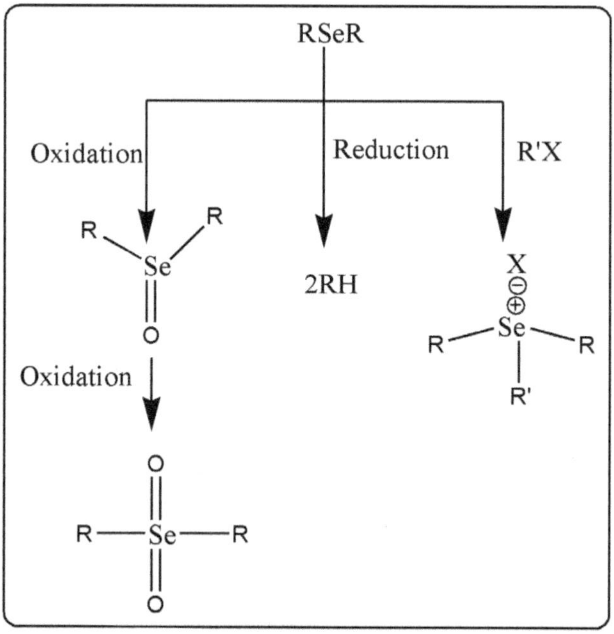

FIGURE 1.27 Chemical reactions of selenides.

to selenoxides by oxidizing agents (hydrogen peroxide, peroxy acids, ozone, and sodium m-periodate). The reaction can be stopped at this stage or further oxidized by strong oxidizing agents to selenones. With various reducing agents (nickel boride, tri-*n*-butyl- or triphenyltin hydride, Raney nickel, lithium triethylborohydride, and lithium in ethylamine) selenides are reduced to hydrocarbon [65,66] (Figure 1.27).

Selenide and sulfide chemistry is quite similar. Metal selenides are used as effective electrodes [67]. Selenides and diselenides have wide applications in the prevention and treatment of cancer [68].

3f. *Selenoxides* (R–Se(O)–R): They are the selenium analog of sulfoxides. *Preparation*: They are prepared easily by oxidizing selenides by oxidizing agents as mentioned above. Selenoxides are further oxidized to selenones $RSe(O)_2R$ under drastic conditions. Selenones are selenium analogues of sulfones. Selenoxides are pyramidal in geometry and with unsymmetrical selenides, chiral selenoxides are obtained that doesn't show pyramidal inversion and their racemization take place in water [69] (Figure 1.28).

The chemistry of selenoxides has been studied extensively [70]. As compared to tosylates or halides, they are good leaving groups [71,72].

Selenoxides are oxidized to selenones $RSe(O)_2R$ under drastic conditions using strong oxidants. Selenoxides with β-hydrogens undergo syn elimination at room temperature or below to form alkene [73,74] (Figure 1.29).

Allylic selenoxides via [2,3] sigmatropic rearrangements are converted to selenenate esters that hydrolyze to allylic alcohols [73–75] (Figure 1.30).

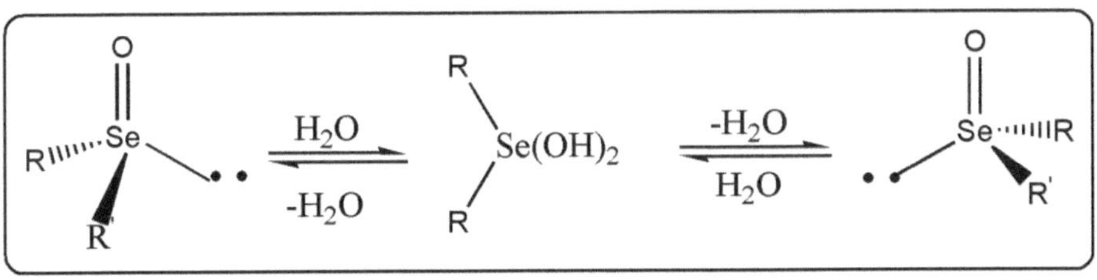

FIGURE 1.28 Racemization of selenoxides in water.

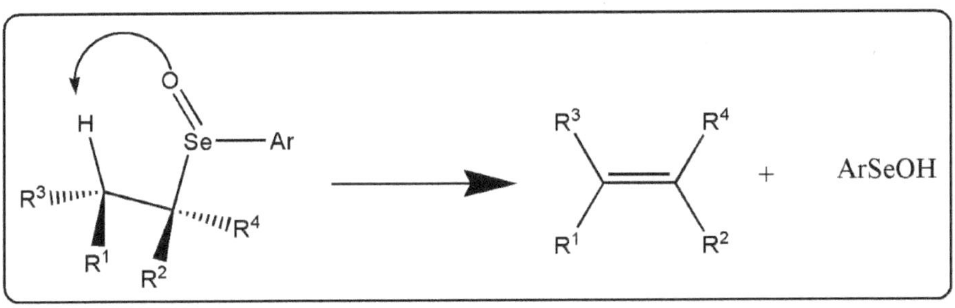

FIGURE 1.29 Syn elimination of selenoxides.

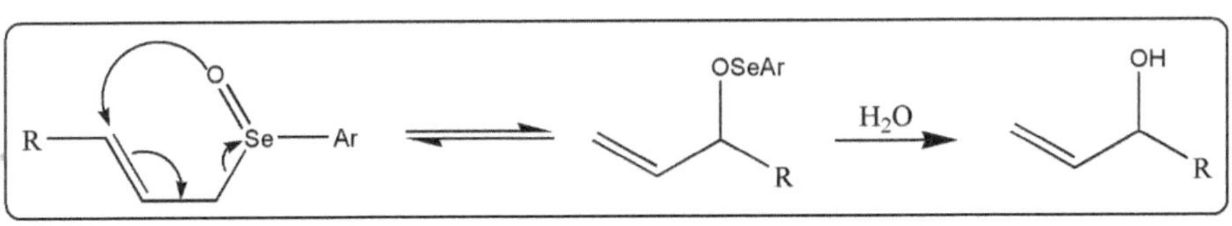

FIGURE 1.30 [2,3] Sigmatropic rearrangements of allylic selenoxides.

3g. *Selenenic acid* (RSeOH) *and seleninic acids* (RSeOOH): They both are an organoselenium oxoacids. Selenenic acids are unstable and form selenoseleninates by self-condensation reaction. They also disproportionate into corresponding seleninic acids and diselenides. The counterparts of selenenic acids are seleninic acids, which are relatively stable so can be isolated (Figure 1.31).

Preparation: Selenenic and seleninic acids are produced by the oxidation of diselenides. Partial oxidation of diselenides with oxidizing agent (hydrogen peroxide) gives corresponding anhydrides, which in the presence of water yield selenenic acid. Further oxidation of anhydrides and selenenic acid produces seleninic compounds anhydrides $(RSeO)_2O$ and seleninic acid (RSeOOH) (Figure 1.32).

The important application of seleninic acids is found upon epoxidation and Baeyer–Villiger reactions. When seleninic acids is oxidized with hydrogen peroxide, peroxyseleninic acids (RSe(O)OOH) formed as an intermediate, which catalyze these reactions [76,77] (Figure 1.33).

$$2RSeOH \longrightarrow RSeOSeR + H_2O$$
$$2RSeOH \longrightarrow RSeOOH + 1/2\ RSeSeR$$

FIGURE 1.31 Reactions of selenenic acid.

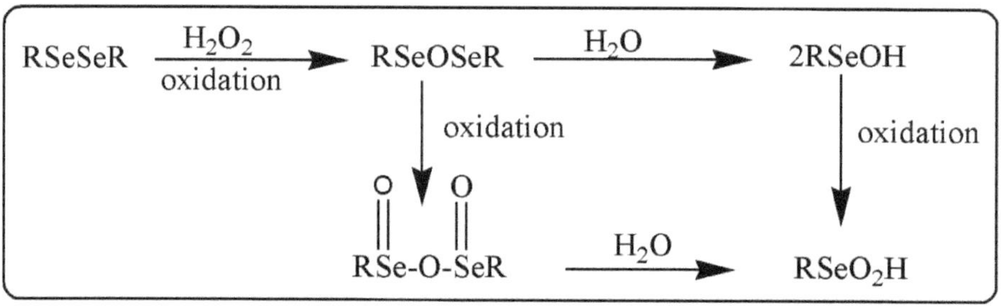

FIGURE 1.32 Oxidation reaction of diselenides.

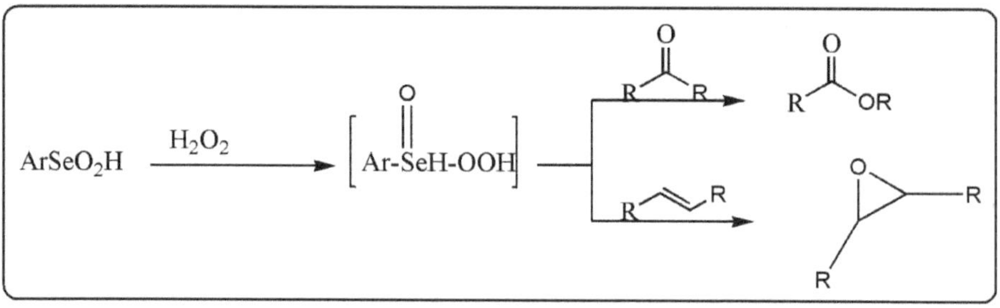

FIGURE 1.33 Oxidation of seleninic acids.

FIGURE 1.34 Preparation of selenonic acid.

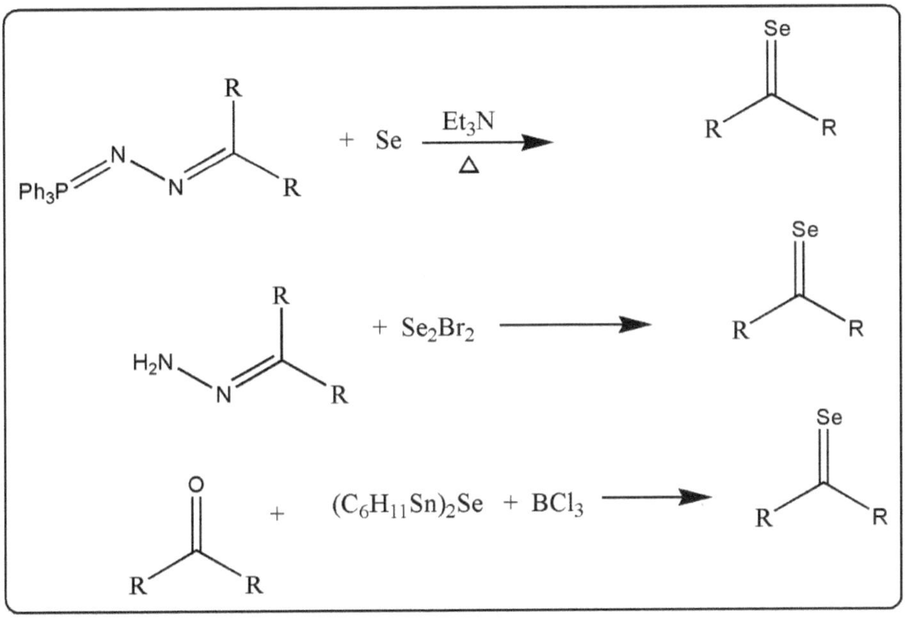

FIGURE 1.35 Synthesis of selones.

3h. *Selenonic acids* ($RSeO_3H$): They are sulfonic acid analogs. Selenonic acids are produced by the oxidation of seleninic acids with potassium permanganate (Figure 1.34).

A detailed study of selenonic acid has not been done. Selenonic acids and its salts are strong oxidants and are hygroscopic. They are isolated by ion exchange chromatography. They have wide synthetic applications.

3i. *Selones* (R2C=Se): Selones are selenocarbonyl compounds, ketone analogues having a C=Se bond. They have a tendency for oligomerization that makes them rare [78]. Nucleophiles attack selones at Se unlike ketones, which are attacked at doubly bonded carbon atom, thereby producing selenium-stabilized carbanion intermediates. They are considered less stable due to the poor 2p and 4p overlap of C=Se bond.

Preparation: Various methods of synthesis of selones are given as shown in Figure 1.35 [79–82].

Chemical properties: Reducing selones to diselenides can be easily done by reducing agents or photolysis. Trapping of less stable selones has been possible by cycloaddition reaction with dienes. They also undergo cycloaddition with diazo compounds to form cycloadducts and substantially lose N_2 and Se to form alkenes.

3j. *Thioselenides* (R–Se–S–R): These compounds are analogous to disulfides having selenium (II)–sulfur(II) bonds.

3k. *Selenones* (R–Se(O)2R): Selenones are the selenium analogues of sulfones. The chemistry of selenones has not been studied.

Chemical properties: Selenones are good leaving groups, comparable to tosylates or halides [83,84]. Vinylic selenones also act as Michael acceptors [85].

Preparation: Selenones are produced by oxidizing selenoxides, which is obtained by oxidizing selenides (Figure 1.36).

Table 1.3 shows the classification of organoselenium compounds [86].

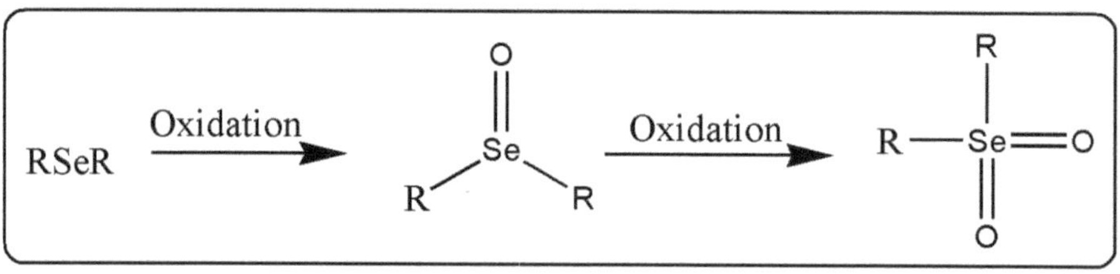

FIGURE 1.36 Preparation of selenones.

TABLE 1.3
Classification of Organoselenium Compounds

S. No.	Organoselenium Compounds		Selenium Analogue
1.	Selenols (RSeH)	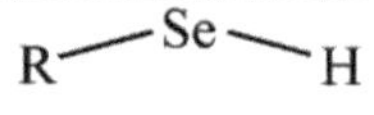	Alcohols and thiols
2.	Selenophenol	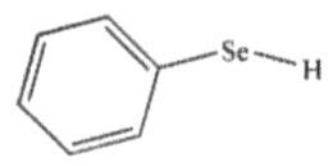	Phenol
3.	Diselenides	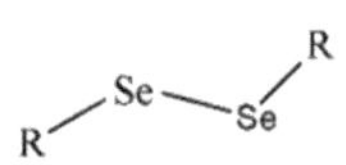	Peroxides and disulfides
4.	Selanyl halides	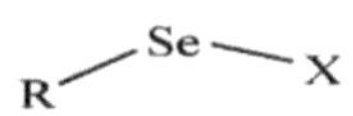	Alky halides
5.	Selenides selenoethers	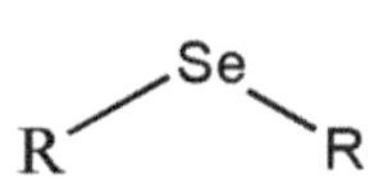	Ethers

TABLE 1.3 (Continued)
Classification of Organoselenium Compounds

S. No.	Organoselenium Compounds		Selenium Analogue
6.	Selenoxides	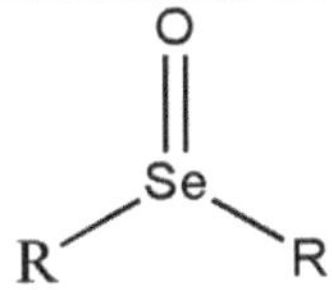	Sulfoxides
7.	Selenenic and seleninic acid	RSeOH and 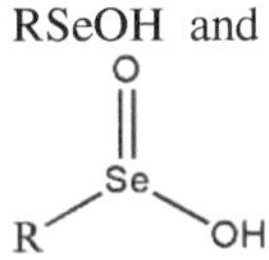	Sulfinic acids
8.	Selenonic acids	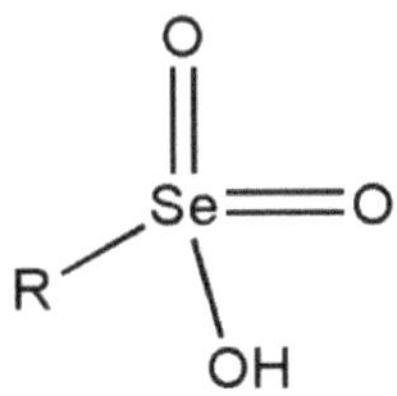	Sulfonic acids
9.	Peroxyseleninic acids	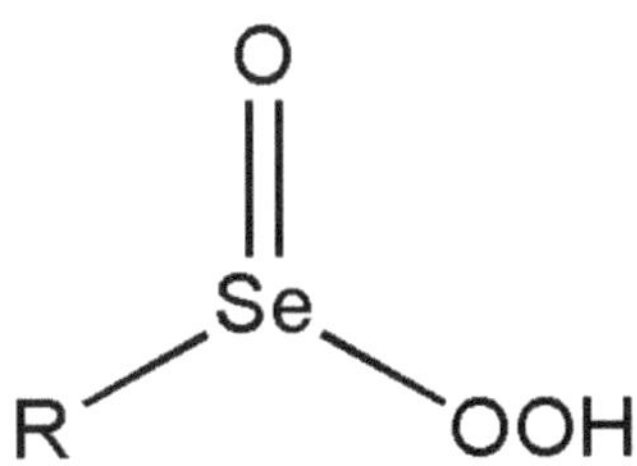	Sulfonic acids
10.	Selones	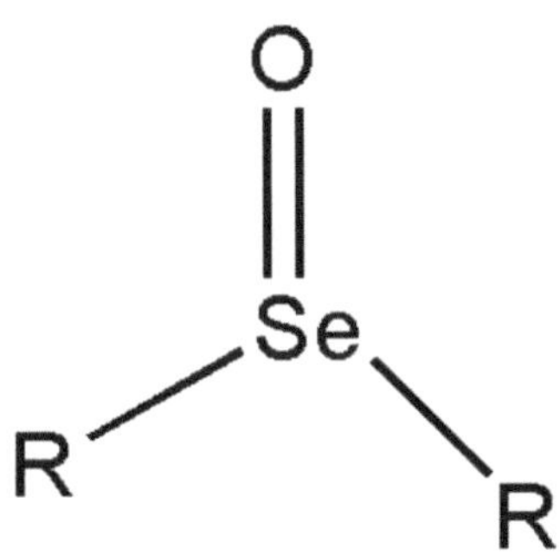	Ketones
11.	Thioselenides	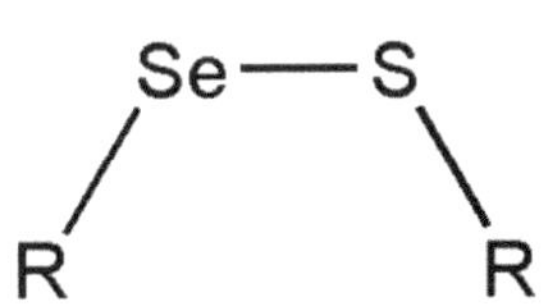	Disulfides

(continued)

TABLE 1.3 (Continued)
Classification of Organoselenium Compounds

S. No.	Organoselenium Compounds		Selenium Analogue
12.	Selenones	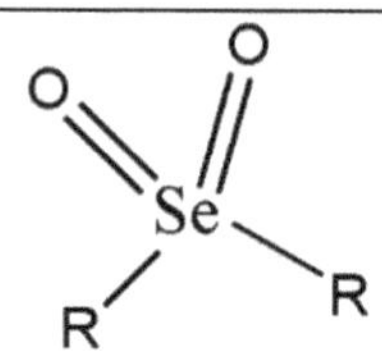	Sulfones

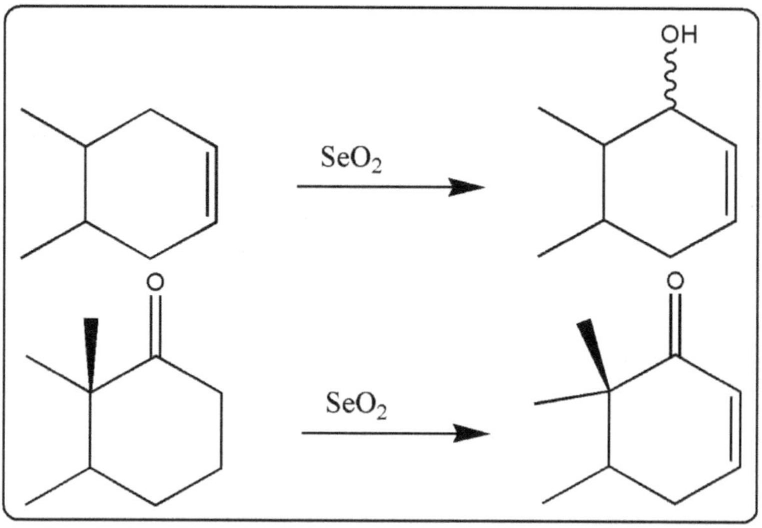

FIGURE 1.37 Oxidation reaction with SeO_2.

1.3.1 Diverse Reactions Induced by Selenium and Its Reagents

The selenium compounds were considered abstruse with limited importance till the 1970s when the perception of chemists toward selenium compounds changed. Selenium dioxide (SeO_2) was the only selenium reagent used by chemists for oxidizing organic compounds like alkenes or ketones till 1970 [87] (Figure 1.37).

Sharpless and Lauer in 1973 reported the utility of sodium phenylselenide as a nucleophile for converting epoxides to allylic alcohols [88] (Figure 1.38).

Soon after the utility of numerous selenium reagents as electrophile were described.

These compounds were then considered valuable and fascinating substances when biochemists confirmed the presence of selenium-containing proteins in the enzyme system of bacterial and mammals. Selenium as a reagent is stable and can be incorporated into various substrates as nucleophiles or electrophiles [89–91]. They

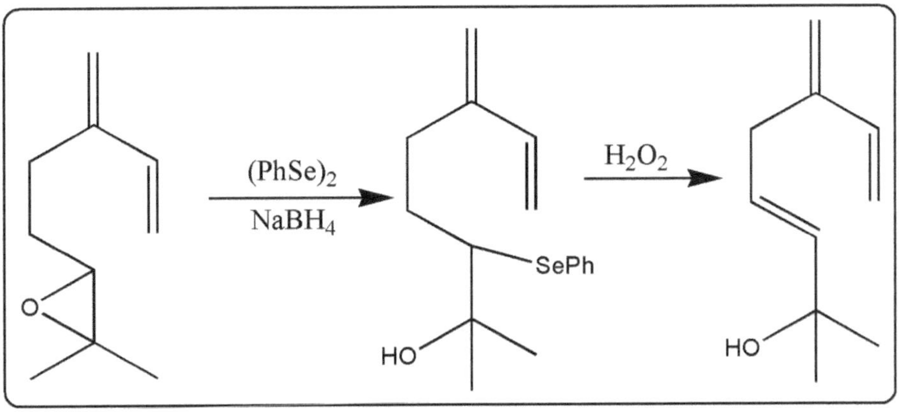

FIGURE 1.38 Utility of sodium phenylselenide reagent.

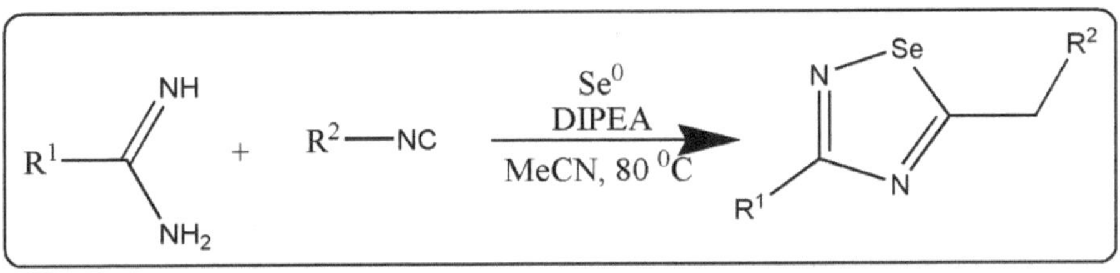

FIGURE 1.39 Selenium metal catalyzed synthesis of 1,2,4-selenadiazol-5-amine derivative.

can be utilized for various organic conversions and can be manipulated at various stages. The time between 1973 and 1975 was a landmark time for selenium chemistry as selenium received immense attention from chemists.

Ji et al. reported the use of selenium metal for the synthesis of 1,2,4-selenadiazol-5-amine derivatives at a 65–95% yield. This novel and facile metal-free procedure occurs through a multi-component reaction (MCR) between elemental Se, imidamides, and isocyanides in the presence of DIPEA and solvents, MeCN [92] (Figure 1.39).

Perin et al. [93] reported a method to synthesize divinyl selenides in single step using acetylene derivatives, selenium (element), $NaBH_4$, and solvent PEG-400 at 60ºC for 5 h. The nucleophile is produced in situ and it has been found that the electron-withdrawing group in acetylene derivatives increases the reactivity and the yield of the product is enhanced [93] (Figure 1.40).

Selenium-based inorganic and organic catalysts have great features of high efficiency and selectivity. They require mild reaction conditions and have excellent compatibility with functional groups. These eco-friendly features have gained widespread attention. In the last decade, there has been a boom in published articles on selenium catalysts [94–104].

Some of the recent applications of selenium catalysts are discussed here:

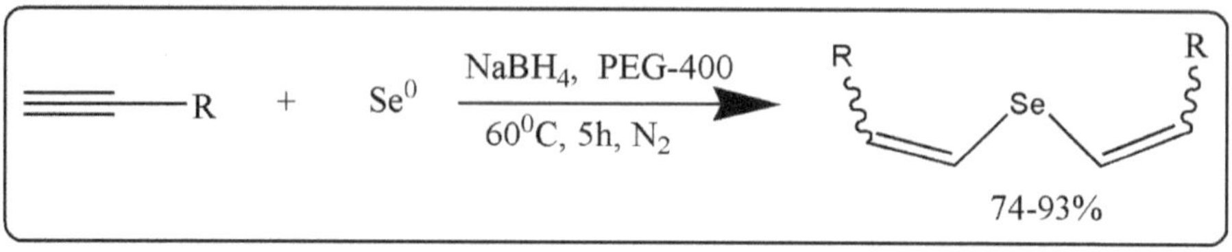

FIGURE 1.40 Synthesis of divinyl selenides by using selenium metal.

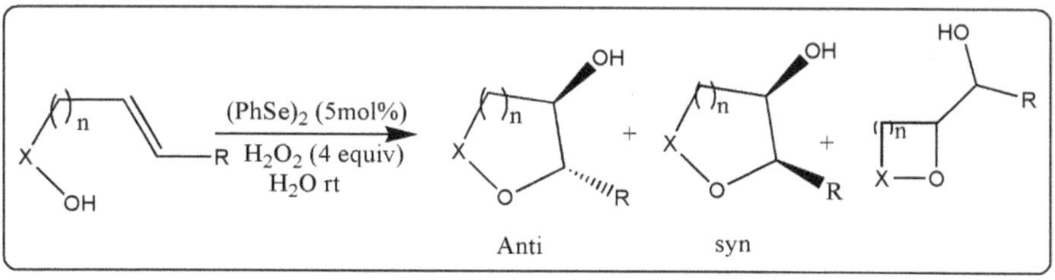

FIGURE 1.41 (PhSe)2-catalyzed synthesis.

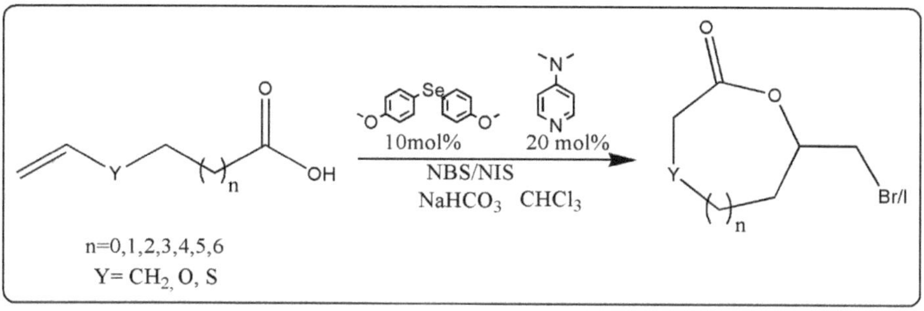

FIGURE 1.42 Organoselenium catalyzed halolactonization reaction of alkenoic acids.

Shanti et al. reported $(PhSe)_2$-catalyzed synthesis of γ- and δ-lactones, or tetrahydropyrans and tetrahydrofurans via oxacyclization reaction of alkenoic acid and alkenols [105] (Figure 1.41).

Kumar et al. reported a methodology to synthesize bromo and iodo lactones of medium size (5- to 11-membered ring size) from alkenoic acid using organoselenium (selenide) and 4-(dimethylamino)pyridine (DMAP) as catalyst and co-catalyst respectively. NBS/NIS is used as a halogenating agent for halolactonization [106] (Figure 1.42).

Yu et al. reported a highly efficient method utilizing heterogeneous Se/C catalyst (developed from glucose) for regiospecific epoxidation of β-ionone [107]. The catalyst has higher TON (turnover number) in comparison to conventional homogeneous organoselenium catalysts (Figure 1.43).

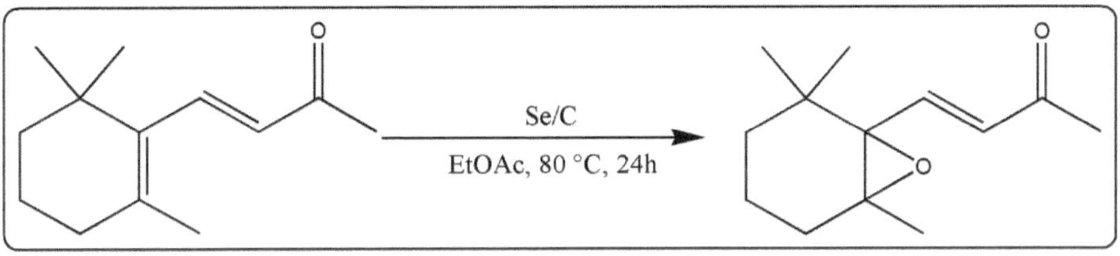

FIGURE 1.43 Regiospecific epoxidation of β-ionone by using Se/C catalyst.

1.4 CONCLUSIONS

Selenium is an essential element of biological importance. It is metabolizable and does not tend to accumulate in the body of an organism. They have wide applications in organic chemistry. Selenium and organoselenium catalysts are gaining attention and the scope of this field is gaining impetus with a manifold increase in researchers along with their publication. This chapter highlighted the chemistry of selenium along with the various reactions catalyzed by selenium and its reagents.

REFERENCES

1. Fordyce, F.M. Selenium deficiency and toxicity in the environment. In *Essentials of Medical Geology*, Springer, Dordrecht, The Netherlands, **2013**, pp. 375–416.
2. Barone, G.; Storelli, A.; Mallamaci, R.; Storelli, M.M. Comparative study on trace metal accumulation in liver of Mediterranean deep-sea fish and their selenium/mercury molar ratios. *Water Air Soil Pollut.* **2017**, *228*, 211. https://doi.org/10.1007/s11270-017-3379-4.
3. Kabata-Pendias, A. Geochemistry of selenium. *J. Environ. Pathol. Toxicol. Oncol.* **1998**, *17*(3–4), 173–177.
4. Fordyce, F. Selenium geochemistry and health. *AMBIO J. Hum. Environ.* **2007**, *36*(1), 94–97. DOI: 10.1579/0044-7447(2007)36[94:SGAH]2.0.CO;2
5. Amouroux, D.; Liss, P. S.; Tessier, E.; Hamren-Larsson, M.; Donard, O. F. X. Role of oceans as biogenic sources of selenium. *Earth Planet. Sci. Lett.* **2001**, *189*(3–4), 277–283. DOI: 10.1016/S0012-821X(01)00370-3
6. Haug, A.; Graham, R. D.; Christophersen, O. A.; Lyons, G. H. How to use the world's scarce selenium resources efficiently to increase the selenium concentration in food. *Microb. Ecol. Health Dis.* **2007**, *19* (4), 209–228. DOI: 10.1080/08910600701698986
7. Imran, M.; Chen, Z.; Mehmood, A.; Rukh, S.; Weixie, W.; Asghar, W.; Iftikhar, F. Distribution of selenium in soils and human health. In *Selenium and Human Health.* Gelen, V.; Kara, A.; Küküirt, A., Eds., IntechOpen, **2023**. doi: 10.5772/intechopen.110636
8. Tan, J. *The Atlas of Endemic Diseases and Their Environments in People's Republic of China*, Science Press, **1989**, p 88.
9. Toxicological Profile for Selenium. Agency for Toxic Substances and Disease Registry (US), **2003** Sep. Available from: https://www.ncbi.nlm.nih.gov/books/NBK600367/
10. Joy, E. J. M.; Broadley, M. R.; Young, S. D.; Black, C. R.; Chilimba, A. D. C.; Ander, E. L.; Barlow, T. S.; Watts, M. J. Soil type influences crop mineral

composition in Malawi. *Sci. Total Environ.* **2015**, *505*, 587–595. DOI: 10.1016/j.scitotenv.2014.10.038

11. Winkel, L. H.; Vriens, B.; Jones, G. D.; Schneider, L. S.; Pilon-Smits, E.; Bañuelos, G. S. Selenium cycling across soil-plant-atmosphere interfaces: A critical review. *Nutrients* **2015**, *7* (6), 4199–4239. DOI: 10.3390/nu7064199
12. Ma, Y.-T.; Liu, M.-C.; Zhou, Y.-B.; Wu, H.-Y. Synthesis of organoselenium compounds with elemental selenium. *Adv. Synth. Catal.* **2021**, *363*(24), 5386–5406. DOI: 10.1002/adsc.202101227
13. Alexander, J., Selenium. In *Handbook on the Toxicology of Metals*, 5th ed., Nordberg, G. F., Costa, M., Eds., Academic Press, **2022**, pp. 729–771.
14. Scalet, B.M.; Slade, S.; Kasper, A.; Lummen, G.; Gitzhofer, K.; van Limpt, H. Selenium emissions from glass melting furnaces: formation, sampling and analysis. A position paper by the Technical Committee 13, "Environment", of the International Commission on Glass (ICG). *Glass Technol. Eur. J. Glass Sci. Technol. A.* **2006**, *47*, 29–38.
15. Wang, C.; Li, S.; Wang, H.; Fu, J. Selenium minerals and the recovery of selenium from copper refinery anode slimes. *J. S. Afr. Inst. Min. Metall.* **2016**, *116*(6), 593–600. ISSN 2411-9717. DOI: 10.17159/2411-9717/2016/v116n6a16
16. Xiao, L.; Wang, Y.; Yu, Y.; Fu, G.; Liu, Y.; Sun, Z.; Ye, S. Enhanced selective recovery of selenium from anode slime using MnO_2 in dilute H_2SO_4 solution as oxidant. *J. Clean. Prod.* **2019**, *209*, 494–504. DOI: 10.1016/j.jclepro.2018.10.144
17. Khanlarian, M.; Rashchi, F.; Saba, M. A Modified sulfation roasting leaching process for recovering se, cu, and ag from copper anode slimes at a lower temperature. *J. Environ. Manage.* **2019**, *235*, 303–309. DOI: 10.1016/j.jenvman.2019.01.079
18. Fthenakis, V.; Wang, W.; Kim, H. C. Life cycle inventory analysis of the production of metals used in photovoltaics. *Renew. Sustain. Energy Rev.* **2009**, *13* (3), 493–517. DOI: 10.1016/j.rser.2007.11.012
19. Gustafsson, A. M.; Foreman, M. R.; Ekberg, C. Recycling of high purity selenium from CIGS solar cell waste materials. *Waste Manag.* **2014**, *34*(10), 1775–1782, ISSN 0956- 053X. DOI: 10.1016/j.wasman.2013.12.021
20. Pearson, R.; Haugen, G. R. Kinetics of the thermal decomposition of H_2Se. *Int. J. Hydrog. Energy* **1981**, *6*(5), 509–519, ISSN 0360-3199. DOI: 10.1016/0360-3199(81)90082-3
21. Nielsen, S.; Heritage, R. J. A method for the purification of selenium. *J. Electrochem. Soc.* **1959**, *106*(1), 39–43. DOI: 10.1149/1.2427260
22. Zhang, J.; Bao, Z.; Xueyi, G.; Danqing, Z.; Chao, S.; Qian, L. . Study on crude selenium purifying by a sulfite sodium leaching method, *Rare Met. Mater. Eng.* **2011**, *40*, 121–125.
23. Xu, N.; He, Xiaocai; Xu, Q.; Diao, W.; Yan, S.; He, Q.; Ren, T. A clean and efficient method for purifying crude selenium based on a selective leaching – directed decomposition purification process. *Mater. Res. Express* **2022**, *9*(10), 106510. DOI: 10.1088/2053-1591/ac977a
24. Burger, A.; Henderson, D. O.; Morgan, S. H.; Feng, J.; Silberman, E. Purification of selenium by zone refining. *J. Cryst. Growth* **1990**, *106*(1), 34–37. DOI: 10.1016/0022- 0248(90)90283-Q
25. Su, C.-H.; Sha, Y. Segregation coefficients of impurities in selenium by zone refining. *J. Cryst. Growth* **1998**, *187*(3–4), 569–572. DOI: 10.1016/S0022-0248(98)00037-2
26. Cole, R. A.; Nathan, L. Skinner process for removing impurities from zone refined materials. US patent (United States Patent 1 1 Cole et al. [451 Sept. so, 1975. Process

for removing impurities from zone refined materials [73] Assignee; Hughes Aircraft Company [22] Filed: Dec. 20, **1972** [21] Appl. No. 317, 505.
27. Badesha, S. S. Ontario; Smith, Thomas W., Penfield, Process for selenium purification, U.S. Patent **1985**, 4548,800.
28. Butterman, W. R.; Brown, R. D. Selenium. Mineral Commodity Profiles. United States Department of the Interior, United States Geological Survey, **2004**.
29. Rosman, K. J. R.; Taylor, P. D. P. Isotopic compositions of the elements 1997 (Technical Report). *Pure Appl. Chem.* **1998**, *70* (1), 217–235. DOI: 10.1351/pac199870010217
30. Audi, G.; Bersillon, O.; Blachot, J.; Wapstra, A. H. The Nubase evaluation of nuclear and decay properties. *Nucl. Phys. A* **2003**, *729* (1), 3–128. DOI: 10.1016/j.nuclphysa.2003.11.001
31. Kondev, F. G.; Wang, M.; Huang, W. J.; Naimi, S.; Audi, G. The NUBASE2020 evaluation of nuclear properties. *Chin. Phys. C* **2021**, *3*, **45**, 030001. DOI: 10.1088/1674-1137/abddae
32. Shaaban, S.; El-Lateef, H.M.A.; Khalaf, M.M.; Gouda, M.; Youssef, I. One-pot multicomponent polymerization, metal-, and non-metal-catalyzed synthesis of organoselenium compounds. *Polymers* **2022**, *14*, 2208. https://doi.org/10.3390/polym14112208.
33. Wallschläger, D.; Feldmann, F. Formation, occurrence, significance, and analysis of organoselenium and organotellurium compounds in the environment. *Met. Ions Life Sci.*; **2010**, *7* , 319–364. DOI: 10.1039/BK9781847551771-00319
34. Krief, A.; Hevesi, L. *Organoselenium Chemistry I. Functional Group Transformations*, Springer-Verlag, **1988**.
35. Burk, R. F., Ed. *Selenium in Biology and Human Health*; Springer-Verlag, 1994.
36. Soda, K.; Tanaka, H.; Esaki, N. Biochemistry of physiologically active selenium compounds. In *Organic Selenium and Tellurium Compounds*, Vol. 2, Patai, S., Ed., John Wiley & Sons, **1987**.
37. Sadeh, T. Biological and biochemical aspects of tellurium derivatives. In *Organic Selenium and Tellurium Compounds*, Vol. 2, Patai, S., Ed., John Wiley & Sons, **1987**. https://doi.org/10.1002/9780470771785.ch8
38. Masukawa, T. Pharmacological and toxicological aspects of inorganic and organic selenium compounds. In *Organic Selenium and Tellurium Compounds*, Vol. 2, Patai, S., Ed., John Wiley & Sons, **1987**. https://doi.org/10.1002/9780470771785.ch9
39. Mugesh, G.; du Mont, W. W.; Sies, H. Chemistry of biologically important synthetic organoselenium compounds. *Chem. Rev.* **2001**, *101*(7), 2125–2179. DOI: 10.1021/cr000426w
40. Sonego, J. M.; de Diego, S.I.; Szajnman, S.H.; Gallo-Rodriguez, C.; Rodriguez J.B. Organoselenium compounds: chemistry and applications in organic synthesis. *Chem. Eur. J.* **2023**, *29*, e202300030. https://doi.org/10.1002/chem.202300030
41. Wessjohann, L. A.; Schneider, A.; Abbas, M.; Brandt, W. Selenium in chemistry and biochemistry in comparison to sulfur. *Biol. Chem.* **2007**, *388(10)*, 997–1006. DOI: 10.1515/BC.2007.138.
42. Birringer, M.; Pilawa, S.; Flohé, L. Trends in selenium biochemistry. *Nat. Prod. Rep.* **2002**, *19* (6), 693–718. DOI: 10.1039/B205802M
43. (a) Krief, A.; Trabelsi, M.; Dumont, W.; Derock, M. Conditions-driven selective synthesis of selenides and selenols from elemental selenium. **2004**, *2004(10)*, 1751–1754. 10.1055/s-2004-829554. (b) Guillemin, J. C.; Bouayad, A.; Vijaykumar, D. First synthesis and characterization of vinylselenols and vinyltellurols *Chem. Commun.* 2000, 1163–1164. (c) Ścianowski, J.; Rafalski, J.; Banach, A.; Czaplewska,

J.; Komoszyńska, A. Synthesis and reactions of the optically active selenols derived from monoterpenes. *Tetrahedron Asymmetry* 2013, *24* (18), 1089–1096. DOI: 10.1016/j.tetasy.2013.07.018. (d) Cooksey, J. P.; Kocieński, P. J.; Blacker, A. J. A scalable process for the synthesis of 1,2-dialkyldiselanes and 1-alkaneselenols. *Org. Process Res. Dev.* **2019**, *23* (11), 2571–2575. DOI: 10.1021/acs.oprd.9b00380. (e) Tanini, D.; Capperucci, A. Synthesis and applications of organic selenols. *Adv. Synth. Catal.* 2021, *363*, 5360.

44. Thapa, B.; Schlegel, H. B. Theoretical calculation of pK_a's of selenols in aqueous solution using an implicit solvation model and explicit water molecules. *J. Phys. Chem. A.* **2016**, *120(44)*, 8916–8922. DOI: 10.1021/acs.jpca.6b09520
45. Singh, R.; Kats, L. Catalysis of reduction of disulfide by selenol. *Anal. Biochem.* 1995, *232(1)*, 86–91. doi: 10.1006/abio.1995.9956.
46. Schmidt, R. L.; Simonović, M. Synthesis and decoding of selenocysteine and human health. *Croat. Med. J.* **2012**, *53*(6), 535–550. DOI: 10.3325/cmj.2012.53.535
47. Papp, L. V.; Lu, J.; Holmgren, A.; Khanna, K. K. From selenium to selenoproteins: synthesis, identity, and their role in human health. *Antioxid. Redox Signal.* **2007**, *9*(7), 775–806. DOI: 10.1089/ars.2007.1528.
48. Back, T. G. Selenium: Organoselenium chemistry. In *Encyclopedia of Inorganic Chemistry* Scott, R. A., Ed., Wiley Interscience, **2006**. https://doi.org/10.1002/9781119951438.eibc0198
49. Tanini, D.; Capperucci, A. Synthesis and application of organic selenols. *Adv. Synth. Catal.* **2021**, *363*(24), 5360–5385. DOI: 10.1002/adsc.202101147
50. Flemer, S. Jr. Selenol protecting groups in organic chemistry: Special emphasis on selenocysteine se-protection in solid phase peptide synthesis. *Molecules* **2011**, *16*(4), 3232–3251. DOI: 10.3390/molecules16043232
51. Santi, C.; Tomassini, C.; Sancineto, L. Organic diselenides: Versatile reagents, precursors, and intriguing biologically active compounds. *Chimia (Aarau)* **2017**, *71*(9), 592–595. DOI: 10.2533/chimia.2017.592
52. Freudendahl, D. M.; Shahzad, S. A.; Wirth, T. Recent advances in organoselenium chemistry. *Eur. J. Org. Chem.* **2009**, *2009*(11), 1649–1664. DOI: 10.1002/ejoc.200801171
53. Bhasin, K. K.; Singh, N.; Kumar, R.; Gupta Deepali, D. G.; Mehta, S. K.; Klapoetke, T. M.; Crawford, M.-J. A convenient synthesis of some symmetrical and unsymmetrical diarylmethyl sulfur and selenium compounds: X-ray crystal structure of diphenylmethylseleno-2-propene and bis[p-chlorophenyl(phenyl)methyl] diselenide. *J. Organomet. Chem.* **2004**, *689*(21), 3327–3334, DOI: 10.1016/j.jorganchem.2004.07.040
54. Taylor, P. C., 2.14. Vinyl and aryl chalcogenides: Sulfur-, selenium- and tellurium-based functional groups. In *Comprehensive Organic Functional Group Transformations*, Katritzky, A. R., Meth-Cohn, O., Rees, C. W., Eds., *Comprehensive Organic Functional Group*, Elsevier Science, **1995**, pp. 705–736. DOI: 10.1016/B0-08-044705-8/00263-6
55. Santi, C.; Tidei, C.; Scalera, C.; Piroddi, M.; Galli, F. Selenium containing compounds from poison to drug candidates: A review on the GPx-like activity. *Curr. Chem. Biol.* **2013**, *7*(1), 25–36. DOI: 10.2174/2212796811307010003
56. Bartolini, D.; Commodi, J.; Piroddi, M.; Incipini, L.; Sancineto, L.; Santi, C.; Galli, F. Glutathione S-transferase pi expression regulates the Nrf2-dependent response to hormetic diselenides. *Free Radic. Biol. Med.* **2015**, *88*(B), 466–480, DOI: 10.1016/j.freeradbiomed.2015.06.039
57. Back, T. G., Ed. *Organoselenium Chemistry: A Practical Approach.* OUP, **1999**.

58. Beaulieu, P. L.; D'eziel, R., Chap. 3. In *Organoselenium Chemistry – A Practical Approach*, Back, T. G., Ed., Oxford University Press, **1999**.
59. Tiecco, M. Electrophilic Selenium, Selenocyclizations.. In *Organoselenium Chemistry. Topics in Current Chemistry*, vol 208, Wirth, T., Ed., Springer-Verlag, **2000**, pp. 7–54. https://doi.org/10.1007/3-540-48171-0_2
60. Kirsch, G.; Christiaens, L., Chap. 11. In *The Chemistry of Organic Selenium and Tellurium Compounds*; Vol. 2, Patai, S., Rappoport, Z., eds., John Wiley & Sons, **1987**.
61. Back, T. G.; Kerr, R. G. The metathesis of N-silylamines and benzeneselenenyl chloride. An efficient selenenamide synthesis. *Can. J. Chem.* **1986**, *64*(2), 308–310. DOI: 10.1139/v86-051
62. Nicolaou, K. C.; Claremon, D. A.; Barnette, W. E.; Seitz, S. P. N-phenylselenophthalimide (N-PSP) and N-phenylselenosuccinimide (N-PSS). Two versatile carriers of the phenylseleno group. oxyselenation of olefins and a selenium-based macrolide synthesis. *J. Am. Chem. Soc.* **1979**, *101*(13), 3704–3706. DOI: 10.1021/ja00507a069
63. Toshimitsu, A.; Uemura, S. Organic selenocyanates and tellurocyanates and related compounds derivatives. In *The Chemistry of Organic Selenium and Tellurium Compounds*, Vol. 2, Patai, S. Ed., John Wiley & Sons, **1987**.
64. Back, T. G. Synthetic and mechanistic aspects of some free-radical and electrophilic organoselenium reactions. *Phosphorus Sulfur Silicon Relat. Elem.* **1992**, *67*(1–4), 203–218. DOI: 10.1080/10426509208045838
65. Grieco, P. A.; Gilman, S.; Nishizawa, M. Organoselenium chemistry. A facile one-step synthesis of alkyl aryl selenides from alcohols. *J. Org. Chem.* **1976**, *41*(8), 1485–1486. DOI: 10.1021/jo00870a052
66. Back, T. G.; Birss, V. I.; Edwards, M.; Krishna, M. V. Study of the reductive cleavage of selenides with nickel boride. A convenient deselenization procedure. *J. Org. Chem.* **1988**, *53*(16), 3815–3822. DOI: 10.1021/jo00251a028
67. Clive, D. L. J.; Chittattu, G. J.; Farina, V.; Kiel, W. A.; Menchen, S. M.; Russell, C. G.; Singh, A.; Wong, C. K.; Curtis, N. J. Organic tellurium and selenium chemistry. Reduction of tellurides, selenides, and selenoacetals with triphenyltin hydride. *J. Am. Chem. Soc.* **1980**, *102*(13), 4438–4447. DOI: 10.1021/ja00533a024
68. Hussain, I.; Sahoo, S.; Lamiel, C.; Nguyen, T. Toan; Ahmed, M.; Xi, C.; Iqbal, S.; Ali, A.; Abbas, N.; Javed, M. S.; Zhang, K. Research progress and future aspects: metal selenides as effective electrodes. *Energy Storage Mater.* **2022**, *47*, 13–43. DOI: 10.1016/j.ensm.2022.01.055
69. Álvarez-Pérez, M.; Ali, W.; Marć, M. A.; Handzlik, J.; Domínguez-Álvarez, E. Selenides and diselenides: A review of their anticancer and chemopreventive activity. *Molecules* **2018**, *23*(3), 628. DOI: 10.3390/molecules23030628
70. Back, T.G.; Ibrahim, N.; McPhee, D.J. Studies of enamidic .DELTA.5-4-azasteroidal selenoxides: preparation, Pummerer reactions, configurational stability, and conversion to carbinol amides. *J. Org. Chem.***1982**, *47*, 3283–3289. https://doi.org/10.1021/jo00138a017
71. Toshimitsu, A.; Uemura, S. Reactions of selenones. In *Organoselenium Chemistry – A Practical Approach*, Back, T.G., Ed., Oxford University Press, 1999.
72. Krief, A.; Dumont, W.; Denis, J.N. Novel functional group transformations involving alkyl phenyl selenones. *J. Chem. Soc., Chem. Commun.*, **1985**, 571–572. https://doi.org/10.1039/C39850000571
73. S. Uemura, S. Fukuzawa, and A. Toshimitsu, A novel, simple method for transformation of C–Se to C–O bonds. *J. Chem. Soc., Chem. Commun.*, **1983**, 1501–1502. DOI https://doi.org/10.1039/C39830001501

74. Nishibayashi, Y.; Uemura, S. Organoselenium chemistry – modern developments in organic synthesis, in *Topics in Current Chemistry*, Wirth, T., Ed., Springer-Verlag, **2000**, *208*, 201–233. 10.1007/3-540-48171-0_8.
75. Reich, H.J. Sigmatropic rearrangements of organoselenium compounds. In *Modern Organoselenium Chemistry*, Liotta, D., Ed., John Wiley & Sons, **1987**, p. 365
76. Nishiyama, Y.; Noboru, S. Reductions with selenium reagents. In *Organoselenium Chemistry – A Practical Approach*, Back, T.G., Ed., Oxford University Press, **1999**; online edn, Oxford Academic, 31 Oct. 2023), https://doi.org/10.1093/oso/9780198501411.003.0006
77. Liotta, D. New organoselenium methodology. *Acc. Chem. Res.*, **1984**, *17*, 28.
78. Back, T.G. Oxidations with selenium reagents. In *Organoselenium Chemistry – A Practical Approach*, Back, T.G., Ed., Oxford University Press, **1999**.
79. Okazaki, R.; Tokitoh, N. Heavy ketones, the heavier element congeners of a ketone. *Acc Chem Res*. **2000**, *33*(9), 625–630. doi: 10.1021/ar980073b.
80. Back, T.G.; Barton, D. H. R., Britten-Kelly, M.R.; Guziec Jr, F.S. Olefin synthesis by two-fold extrusion processes. Part 3. Synthesis and properties of hindered selenoketones (selones) *J. Chem. Soc., Perkin Trans. 1*, **1976**, 2079–2089. https://doi.org/10.1039/P19760002079
81. Silks, L. A. P.; Wu, R.; Dunlap, R. B.; Odom, J. D. Synthesis and applications of chiral selones. *Phosphorus, Sulfur, and Silicon and the Related Elements*, **1998**, *136*(1), 209–214. https://doi.org/10.1080/10426509808545945.
82. Guziec F. Seleno and telluro carbonyl compounds . In *The Chemistry of Organic Selenium and Tellurium Compounds*, Patai, S.; , Eds., John Wiley & Sons, **1987**, Vol. 2.
83. Murai, T.; Kato, S. Organoselenium chemistry – Modern developments in organic synthesis. In *Topics in Current Chemistry*, T. Wirth, Ed., Springer-Verlag, Berlin, **2000**, 208, pp. 177–199.
84. Sayama, S.; Onami, T. Favorable effect of ortho substituents on selenoxide and elimination of n-alkyl aryl selenoxides. *Tetrahedron Lett.*, **2000**, *41*(29), 5557–5560, https://doi.org/10.1016/S0040-4039(00)00879-0
85. Uemura, S.; Fukuzawa, S.; Toshimitsu, A. A novel, simple method for transformation of C– Se to C–O bonds. *J. Chem. Soc., Chem. Commun.*, **1983**, 1501–1502. https://doi.org/10.1039/C39830001501
86. Shimizu, M.; Ando, R.; Kuwajima, I. Highly efficient method for ethylenic or acetylenic ketones via fragmentation of hydroxy vinyl selenones. *J. Org. Chem.*, **1981**, *46*, 5246–5248. https://doi.org/10.1021/jo00338a048
87. Mukherjee, A.J.; Zade, S. S.; Singh, H.B.; Sunoj, R.B. Organoselenium chemistry: Role of intramolecular interactions. *Chem. Rev.* **2010**, *110* (7), 4357–4416. https://doi.org/10.1021/cr900352j
88. Sharpless, K.B.; Lauer, R.F. Mild procedure for the conversion of epoxides to allylic alcohols. First organoselenium reagent. *J. Am.Chem.Soc.* **1973**, *95*, 2697–2699. https://doi.org/10.1021/ja00789a055
89. Clive, D.L.J. Modern organoselenium chemistry, *Tetrahedron*, **1978**, *34*, 1049–1132.
90. (a) Reich, H.J. Functional group manipulation using organoselenium reagents. *Acc. Chem. Res.* 1979, *12*, 22–30. (b) Liotta, D.; Monahan, R. Selenium in organic synthesis. *Science*, **1986**, *231*(4736), 356–361. http://www.jstor.org/stable/1696597
91. Fang, Y.; Zhu, Z-L.; Xu, P.; Wang, S-Y.; Ji, S-J. Aerobic radical-cascade cycloaddition of isocyanides, selenium and imidamides: facile access to 1,2,4-selenadiazoles under metal-free conditions. *Green Chem.*, **2017**, *19*, 1613–1618. http://dx.doi.org/10.1039/C6GC03521C

92. Perin, G.; Barcellos, A.M.; Luz, E.Q.; Borges, E.L.; Jacob, R.G.; Lenardão, E.J.; Sancineto, L.; Santi, C. Green hydroselenation of aryl alkynes: Divinyl selenides as a precursor of resveratrol. *Molecules*, **2017**, *22*(2), 327. http://dx.doi.org/10.3390/molecules22020327 PMID: 28230754
93. Singh, F.V.; Wirth, T. Selenium reagents as catalysts. *Catal. Sci. Technol.*, **2019**, *9*, 1073–1091.
94. Lenardão, E.J.; Santi, C.; Sancineto, L. *New Frontiers in Organoselenium Compounds*. Springer, **2018**. https://doi.org/10.1007/978-3-319-92405-2
95. Ortgies, S.; Breder, A. Oxidative alkene functionalizations via selenium-π-acid catalysis. *ACS Catalysis*, **2017**, *7*. 5828–5840. 10.1021/acscatal.7b01216.
96. Guo, R.; Liao, L.; Zhao, X. Electrophilic selenium catalysis with electrophilic N-F reagents as the oxidants. *Molecules* **2017**, *22*, 835. https://doi.org/10.3390/molecules22050835
97. Back, T.G. Oxidations catalyzed by seleninic acids and anhydrides, their precursors and congeners, *Curr. Green Chem.* **2016**, *3(1)*, 76–91. doi 10.2174/2213346103666160127003954
98. Breder, A.; Ortgies, S. Recent developments in sulfur- and selenium-catalyzed oxidative and isohypsic functionalization reactions of alkenes. *Tetrahedron Lett.*, **2015**, *56(22)*, 2843–2852.
99. Młochowski, J.; Wójtowicz-Młochowska, H. Developments in synthetic application of selenium(IV) oxide and organoselenium compounds as oxygen donors and oxygen-transfer agents. *Molecules,* **2015**, *20*, 10205–10243. https://doi.org/10.3390/molecules200610205
100. Santoro, S.; Azeredo, J.B.; Nascimento, V.; Sancineto, L.; Braga, A.L.; Santi, C. The green side of the moon: Ecofriendly aspects of organoselenium chemistry, *RSC Adv.*, **2014**, *4*, 31521–31535.
101. Breder, A. Oxidative allylic amination reactions of unactivated olefins – at the frontiers of palladium and selenium catalysis. *Synlett*, **2014**, *25(7)*, 899–904.
102. Santi, C.; Santoro, S.; Battistelli, B. Organoselenium compounds as catalysts in nature and laboratory, *Curr. Org. Chem.*, **2010**, *14*(20), 2442–2462, doi 10.2174/138527210793358231
103. Santi, C. *Organoselenium Chemistry: Between Synthesis and Biochemistry*, Bentham Science Publishers, **2014**. https://doi.org/10.2174/97816080583891140101
104. Sancineto, L.; Mangiavacchi, F.; Tidei, C.; Bagnoli, L.; Marini, F.; Gioiello, A.; Scianowski, J.; Santi, C. Selenium-catalyzed oxacyclization of alkenoic acids and alkenols. *Asian J. Org. Chem.* **2017**, *6*, 988.
105. Verma, A.; Jana, S.; Prasad, C.D.; Yadav, A.; Kumar, S. Organoselenium and DMAP co-catalysis: Regioselective synthesis of medium-sized halolactones and bromooxepanes from unactivated alkenes *Chem. Commum.*, **2016**, *52*, 4179–4182.
106. Yang, Y.; Fan, X.; Cao, H.; Chu, S.; Zhang, X.; Xu, Q.; Yu, L. Fabrication of Se/C using carbohydrates as biomass starting materials: an efficient catalyst for regiospecific epoxidation of β-ionone with ultrahigh turnover numbers. *Catal. Sci. Technol.*, **2018**, *8*, 5017–5023.

2 Chemistry of Vanadium

Sangeeta Bajpai, Monika Kamboj, Garima Pandey and Bimal Krishan Banik

2.1 INTRODUCTION

Vanadium is a well-known chemical element with a symbol 'V' and an atomic number 23. It is a member of first row transition metal series. It was discovered twice. Firstly, in the year 1801 Spanish mineralogist Andrés Manuel del Río discovered it, in Mexico, from a specimen of vanadinite, $Pb_5(VO_4)_3Cl$ (earlier known as Mexican 'blue lead' ore) [1] (Figure 2.1). He found its salts possessing wide range of colors and named it as *panchromium* (a Greek word means "all colors"). Later, del Río found that most of its salts turned red on heating, so he renamed the element as *erythronium* (Greek word meaning "red"). However, this discovery was dismissed by French chemist Hippolyte-Victor Collet-Descotils, concluding the ore to be chromium. Then after in 1830, Swedish chemist Nils Gabriel Sefstrom rediscovered it from the samples of cast iron found in mines in Sweden. The name vanadium was coined from Vanadis (a Scandinavian goddess of beauty and fertility) due to the beautiful colors of vanadium compounds in solution. Pure vanadium metal was first isolated in 1867 by the English chemist Henry Enfield Roscoe during the hydrogen reduction reaction of vanadium dichloride (VCl_2) and showed that previously isolated samples were vanadium nitride (VN).

In 1925, around 99.7 percent pure vanadium was obtained by the American chemists John Wesley Marden and Malcolm N. Rich by the reduction of vanadium pentoxide (V_2O_5) with calcium metal through a variety of smelting, leaching and roasting processes.

2.2 OCCURRENCE AND GEOCHEMISTRY

Vanadium holds 22nd position among the category of the most abundant metallic elements found in the Earth's crust [2] It is the fifth most frequently occurring transition element [3].

This "rarely" found pure vanadium exists in its almost 65 naturally occurring compounds. It is a significant constituent of some fossil fuels. Both vanadium (V) and zinc (Zn) possess identical concentration on the Earth's crust. It is nearly ten times greater than lead (Pb) and twice more than copper (Cu). Crude oil, coal, tar sands

DOI: 10.1201/9781032615110-3

FIGURE 2.1 Crystalline vanadium (*Source*: photo by Heinrich Pniok license FAL (https://artlibre.org/licence/lal/de/).

deposits, and bauxite ore are also some of the important sources of vanadium. It is also obtained as a by-product of uranium mining. In the petroleum sector, it is considered as an abundant component. Commercial sources of vanadium are the minerals, viz. Patronite, Carnotite, Vanadinite, Roscoelite and vanadium-bearing magnetite [4]. Vanadium is mostly obtained by extraction of Patronite (VS_4) ores. Vanadium's biogeochemical cycle is represented in Figure 2.2. It depicts the natural pathway of vanadium showing the circulation of its concentration via chemical, geological, and biological perspectives.

Table 2.1 shows an overview of the annual fluxes in the global biogeochemical cycle of vanadium. It is simplified as per recent compilation by Schlesinger et al. [6], and in turn is based in part on the earlier works of Hope et al. [7]. As per reports of Monakhov et al. [8], approximately 130,000 tons of vanadium is mined every year. The most common use of vanadium is as alloy steel. It is used in combination with chromium, manganese, nickel, boron, and tungsten. Due to increased demand for high-grade steel, the growth in global production of vanadium has nearly doubled in the past 15 years, thereby leading to 100 percent price increase in mined vanadium during 2017 [9]. Vanadium is also a trace contaminant in coal and petroleum products (in the heavy fractions of raw oil), which explains the high annual figure resulting from coal exploration (Table 2.1). A large annual release of vanadium as aerosols (100 and 280 Gg yr^{-1}) is seen due to the combustion of fossil fuels, particularly oil. A large part of the mined vanadium is stored in landfills and in products. The "natural" biogeochemical vanadium cycle is dominated by the riverine transport of mechanically weathered, particulate vanadium, and the sedimentation of this vanadium in

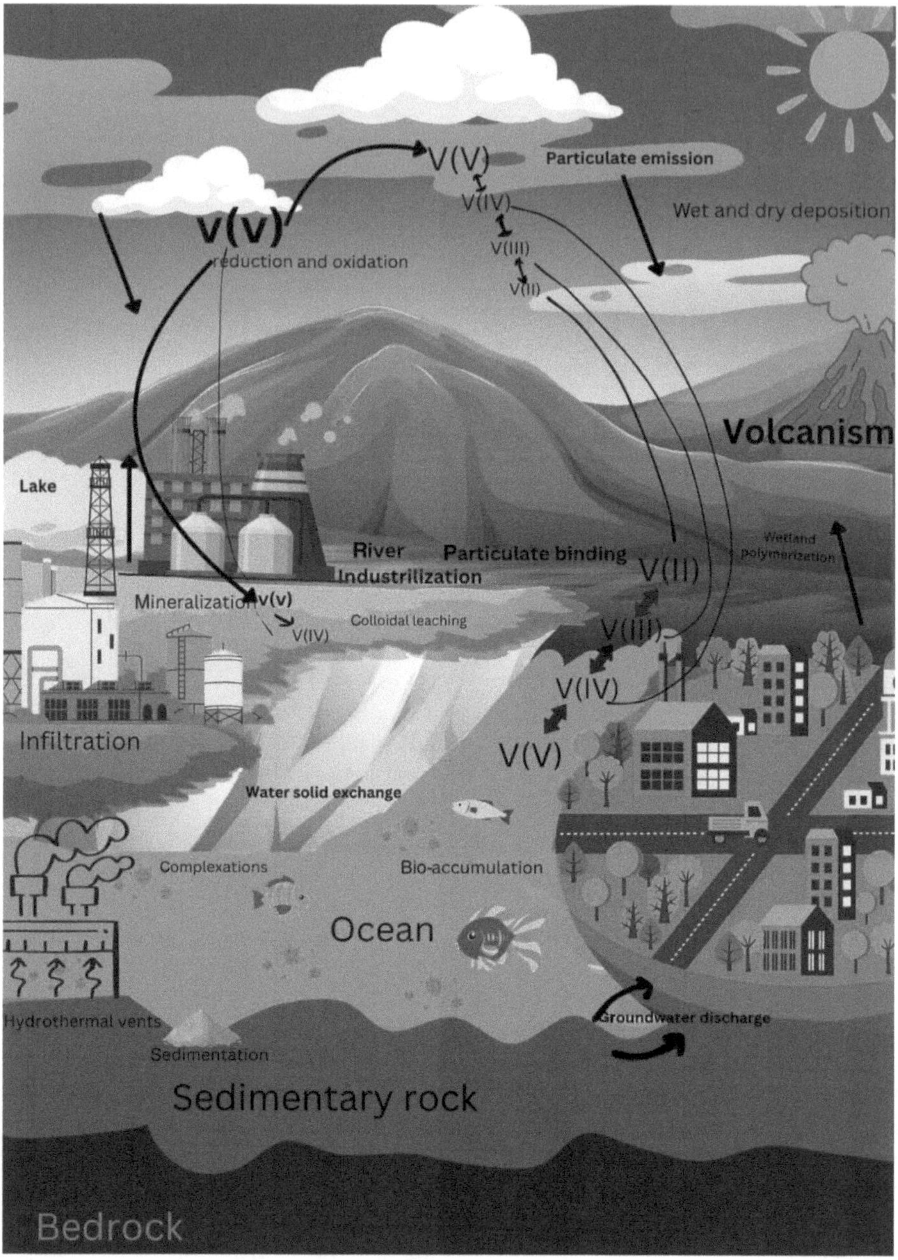

FIGURE 2.2 Biogeochemical cycle of vanadium [5] (adapted and modified).

TABLE 2.1
An overview of the annual fluxes in the global biogeochemical cycle of vanadium [11]

Source	Vanadium in $Gg\ V yr^{-1}$
Rivers	Dissolved V: 435
	Particulate V: 1600
Weathering	Chemical: 20
	Mechanical: 1600
Sedimentation	1600
Landfill products	435
Petroleum exploration	140–410
Fossil fuel aerosols	100–280
Coal mining	180–190
Subduction	150
Dust	150
Ore mining	130
Atmospheric deposition	72
Hydrothermal vents	22–28
Suboxic sediments	22
Oxic sediments	15–41
Volcanoes	5–29

the oceans. A mean residence time of 100,000 years has been estimated for vanadium in oceans [10].

2.2.1 Mode of Vanadium Occurrence

Vanadium enters the hydrosphere through continental weathering, volcanic ash and aerosols, and from hydrothermal vent systems [12–14]. It naturally forms a variety of metal complexes in terrestrial and marine environments [15] that are mediated by its valence states. Vanadium shows a wide range of oxidation states from −3 to +5 in many of its chemical compounds. Some vanadium compounds, exhibiting higher oxidation states, shows good "Lewis's acidity" property. "Oxophilic" property (the tendency of certain chemical compounds to form oxides by hydrolysis or abstraction of an oxygen atom from another molecule) is also shown by various vanadium compounds. Due to some of its valuable properties, viz. existence of multiple oxidation states, its oxophilicity behavior, Lewis's acid character, ease of hydrolysis, and polymerization, vanadium metal is richer chemical element than that of many other elements [16]. Vanadium exhibits different oxidation states in surface or near surface environments (Figure 2.2). It mainly occurs as a quadrivalent metal in the form of a vanadyl (VO^{2+}) ion in moderately reducing (0.5 to –0.2 Eh Volts) and within normal pH (4–7) conditions, as typical as for environments where carbonaceous sediments accumulate. Under these conditions, the cation is readily hydrolyzed and strongly adsorbs onto particles to form stable complexes with humic acids and porphyrins

[15]. Commonly accepted pre-oil forming metalation pathways involve the sequestration of transition metals into specific classes of compounds called metallo-porphyrins, the most well recognized being vanadyl and Ni porphyrins [17–20]. The initial source of these structurally labile geoporphyrins are bacterial and plant pigments such as chlorophyll a [17,18, 20,21]. The Mg^{2+} ion of chlorophyll a is replaced by hydrogen during transport through the water column. During early to late diagenesis, vanadium becomes ligated to the tetrapyrrole structure [20], resulting in locking of around 50 percent of the total vanadium content within porphyrins [22,23] with the remaining chelates likely being comprised of cracked porphyrins, metal-binding non-porphyrin species [24–27], and naphthenic acid salts [28,29].

The initial metal source of porphyrins is unclear. Amalgamation of VO^{2+} into porphyrins may result either from diffusion into organic-rich sediments from the overlying water column [30] or through oil–rock and/or water–rock interactions within the subsurface [31–33]. Potential sources may also expand the deposition of sedimentary organic matter enriched in organisms which preferentially extract these metals from the environment. To code with example, vanadyl may initially come from the vacuoles of certain marine urochord, which accumulate vanadium more than 100 times greater than what is naturally present in seawater [34,35].

Based on above studies, vanadium may have been divided into three primary groups in naturally occurring Metallo-organic compounds [5]:

(i) Mixed ligands tetradentate
(ii) Humate complexes
(iii) Tetrapyrrole complexes

This basic categorization is based on the type of the ligands that are involved in the interaction of metal bonding with the metallo-organic compounds [5].

In mixed ligands tetradentate, four ligands can be bonded with the metals containing various combinations of nitrogen, oxygen, and sulfur (Figure 2.3). For this type of complex, usually, bivalent metallic cations are required. Their coordination is not an octahedra form of stable square planer [36]. These ligands show distorted arrangement in less stable coordination sphere of square pyramidal or tetrahedral

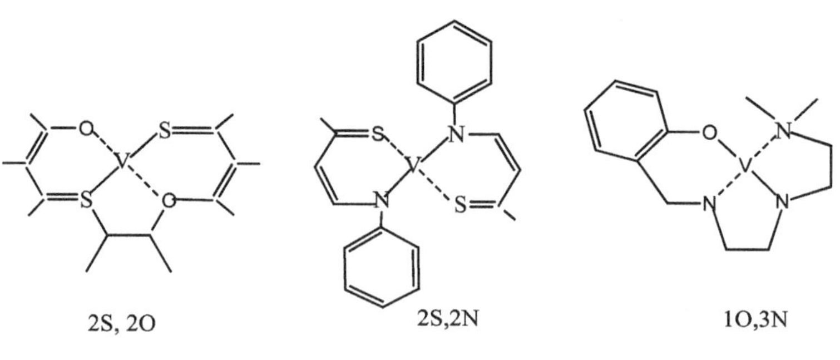

FIGURE 2.3 Mixed ligand tetradentate complexes.

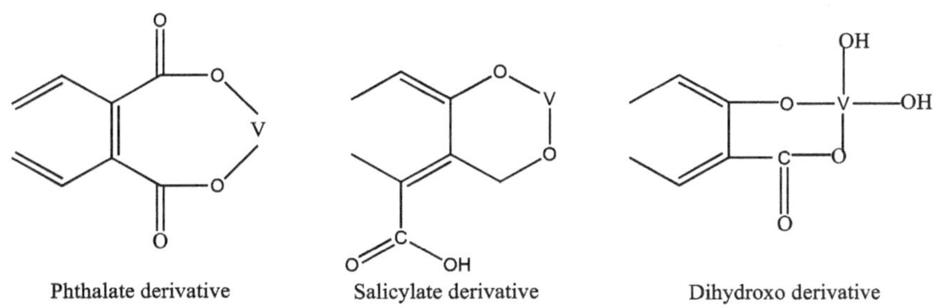

FIGURE 2.4 Humate complexes.

geometry. In dilute acids, demetallation of vanadium may occur, and it may have low stability in mixed ligand tridentate.

Humate complexes are the metallo-organic compounds containing oxygen as bonding ligands (Figure 2.4) [5]. This kind of compound naturally occurs in humic complexes. Typically, they contain oxygen atoms in phenolic hydroxide or carboxyl functional groups. In these complexes, researchers suggested the bonding of phenolic hydroxide (functional group) with metals, up to some extent. In humate compounds, the primary ligands as a carboxyl functional group contain an oxygen atom. Reports suggested that at low pH, the humate compounds may form by the cations of vanadyl (VO^{2+}). In naturally occurring humic compounds, the stability of vanadium is significantly lower than the desired trivalent cations of ferric and aluminium, supported by the higher amount of Fe and Al than vanadium. In humic compounds of marine origin, greater number of divalent cations are present relative to the trivalent ones. In some marine humic compounds, it is due to the increasing sulfur and nitrogen functional groups that act as a ligand, which contains oxygen atoms as a functional group. These kinds of ligands might preferably form the divalent mixed ligands, that is, tetradenates. Moreover, in humate compounds, vanadium is not preferred due to its unstable behavior, which suggests they may not survive in diagenetic settings.

Tetrapyrrole complexes are the metallo-organic compounds containing four atoms of pyrrole–nitrogen serving as a ligand (Figure 2.5) [5]. The present definition includes the higher aromatic and the pseudo-aromatic tetrapyrroles based on the pervasive research of metalated porphyrins; this kind of structure may present in the subunits of macromolecules or as a free molecule of organic matter. Tetrapyrrole complexes form strong bonds with vanadyl cation, in contrast to other metallo-organic compounds showing higher thermal stability, resistance to strong acid, and having inert behavior to cation exchange reaction [30]. Even though, identification of Metallo-organic compounds containing vanadium is difficult. Vanadium compounds in crude oil, asphalt, and bitumen, is thought to be associated with the tetrapyrrole compounds due to strong bonding.

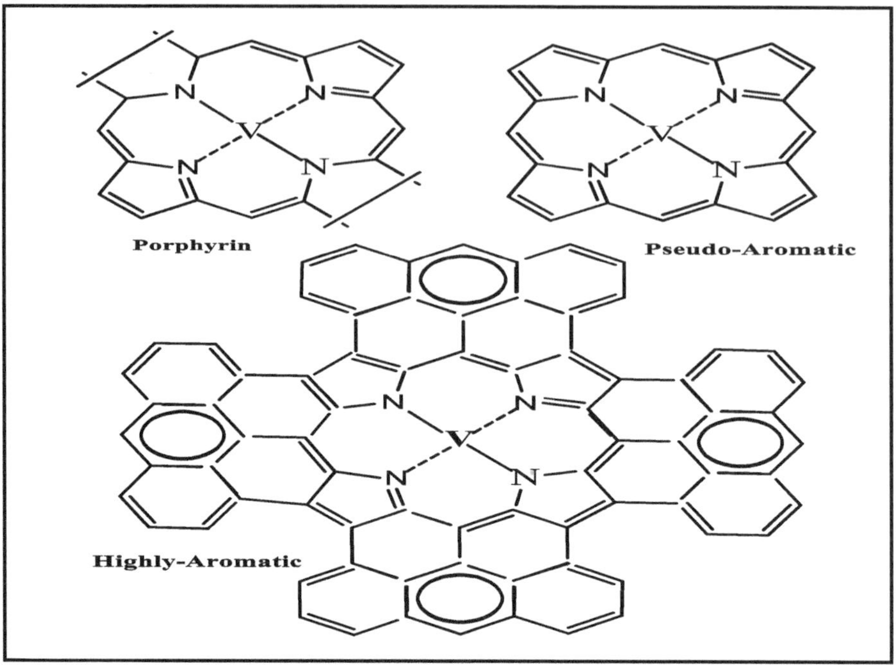

FIGURE 2.5 Tetrapyrrole complexes. (Adapted from Ref. [5].)

2.2.2 Global Occurrence of Vanadium Isotopes [37]

Only two naturally occurring isotopes are available of vanadium that are stable in nature. Apart from these, it has only two synthetic isotopes that are stable. Out of its synthetic isotopes, ^{49}V is the most stable with a half-life of 330 days. Its two naturally occurring isotopes are ^{50}V and ^{51}V. Out of its naturally occurring isotopes, ^{51}V is the most abundant isotope. Its alloys are also available with other elements such as Fe, Al, Ti, and Mo.

Stable isotopes of vanadium ($^{51}V/^{50}V$ isotope ratios) play an important role and act as a potentially powerful tool for studying the biogeochemical cycle of vanadium. However, sufficiently precise analytical methods have been developed only recently [38], and this research field is therefore still in its infancy. So far, only rather small fractionations (around 2 percent for crude oils) have been observed [37,39]. A relatively broad range of $\delta^{51}V$ values of crude oils ranging from –1.64 to –0.22 % has been reported by Ventura et al. [37]. This may reflect different redox conditions when the oil was formed. Seawater has a homogeneous isotopic composition with $\delta^{51}V = 0.20 \pm 0.15\%$ [40]. Several minerals have also been investigated, with a variation in $\delta\ ^{51}V$ values of ~1.5 percent [40,41]. Theoretical calculations predicted that the lighter ^{50}V isotope is preferentially adsorbed to goethite, resulting in substantial isotopic fractionations [42], but this still needs to be explored.

More recently, radiogenic isotopes techniques including Sm–Nd and Re–Os have also been used to determine migration of oil, geochronology, and/or fingerprinting

[43,44]. Crude oils are found to be commonly enriched in transition metals such as V, Ni, and Mo [45,46]. Technical advances at present allow the exploration of stable isotope compositions of vanadium in crude oils. To investigate transition metal isotope biogeochemistry, crude oils appear as an ideal substrate due to their accessibility and extensive global occurrence. Ventura et al. [37] reported the first high-precision measurements study of V, Ni, and Mo stable isotopes. This was determined for a set of globally distributed crude oils as a first examination of the magnitude of potential stable isotope fractionation. Several reports depicting the long research history of elemental analytical composition of Vanadium and other transition elements in crude oils are available [17–20,22,31,32,46,47].

Major segments of universe, viz. rocks and minerals, soil, and water bodies contain vanadium to the greater extent than found in other segments. The related precise information is discussed here.

2.2.3 Vanadium in Rocks

Vanadium, occurring in black rocks, has environmental and economic impacts. Black rock sediments are distributed globally in the form of silty, carbonaceous, and siliceous shales, sandstones, mudstones, etc. [48]. These black rocks are a new type of mineral resource due to the presence of some precious elements, viz. uranium, molybdenum, and vanadium. Vanadium is an extensive metal element in the environment. It is widely used in all fields of geochemistry.

Reports [5] reveal the average concentration of vanadium to be approximately 240 ppm in the black organic content, found in these black rock sediments. The mechanism and conditions of vanadium accumulation in these sediments involves the deposition of vanadyl V ions. These vanadyl ions are formed by the reduction of vanadate V(V) ions through the action of organic matters or H_2S found in the oxygenated ocean water. Oxygenated ocean water is suitable for the dissolved vanadate V(V) species. Most favorable vanadium-enriched minerals are clays and pyrite. Vanadium is mainly derived from the metamorphic transformation of its precursor (porphyrin pigments and chlorophyll) from the organism. Ages depicts that the massive transgression in the sea level created a favorable and survival environment for organisms. In addition, massive nutrient supply in the form of vanadium and other metal elements from the deep earth was bought by the hydrothermal activities. These creatures consumed the vanadium-rich nutrients, due to which their whole body was found to be vanadium enriched. Later, dead bodies of these organisms were submerged in the sediments, which on diagenetic actions became a part of these black sediments along with organic carbon. Therefore, these black rocks are enriched in vanadium and organic carbon. Adsorption, reductions, and complexation are some of the active factors responsible for the precipitation of dissolved vanadium into the organically rich sediments.

In sediments, presence of mainly three forms of vanadium, viz. trivalent V (III), tetravalent V (IV), and pentavalent V (V) is reported by Larson [5,49]. A good concentration of vanadium has been found in both types of sediment (soils and rocks) of the volcanic regions, peat, crude oils, oil shales, bitumen, and asphalt. Aydin et al.

TABLE 2.2
Chemical Composition of Some Select Vanadium Minerals [11,52–54]

Chemical Group	Chemical Composition	Name of the Mineral
Sulfides	VS_4	Patronite
	Cu_3VS_4	Sulvanite
Sulfates	$VOSO_4.5H_2O$	Minisragrite
	Pb3Zn3TeO6(VO4)2	Cheremnykhite
Silicates	K(V,Al,Mg)2AlSi3O10(OH)2	Roscoelite[a]
		Kaolinite and Sepiolite[b]
Oxides	V2O5.3H2O	Navajoite
	(V3+, Fe3+) OOH	Montroseite
Phosphates[c]	Pb5(VO4)3Cl	Vandinite
Vanadates	Pb2V2O7	Chervetite
	Ca(UO2)2V2O8×5-8H2O	Tyuyamunite
	CaV2O6×4-5H2O	Calcium vanadate hydrate
	K2(UO2)2(VO4)2×3H2O	Carnotite
	Cu3V2O7(OH)2×2H2O	Volborthite

[a] Vanadium present as V^{3+} in the octahedral layer of silicate minerals.
[b] Vanadium present as V^{4+} in the octahedral layer of silicate minerals [52, 53].
[c] Vanadate can enter solid solutions with apatite, due to its ability to substitute for the *o*-phosphate anion [54]. An example of an apatite-type mineral is vanadinite, Pb5(VO4)3Cl. In vanadinite, vanadate (V) has replaced phosphate because of the structural similarity between vanadate and phosphate.

[50] reveal that in crude oils, ~0.12 percent of vanadium is present in the form of VO-porphyrins. In the sediments, the concentration of vanadium is used for mineralogy study of the parent rock. Almost 65 minerals in nature contained vanadium. The notable minerals containing vanadium are vanadinite, mottramite, patronite, roscoelite, and carnotite [4].

Nriagu in1998 [51] reported more than 80 minerals containing vanadium in the bedrocks. The chemical groups and chemical composition of some select minerals are listed in Table 2.2 [11].

A good relationship is observed between vanadium and organic matter due to (i) the formation of complex aggregates between them or (ii) reduction of vanadium to its lower valence state, particularly at a lower pH level [55]. The adsorption property of vanadium creates compatible behavior with clay. Vanadium easily integrates with it, and as a result high concentration of vanadium is found in organically rich sediments. The concentration of vanadium is also high in peats as well.

2.2.4 Vanadium in Soil

Vanadium in soils is present in metallic form. Commonly, it could occur with oxides of iron, aluminium, manganese, lead, calcium, zinc, and copper. Vanadates (the most typical form of vanadium) is present in sediments [5,56]. The vanadium content in the soil is driven by the mineralogical composition of the parent material. Vanadium in the soils is primarily derived from the parental rocks [5,57]. However, recent studies showed that vanadium in soils could be enriched by anthropogenic activities such as burning fossil fuels, mining, pesticides, industries, fertilizers, and recycling domestic wastes [5,51,58]. Titanomagnetite deposits are the fundamental vanadium source in which vanadium minorly substitutes the iron [59]. Similarly, organic fractions, oxides of iron, and argillaceous minerals are also comprised of vanadium [5,60].

Development of high-temperature vanadium industrial activities, viz. electronic printing and dyeing has increased the concentration of anthropogenic vanadium in the environment [61,62]. China is rich in vanadium mineral resources, with vanadium titanium magnetite reserves up to 10 billion tons. It is the third-largest vanadium reserves in the world [63]. The Panzhihua region of China is famous for its vanadium titanomagnetite resources. In this region, the ranges of vanadium contents in soil from a smelting area and a mining area were found to be 208–938 mg/kg and 112–591 mg/kg, respectively [64].

The mean content of vanadium in the soil has been estimated to 108 mg kg^{-1} in contrast to the most regional surveys, which suggest lower values. According to the FOREGS database, in European topsoils, the median total vanadium concentration was found to be of 60.4 mg kg^{-1}, with a range from 1.28 to 537 mg kg^{-1}. Subsoils were slightly higher in vanadium, where the median total vanadium concentration was found to be of 62.8 mg kg^{-1} [11]. For American soils, vanadium concentration value of 84 mg kg^{-1} was reported by Shacklette and Boerngen [65]. Studies [66] depicted that a multiple regression model containing total amount of Fe, Al, and Mg could explain 84 percent of the variation in total vanadium for floodplain soils in the FOREGS data set. Soils engineered in black shales usually have comparably high vanadium concentrations due to its enrichment under the euxinic conditions in which they were formed [67]. Reports by Chermak et al. [68] envisages the worldwide average vanadium concentration to be around 790 mg kg^{-1} in black shales with maximum values of up to 2,540 mg kg^{-1} [69].

2.2.5 Vanadium in Water Resources

Earth is unique and distinct among all the planets. It is known as "blue planet" due to the massive presence of water on the earth where 97 percent of the water is salty, 2 percent is frozen, and only 1 percent is usable. Vanadium is an important heavy metal pollutant for surface water and ground waters, impacted mostly by mining [70,71]. Vanadium is the most abundant transition metal in the aquasphere with an average content same as that of zinc [70,72] .The concentration of dissolved vanadium in rivers is found to be of 0.78 ug/L. Fluctuation in the relationship of various environmental factors and vanadium in solution in different environments is seen. For example, low content of dissolved vanadium in streams might be due to a strong relationship with particulate matter and colloidal form of Fe [73].

However, in major rivers, the concentration of dissolved vanadium is closely associated with the origin of rock and the weathering. For example, the dissolved vanadium in the Wyoming River of the United States is 30–220 mg/L [5]. In aquasphere, vanadium concentration is chiefly dependent on seasonal variations. During hot season, the concentration of dissolved vanadium in marine and freshwater systems has been reported at a higher level in Florida Bay [74], Biwa Lake, and Toyko Bay [5], and Long Island Sound. This seasonal mobility of vanadium could be affected by various factors. The compatible relationship between the water particulates and the dissolved vanadium could be due to the substantial dependency of Eh and pH settings, which can vary seasonally. Fluctuations in vanadium concentrations in acidic streams are due to the impact of photoreduction property of light [5].

In the open ocean, the average dissolved vanadium concentration is found to be 35–45 nmol/L [75]. The presence of stable and high concentration of vanadium in oceans is reported whereas it shows lower concentration and a non-conservative behavior in coastal waters [76]. In the global oceans, comparative study between the suspended particulate form of vanadium reservoirs and the dissolved vanadium reservoirs is reported to be relatively smaller (1.8×10^{10} g) in the former than (2.7×10^{15} g) in the latter [77]. Oceanic sediments are the biggest vanadium pool (1.7×10^{20} g) on the Earth's surface [77].

In marine sediments, vanadium accumulation with a higher concentration (>400 mg/kg) is observed due to the significant role of volcanic activities [78].

The groundwater specimens (around 10 percent) from California and some other states of United States contain more than 25 lg/dm^3 of vanadium. It is because of vanadium washout from water-bearing sediments [79]. In oxic water region, the most dominant form of vanadium is V(V). In contrast, the vanadyl ion VO^{2+} (IV) is more stable in reducing conditions and tends to hydrolyze. Yangtze River of China (the most significant freshwater source in China) contains vanadium from 0.24 to 64.5 lg/L [80]. In water, the geochemical attributes of vanadium control the biogeochemical cycle of vanadium in the surface environment (Figure 2.2). The biological evolution of vanadium is a composite function of solution chemistry, biology, pH, Eh, and several surface environments.

In natural water resources, vanadium exists in three different states +5, +4, and +3. Various vanadate species V(III), V(IV), and V(V) are stable in anoxic, oxic, and suboxic environments, respectively [81]. Different spectroscopic studies and chemical techniques depicts presence of V(IV) and V(V) species in a natural aqueous system [5,82,83]. Vanadium complexation property, with both inorganic and organic ligands, can enhance its mobility at the surface of the interface (solid water) [84]. Vanadium (IV) easily forms complexes with a broader range of inorganic and organic ligands relative to vanadium V(V). The typical compounds formed with the inorganic ligands are VSO_4^- for V(III); $VO(OH)CO_3^-$, $VOCl^+$, VOF_4^{2-}, VOF_3^-, $VO(CH_3COO)$, $VO(CH_3COO)_2$ VOF^+ VOF_2, $VOCO_3$, $VOSO_4$, $VO(OH)C_2O_4^-$, $VO(C_2O_4)_2^{2-}$ for V(IV); $H_4VO_4C_2O_4^-$ and $H_4VO_4(C_2O_4)_2^{3-}$ for V(V) [84]. A wide range of organic substances shows good affinity with vanadium comprising humic substances and fulvic acid through oxygen groups. Besides, the organic substances form comparatively weaker complexes with vanadate oxyanion than oxycations V(IV) [5]. Hence, the V(V) organic compounds are found significant to reduce V(V) and stability of vanadium against reductions can be achieved by binding with humic compounds at pH values [85]. The reduction of abiotic and biotic V(V) can result in both V(III) and V(IV) compounds [86].

A variable concentration of vanadium is observed in sea water. In the English Channel, the concentrations are between 0.5 and 1 μg V L^{-1}. In the Landsort Deep of the Baltic Sea, the dissolved vanadium was only about 0.15 μg V L^{-1}. In a large area of the East Pacific region, dissolved vanadium was between 1.3 and 1.4 μg V L^{-1}, but somewhat higher (1.8 μg V L^{-1}) near the Peru Margin. The surface water of the East Pacific is ~5 percent lower in vanadium than the deeper waters, because of biological uptake and incorporation into particulate vanadium, which is higher at the surface. Vanadium mean concentration has been predicted to 1.8 μg V L^{-1} [11].

2.3 ATOMIC, CHEMICAL, AND PHYSICAL PROPERTIES

Vanadium is the member of the first transition series. In the periodic, table this chemical element is placed in the Vth group and IVth period. The symbol for the vanadium element is V with the atomic number 23. Vanadium does not show any similarity in physical and chemical properties with niobium although they are present in the same group. The atomic and physical properties of Vanadium are listed in Table 2.3. As vanadium is found in the 4th group so, it is a d-block element. Vanadium has a blue-silvery-gray metallic appearance. In highly pure state, it appears as white in color, soft and ductile. It possesses high melting point, due to which it is referred to as a refractory metal [4]. It has a low fission neutron section and good structural strength.

Bulk samples of vanadium are inert to air, moisture, alkalis, and other nonoxidizing acids other than HF at 25°C. However, the metal is attacked by concentrated sulfuric acid, nitric acid, and aqua regia. At elevated temperatures, vanadium combines with most nonmetals including nitrogen, carbon, silicon, and boron to give a variety of products such as VN, VC, VSi_2, V_3Si, VB, and VB_2. It reacts with chlorine at 180°C

TABLE 2.3
Atomic and physical properties of vanadium [88]

Properties	
Atomic number	23
Atomic mass	50.942 g.mol^{-1}
Density	6.11 g.cm^{-3} at 20°C
Electronic configuration	$[Ar]3d^34s^2$
Electronegativity	1.63 (Pauling Scale)
Van der Waals radius	179pm
Atomic radius	135pm
Covalent radius	153(8) pm
Common oxidation states	+5, +4, +3, +2
Electron affinity	0.525 eV
Ionization potential I^{st}	6.74 eV
Melting point	1910 °C
Boiling point	3407 °C

(which is low temperature) forming vanadium tetrachloride (VCl_4). Most of these compounds are nonstoichiometric and interstitial. In air, V_2O_5 is formed along with lower oxides [87].

Vanadium exhibits good corrosion resistance property to salt water, alkalies and acids (HCl and H_2SO_4). However, it oxidizes above 660°C. Oxidation states of vanadium govern the chemistry of vanadium compounds. On heating in air at variable temperatures, it oxidizes to trioxide (brownish black), tetraoxide (blue black), or pentoxide (reddish orange) in color. Vanadium pentoxide (V_2O_5) is a well-known industrially important vanadium compound.

The electronic configuration of vanadium is $1s^2\ 2s^2\ 2p^6\ 3s^2\ 3p^6\ 3d^3\ 4s^2$ or it can be written as [Ar] $3d^3\ 4s^2$. It has two electrons in K shell, eight electrons in L shell, 11 electrons in M shell and two electrons in its outermost shell N.

2.3.1 Chemical Properties of Vanadium

The oxidation states of vanadium govern its chemistry. Oxides of vanadium show different acidic and basic properties. As vanadium dioxide (VO_2) is basic and weakly acidic in nature, vanadium trioxide (V_2O_3) and vanadium monoxide (VO) have basic characters whereas vanadium pentoxide (V_2O_5) possesses acidic and weakly basic property. In aqueous medium, vanadium occurs as $(VO_3)^-$ or $(V_3O_9)^{3-}$ anion and as $(VO_2)^+$ cation in strongly acid solution. Vanadium in +4 oxidation state [V(IV)] forms oxycations $(VO)^{2+}$ and oxyanions $(V_4O_9)^{2-}$; V (III) and V(II) compounds exist as hydrated ions $[V(H_2O)_6]^{3+}$ and $[V(H_2O)_6]^{2+}$, respectively, in aqueous medium.

2.4 VANADIUM COMPOUNDS

Vanadium forms a variety of compounds, viz. hydrides, halides, oxyhalides, sulfates, sulfides, and phosphates (Figure 2.6).

2.4.1 Hydrides

Vanadium metal is almost inert to hydrogen at both room temperature and higher temperatures. A review on vanadium hydrides is available [89]. Vanadium hydrides are obtained by absorbing hydrogen reversibly, upto the final composition $VH_{0.94}$ when heated at 300°C in a 48.4 percent hydrogen atmosphere at 1 atm pressure. The product obtained is gray metallic material whose brittleness increases with increasing hydrogen content. The structure has a single phase having a tetragonal unit cell and lattice parameters $a = 3.02$, $c = 3.36$ [87].

A nonstoichiometric dihydride of vanadium having the composition $VH_{1.64}$ was reported by crushing vanadium monohydride under high hydrogen pressures at 70 atm. X-ray powder diffraction studies envisage the presence of an fcc phase along with bct (body centered tetragonal) monohydride. The same authors also reported that on treatment with hydrofluoric acid, vanadium hydride having the composition $VH_{1.77}$ was obtained. Vanadium metal has a bcc structure with a lattice parameter $a_0 = 3.04$

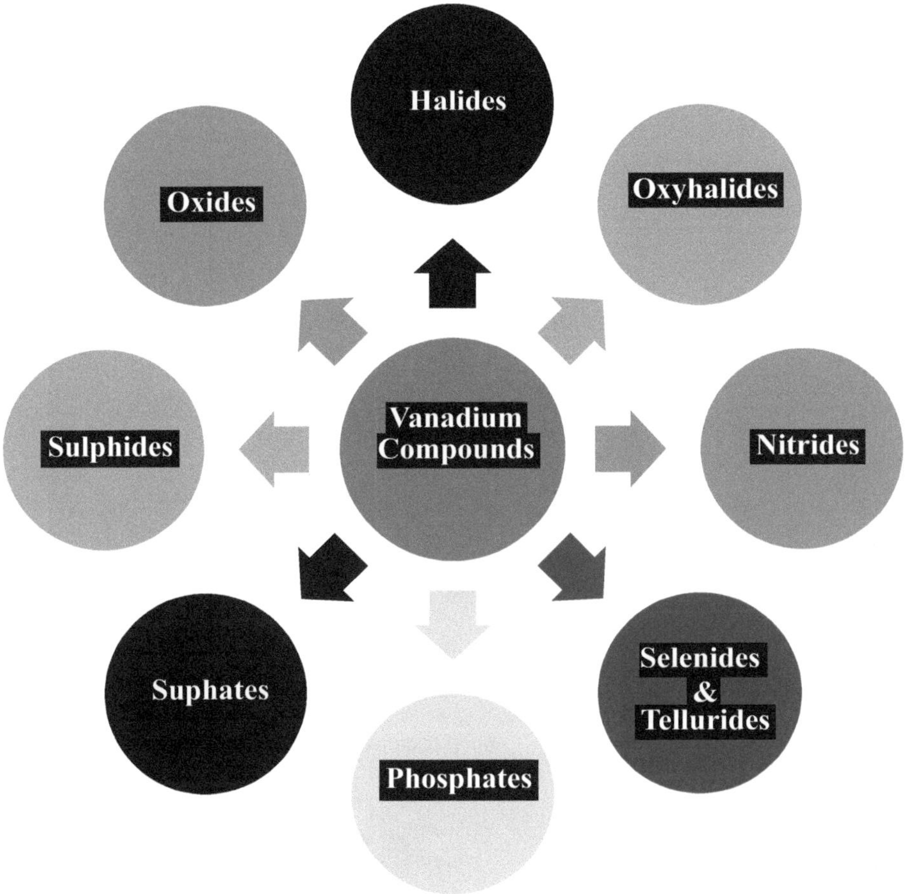

FIGURE 2.6 Types of vanadium compounds.

Å. Stochiometric vanadium hydrides, viz. VH_4, have been observed spectroscopically in a hydrogen matrix at 2–4°K [89].

2.4.2 Halides

Vanadium halides are very important class of compounds providing convenient starting materials for other species. Their synthesis and properties have been comprehensively reviewed and references to individual synthetic methods can be found in these works. Only vanadium tetrachloride (VCl_4) is commercially important whereas vanadium tetrachloride (VCl_3) is of less commercial interest. F–F bond is comparatively weaker than analogous bonds in other halides, which is the main cause for the absence of pentavalent vanadium halides other than VF_5. Among vanadium tetrahalides, VCl_4 is unstable and VBr_4 decomposes at –23°C [87]. Known halides are listed in Table 2.4.

TABLE 2.4
Vanadium halides

Oxidation State	Compound	Preparation and Properties
II	VF_2 [87]	i. It appears as blue crystals. ii. It can be prepared by reduction of VF_3 by H_2 in the presence of HF at 1150°C. $2VF_3 + H_2 \rightarrow 2\ VF_2 + 2\ HF$ iii. It acts as a strong reducing agent . In the presence of Mg $(OH)_2$ it can reduce N_2 to N_2H_4 . iv. It gives $[V(H_2O)_6]^{2+}$ ions on reacting with water. $V^{2+} + 6\ H_2O \rightarrow [V(H_2O)_6]$
	VCl_2 [90,91]	i. It appears as apple green solid. ii. It can be prepared by thermal decomposition of VCl_3 $VCl_3 \rightarrow VCl_2 + VCl_4$ iii. With water it gives purple colored hexa aquo ions $[V\ (H_2O)_6]^{2+}$which on evaporation, produces crystals of $[V(H_2O)_6]Cl_2$
	VBr_2	i. It appears as orange brown solid. ii. The compound is produced by the reduction of vanadium (III) bromide. $2VBr_3 + H_2 \rightarrow 2VBr_2 + 2HBr$
	VI_2 [87,92]	i. It appears as red solid. ii. It can be prepared by the direct reaction of the elements. iii. By reducing vanadium (III) chloride with iv. trimethylsilyl iodide, solvated vanadium (II) iodides can be prepared
III	VF_3 [87]	i. It is a yellow green refractory solid. ii. It is prepared by passing anhydrous HF in N_2 over VCl_3 or VCl_2 at 600°C in a platinum crucible. iii. It is slightly soluble in water, giving acidic solutions, and is virtually insoluble in organic solvents. It sublimes without decomposition at 800°C. v. It is obtained in a two-step procedure from V_2O_3: The first step entails conversion to the hexafluoro vanadate (III) salt using $(NH_4)HF_2$ (ammonium bifluoride): $V_2O_3 + 6\ (NH_4)HF_2 \rightarrow 2\ (NH_4)_3VF_6 + 3\ H_2O$ In the second step, the hexafluoro vanadate is thermally decomposed. $(NH_4)_3VF_6 \rightarrow 3\ NH_3 + 3HF + VF_3$

TABLE 2.4 (Continued)
Vanadium halides

Oxidation State	Compound	Preparation and Properties
	VCl_3 [87]	i. It appears as anhydrous red violet crystals. ii. It is prepared by heating VCl_4 at 160–170°C in the presence of flowing stream of inert gas thereby removing Cl_2. Purple solid obtained. iii. VCl_3 can also be prepared by the reaction between vanadium (III) oxide and thionyl chloride at 200°C. $V_2O_3 + 3\ SOCl_2 \rightarrow 2\ VCl_3 + 3\ SO_2$
	VBr_3 [87]	i. It appears as grayish crystalline material. ii. VBr_3 has been prepared by treatment of vanadium tetrachloride with HBr: $2VCl_4 + 8\ HBr \rightarrow 2\ VBr_3 + 8\ HCl + Br_2$ iii. It can also be prepared by the reaction between bromine and ferrovandine $2\ V + 3\ Br_2 \rightarrow 2\ VBr_3$ $2VFe + 6\ Br_2 \rightarrow 2\ VBr_3 + FeBr_3$ iv. It is soluble in water giving green color. v. Pure VB3 is obtained on sublimation at 350°C in a bromine atmosphere. vi. On heating around 500°C, a red vanadium (IV) bromide can be separated by rapid cooling, which decomposes at –23°C.
	VI_3 [93]	i. It is a brownish black crystalline powder. ii. It can be prepared by the reaction of vanadium metal powder and iodine at 500°C. iii. The purification of vanadium metal via chemical transport reaction involves reversible formation of VI_3 in the presence of I_2 and its subsequent decomposition to yield pure metal: $2\ V + 3\ I2 \rightleftharpoons 2\ VI_3$
IV	VF_4 [87]	i. It is a bright green hygroscopic solid. ii. It can be prepared by treating VCl_4 with anhydrous HF in an inert solvent CCl_3F at –78°C. $VCl_4 + 4\ HF \rightarrow VF_4 + 4\ HCl$ iv. It decomposes at 325°C, undergoing disproportionation to VF_3 and VF_5 $2\ VF_4 \rightarrow VF_3 + VF_5$

(continued)

TABLE 2.4 (Continued)
Vanadium halides

Oxidation State	Compound	Preparation and Properties
	VCl_4	i. It appears as reddish-brown viscous liquid. ii. VCl_4 is produced in the reaction between dry chlorine and vanadium or by the action of chlorinating agents such as $COCl_2$, $SOCl_2$, S_2Cl_2, or CCl_4 on V_2O_5 or the metal itself.
	VBr_4	It has been isolated by quenching (−78°C) the vapor obtained from heating the solid tribromide to 325°C.
	VI_4	It is unstable and has not been isolated.
V	VF_5	i. It is white solid in appearance. ii. It can be prepared by fluorination of vanadium metal: $2\,V + 5\,F_2 \rightarrow 2\,VF_5$ iii. On disproportionation at 650°C, equal amounts of the solid VF_3 and volatile VF_5 are obtained. $2\,VF_4 \rightarrow VF_3 + VF_5$

2.4.3 Oxohalides

The synthesis and properties of vanadium oxo halides are well documented [87]. These compounds form an important set of chemical compounds. Based on oxidation states they can be classified as:

- Pentavalent oxohalides [VOX_3] (X = F, Cl, Br) & VO_2X (X = F, Cl). Vanadium(V) oxy trichloride ($VOCl_3$) (a pale-yellow liquid) and vanadium(V) oxy tribromide ($VOBr_3$) have somewhat similar physical properties.
  (i) Both are extremely unstable toward moist air.
  (i) The gas-phase structures of [VOC_{l3}] and [$VOBr_3$] show trigonal distorted tetrahedral geometry as envisioned for [VOF_3].
  (ii) $VOCl_3$ is readily hydrolyzed and forms coordination compounds with simple donor molecules, for example, ethers. But it is reduced by reaction with sulfur-containing ligands and molecules. It dissolves sulfur and is completely miscible with many hydrocarbons and nonpolar metal halides.
- Tetravalent oxohalides [VOX_2] (X = F, Cl, Br). All the oxohalides of vanadium (IV), [VOX_2] (X = F, Cl, Br), are known.
  (i) The complexes are green and monomeric.

(ii) The presence of strong vanadium–oxygen bond of the VO^{2+} group enriches its chemistry.
(iii) $[VOCl_2]\cdot2.5H_2O$ is the best-known hydrate of vanadium. The known hydrates $[VOF_2]\cdot4H_2O$ and $[VOF_2]\cdot2H_2O$ and several hydrates of $[VOCl_2]$ have also been reported, but many of their stoichiometries are uncertain.

- Trivalent oxohalides [VOX] (X = Cl)
  (i) Crystalline VOCl has a layered structure. It is prepared by reaction between V_2O_5 and VCl_3.
  (ii) Geometrically, each oxygen atom is surrounded by a tetrahedron of vanadium atoms. Each vanadium atom is surrounded by four oxygen and two chlorine atoms in a distorted octahedral arrangement [87].

2.4.4 Oxides

Reports on vanadium oxides are available [87]. The behavior and stability of vanadium pentoxide (V_2O_5) lies between TiO_2 and CrO_2 (highest oxides). V_2O_5 readily dissolves in both acids and alkalies. Vanadate ions, in their simpler mononuclear form, are present in strongly alkaline solution. Dioxyvanadium (V) cation $(VO_2)^+$ is present as the main species in strongly acidic solutions. Oxides in the form of $(VO_3)^-$, $(HVO_4)^{2-}$, $(V_3O_9)^{3-}$, $(V_4O_{12})^{4-}$, $(V_{10}O_{28})^{6-}$ complex ions are present in alkaline /acidic solutions of intermediate strength. Known oxides are listed in Table 2.5.

2.4.5 Sulfates and Sulfides [87]

The +2 and +3 oxidation states of vanadium are known to form vanadium sulfates. Trivalent vanadium sulfate can be crystallized as a hydrate from the aqueous solution. It shows strong reducing character and produces a series of alums of the type $M^IV(SO_4)_2\cdot12H_2O$, which have been shown to contain the $[V(H_2O_6)]^{3+}$ ion. The electrolytic reduction of $NH_4[VO_3]$ in sulfuric acid gives air-stable blue-violet crystals of ammonium alum, $(NH_4)V(SO_4)_2\cdot12H_2O$. The violet vanadium (II) sulfate heptahydrate [$VSO_4\cdot$7H2O] may be prepared either by electrolytic sodium amalgam or by reduction of V_2O_5 solution in sulfuric acid using zinc as reducing agent. Electrolytic reduction of vanadyl (IV) sulfate solution gives vanadium hexahydrate $[VSO_4\cdot6H_2O]$. Aqueous solutions of both salts, on treating with respective alkali metal sulfates, produces reddish violet Tutton's salts, $[M_2\{V(H_2O)\ 6\}\cdot(SO_4)_2$ (M = NH_4, K, Rb, or Cs)]. These vanadium alums show more stability than the simple vanadium sulfate, especially in the presence of excess acid. The properties of vanadium (IV) oxysulfate. [($VOSO_4.5H_2O$), vanadyl sulfate] is also reported. Vanadium sulfides, viz. VS, VS_2, VS_4, V_2S_3, and V_2S_5, have been reported. Some may not exist whereas some have been characterized thoroughly.

TABLE 2.5
Vanadium oxides

Oxidation States	Compound	Preparation, Properties and Applications	Structure
II	Vanadium (II) oxide (VO) [87,94]	(i) Nonstoichiometric material. (ii) Grayish black metallic luster. (iii) Electrical conductivity similar as metals. (iv) Metal–metal bonding increases as oxygen content decreases. (v) Its potential use is as a molecular vapor in chemical synthesis at low or relatively low temperatures.	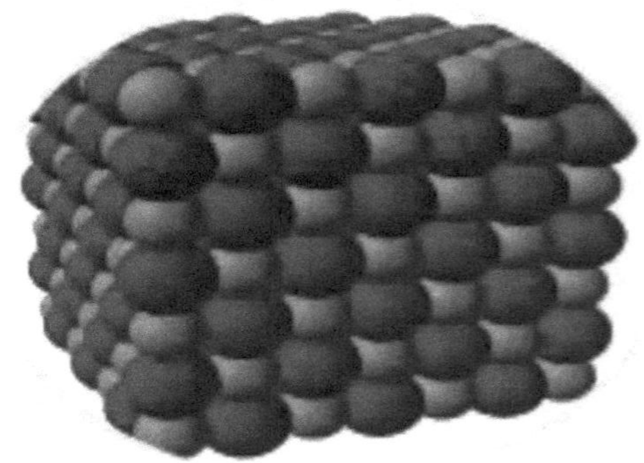 It adopts the distorted NaCl structure containing weak V–V bonds.

III	Vanadium (III) oxide known as vanadium sesquioxide (V_2O_3) [95]	(i) It can be obtained by the reduction of vanadium pentoxide or ammonium metavanadate by natural gas /hydrogen / carbon monoxide. (ii) Appears as a black solid. (iii) Oxidation by air proceeds slowly at ambient temperatures, but oxidation by chlorine at elevated temperatures is rapid giving $VOCl_3$ and V_2O_5. (iv) It is antiferromagnetic with a critical temperature of 160 K and at this temperature its conductivity changes from metallic to an insulating one.	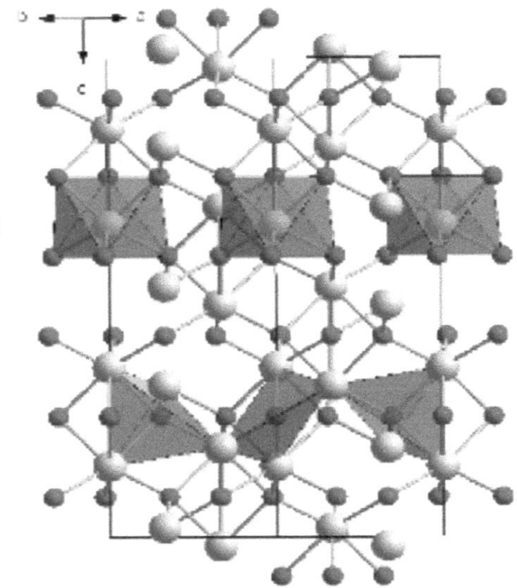 It has rhombohedral corundum type structure at ambient temperature, in which the vanadium atoms form 3D V–V chains, and the oxygen atoms around the vanadium sites form distorted octahedron.

(continued)

TABLE 2.5 (Continued)
Vanadium oxides

Oxidation States	Compound	Preparation, Properties and Applications	Structure
IV	Vanadium (IV) oxide known as vanadium dioxide (VO_2) [96]	(i) It is prepared by the reaction of V_2O_3 and V_2O_5: $V_2O_3 + V_2O_5 \rightarrow 4VO_2$ (ii) It slowly oxidizes in air. (iii) In acidic medium it dissolves to give VO^{2+} ions and in hot alkaline medium it gives $HV_2O_5^-$ (iv) In air, it gradually converts to V_2O_4(Indigo blue). (v) Naturally it occurs as one of the rare minerals *karelianite*.	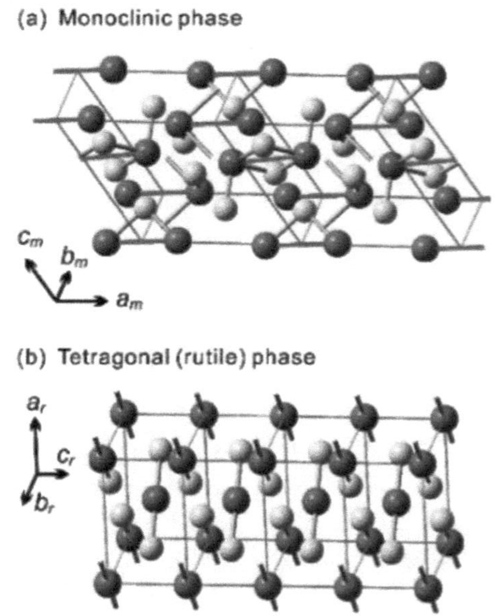 At $T_c < 340$ K (67°C), it has a monoclinic structure. Here the V^{4+} ions form pairs along the c-axis, leading to alternate short and long V–V distances of 2.65 Å and 3.12 Å respectively. At $T_c > 340$ K (67°C), it has a rutile structure. Here the V^{4+} ions are separated by a fixed distance of 2.96 Å

V	Vanadium (V) Oxide known as Vanadium pentoxide (V_2O_5) [97]	(i) It is a brownish yellow solid in appearance. (ii) It is prepared by decomposition of metavanadate at 500–550°C. $NH_4VO_3 \rightarrow V_2O_5 + 2NH_3 + H_2O$ (iii) It acts both as amphoteric oxide and oxidizing agent due to having high oxidation state. (iv) It has a wide range of industrial applications where it acts as catalyst, viz. in the production of maleic anhydride and phthalic anhydride which are potential agents for polymer industries. (v) It is predominantly used in production of ferrovanadium. (alloy formation). (vi) In sulfuric acid production, it acts as a very potential catalyst. It catalyzes the mildly exothermic oxidation of SO_2 to SO_3 by air (a well-known crucial reaction) $2SO_2 + O_2 \Leftrightarrow SO_3$ (vii) Today, sulfuric acid has become the cheap commodity chemical because of the effective catalytic role of V_2O_5. The reaction is carried out between 400–620°C. V_2O_5 acts as inactive catalyst below 400°C and it begins to break down above 620°C. Prompted by the reaction where reduction of V_2O_5 to VO_2 takes place via SO_2, following catalytic cycle is as: $SO_2 + V_2O_5 \rightarrow SO_3 + 2VO_2$ $2VO_2 + ½O_2 \rightarrow V_2O_5$ (viii) In power plants and diesel engines, it is used as catalyst in the selective catalytic reduction (SCR) of nitrogen oxide emissions.	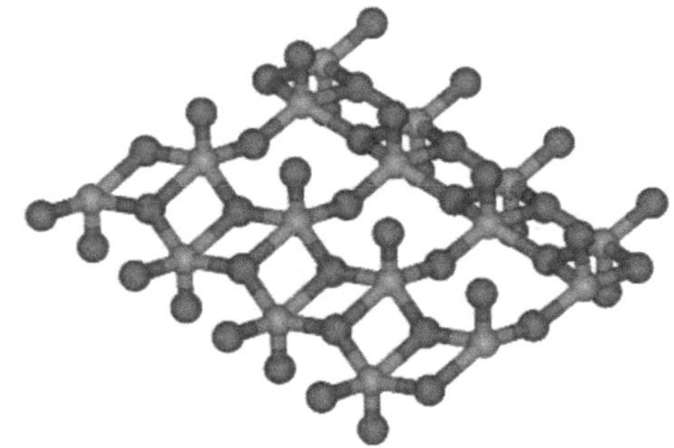 It possesses orthorhombic layered crystal structure. The distorted VO_5 square pyramids are connected by sharing edges and corners with periodic up–up–down–down sequence to form the 2D structure, and the layers are stacked along c-axis with weak van der Waals interaction.

2.4.6 Vanadates

Sodium, potassium, and ammonium metavanadates are of commercial importance. Three forms of vanadate—(i) meta vanadate (MVO_3), (ii) ortho vanadate (M_3VO_4), and (iii) pyro vanadate ($M_4V_2O_7$)—are identified. Their nomenclature resembles phosphates. Discrete tetrahedral $(VO_4)^{3-}$ and $(V2O_7)^{4-}$ anions are present in ortho and pyro vanadates, respectively. Metavanadates show different structures for the anhydrous and hydrated salts. In the anhydrous salts, viz. KVO_3 and $NH4VO_3$, the vanadium atoms are four-coordinate, with VO_4 tetrahedra linking through two oxygen atoms. In the hydrated form, $KVO_3.H_2O$, the vanadium atoms is five coordinated, with three shared oxygen atoms per vanadium and two terminal oxygen atoms.

2.4.7 Other Compounds [87]

Other remaining known compounds of vanadium, for example, vanadium selenides, tellurides,-nitrides, phosphides, carbides, silicides, borides, and phosphates are very well documented.

2.5 COORDINATION CHEMISTRY OF VANADIUM [87]

The formation of coordination complexes by vanadium is based on its variable oxidation states, which ranges from –1 to +5. Earlier reports discuss about the large number of vanadium coordination complexes formed by its higher oxidation states. Vanadium coordination complexes have biological significance where it plays important role in both oxidative and reductive catalytic transformations in biological systems. Reports on vanadium porphyrin complexes present in petroleum, shale deposits, bitumen, and coal are also available. The synthesis and reactions of some select coordination complexes of vanadium, based on the oxidation states of vanadium, are discussed here.

2.5.1 Coordination Complexes with Low Oxidation States of Vanadium (V^0 and V^{-1}) [87]

Stable complexes of vanadium, with low oxidation states, are known. Their stability is governed by some selective ligands such as bipyridyls, 1,10-phenanthroline, nitric oxide, and tert-phosphines, which stabilize such oxidation states. The nuclear magnetic resonance (NMR) spectroscopy reveals diamagnetic behavior of these complexes based on the electronic structures of vanadium (I) and vanadium (–I) complexes. V^0 complexes show paramagnetic behavior based on EPR spectroscopic studies. Some examples of vanadium complexes with V^0 and V^{-1} oxidation states are discussed here.

V^{-1} complexes via reduction process:

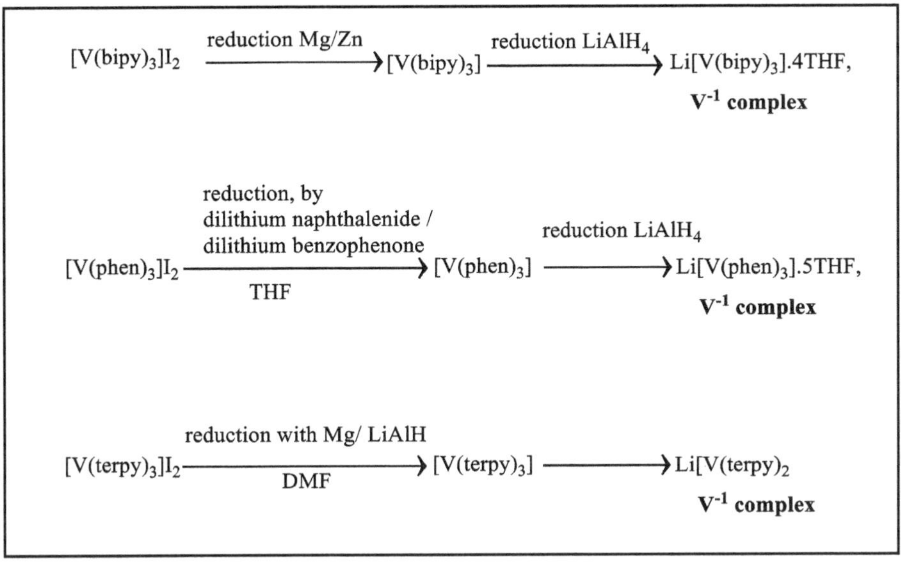

V^{-1} complexes via stabilization:

X-ray structures of trans-$[V(N_2)_2(Ph_2PCH_2CH_2PPh_2)_2]^-$ and *cis*-$[V(N_2)_2\{(Ph_2P)(CH_2CH_2 P Ph_2)_3\}]^-$ ions stabilized by tertiary phosphines along with dinitrogen ligands are reported. Some unstable mono dinitrogen compounds and complexes with monodentate phosphines have also been obtained. Nitrosyl ligand stabilized complexes $[V(NO)_2L_2]$ [CN] (L = bipy or phen) is also reported . Nitrosyl ligands, like dinitrogen, are good π-acceptors that serve to remove electron density from electron-rich metals.

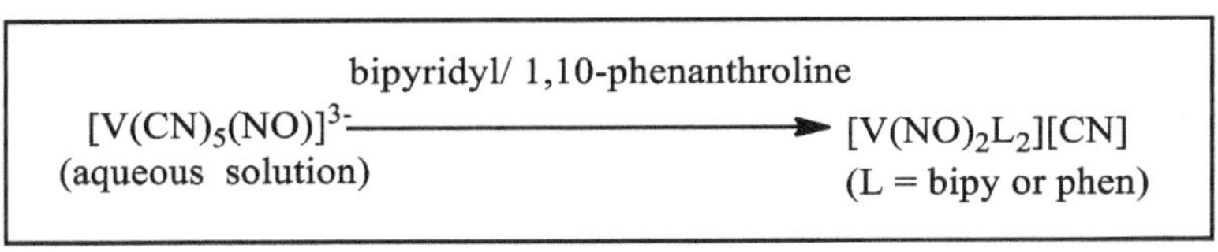

V^0 Complexes

The vanadium complex $[V(bpy)_3]$ has been studied by EPR spectroscopy, which demonstrated the extensive delocalization of single unpaired electron on the bipyridyl ligands. [V(PF3)6] (a V0 complex) has been prepared by the oxidation of [Na(diglyme)2][V(PF3)6] with $[C_6H_5N_2]BF_4$ (benzene diazonium tetrafluoroborate).

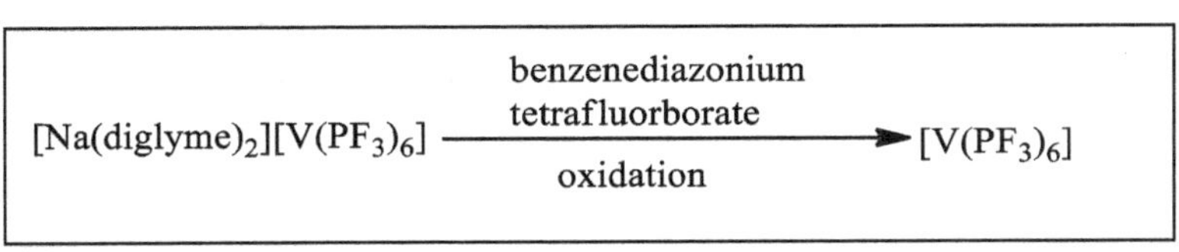

V^I Complexes

Few nitrosyl complex compounds have been prepared where vanadium center is V^I. The co-ligand involved in these complexes is triethanolamine (tea). For example, structurally characterized $[V(NO)(tea)]^-$ shows ΨTBP (distorted trigonal-bipyramidal) environment where amine and nitrosyl nitrogen donors occupy the axial positions.

2.5.2 Coordination Complexes with Vanadium (II) [87]

Vanadium II complexes have been well documented by different research groups. Though their synthesis and characterization were hampered earlier, due to air- and moisture-sensitive behavior, but today with the advancement of various techniques of handling moisture- and air-sensitive compounds and studies on compatible O, N, S, P, and halide ligation systems, various complexes are reported. Reports on sulfate salts of vanadium, viz. $VSO_4{\cdot}6H_2O$, Tutton's salt $[M_2\{V(H_2O)_6\}(SO_4)_2]$ (M = NH_4, K, Rb, or Cs) are available.

Vanadium complexes with compatible O, N, S, P and halide ligation systems
Vanadium complexes with oxygen ligands are known and reported by various research groups. They have vast chemistry and are of interest due to their structural and complexing properties. The related compounds are listed in Table 2.6.

TABLE 2.6
Vanadium (II) complexes

Ligand	Examples
Oxygen	$[V(py)_3(C_6H_3Ph_2\text{-}2,6)_2]$
	$[V(C_6H_3Pr^i_2-2,6)_4\{Li(thf)\}_2]$
Nitrogen and Phosphorous	*trans*-$[VCl_2(Me_2NCH_2CH_2NMe_2)_2]$
	$[V\{(SiMe_3)NC(Ph)N(SiMe_3)\}_2(thf)_2]$
	$[VX_2(R_2PCH_2CH_2PR_2)_2]$ (X = halide or BH_4^-, R = Me or Et)
	$[\{V(PMe_3)_3\}2(\mu\text{-}H)_3]^+$
	$[VX_2L_4]$ and $[VL_6]X_2$ (L = heterocyclic amine, e.g. pyrazole, imidazole, 2-methylimidazole, benzimidazole, isoquinoline; X = anionic ligand)
	$[V(terpy)(bipy)L]^{n+}$
	$[V^{II}Y_2L]$ (Y = thf, $PPhMe_2$).
Others	$[L_3V(\mu\text{-}Cl)_3VL_3]$ (L = thf or 3-Me-thf)
	$[L_3V(\mu\text{-}Cl)_3VL_2(PPh_3)]$
	trans-$[VX_2L_4]$ (X= Cl, Br, I, SCN, N_3, PhS, EtS or BH_4; L = py)
	triangulo-$[V_3(\mu\text{-}Cl)_3(\mu^3\text{-}Cl)L_3]$+ (L = MeOH, EtOH, etc.)
	tetranuclear species $[V_4(\mu^3\text{-}Cl)_2(\mu\text{-}Cl)_2(\mu\text{-}CF_3CO_2)_2(thf)_6]$
	$[V(2\text{-}SC_6H_4SCH_2CH_2SCH_2CH_2C_6H_4S\text{-}2)(Me_2NCH_2CH_2NMe_2)]$.
	$[\{V(9[ane]S_3)\}_2(\mu\text{-}Cl)_3]Cl$, $[VI_2(thf)(9[ane]S_3)]$

TABLE 2.7
Vanadium (III) coordination complexes [98]

Ligand	Examples
Oxygen	• $[V(H_2O)_6]^{3+}$, • $[VCl(OC_6H_3Pr^i_2\text{-}2,6)_2(thf)_2]$ • $[\{V(OPh)_2(thf)\}_2(\mu\text{-}OPh)_2]$. • $[\{VCl_2[CH_3C(O)OC_2H_5]_2\}_2(\mu\text{-}Cl)_2]$. • $[V(acac)_3]$, $[V(oxalate)_2(H_2O)_2]^-$, $[\{V(catecholate)_2(L)\}_2(\mu\text{-}O)]$.
Nitrogen	• $[VX_3(RCN)_3]$ (X = halide; R = Me, Et. or Pr.) and $[VCl_3L_3]$ (L= pyridine and γ -picoline). • $[VCl_3(MeCN)_3]$ (*fac*- and *mer* isomers in solution) • $[VCl_3(Py)_3]$ (*mer*-structure in the solid state). • $[\{V[Me_3SiN(CH_2CH_2NSiMe_3)_2]\}_2(\mu\text{-}H)_2]$ and $[\{V[CH_3C(CH_2NPr\textit{i})_3]\}_2(\mu\text{-}N_2)]$. • $[(C_6H_5NCH_2CH_2)_3N]V$. • $(TTP)V^{III}Cl.THF\ (2C_6H_6)$. • $[V_2L(H_2O)_4][C1O_4]_2C1_2$, $[V_2L(H_2O)_4][ClO_4].H_2O$, $[V_2L(H_2O)_4][SO_4].3H_2O$
Sulfur	Refer Chapter 3 for compounds containing sulfur ligands.
Others	$[\{VCl_3\}_2(\mu\text{-}Cl)_3]_3^-$ and $[V_3OCl_4\{O_2C(C_6H_4SO_2^{2-})\}5]-$.

2.5.3 Coordination Complexes with Vanadium (III) [87]

For these coordination complexes, VCl_3 (a well-known vanadium (III) chloride) acts as a key source. The electrolytic reduction of V^{IV} and V^{V} solutions also forms this class of compounds. These compounds are unstable in air and moisture. Vanadium (III) complexes, with monodentate-, bidentate, and multidentate ligands having N, O, and S donor atoms, are synthesized and structurally characterized. Their geometrical, spectral, and biological catalytical studies and toxic effects are very well documented. The related compounds are listed in Table 2.7.

2.5.4 Coordination Complexes with Vanadium (IV) [87]

This is the most important stable oxidation state of vanadium. It is obtained by both V^{III} and V^{V} ions by the following general reaction.

$$V(III) \xrightarrow{\text{air oxidized}} V(IV) \xleftarrow{\text{mild reducing agents}} V(V)$$

The blue $[VO(H_2O)_5]^{2+}$ (oxovanadium (IV) cation) is used as a precursor for variety of complexes and is a common species in aqueous solution and most of the salts. The vanadium chemistry depicting the dominating character of VO^{2+} ions (vanadyl ions) over aqueous chemistry of V^{IV} has been very well studied by various scientific

TABLE 2.8
Vanadium (IV) coordination complexes

Ligand	Examples
Oxygen	• $[VO(acac)_2]$ • [VO(salicylaldimine)] and VO(phthalocyanine) • [VO(SALIMH)(acac)(MeOH)], • [VO(SALIMH)SAL], and [VO(SALIMH)2(EtOH)]. • $Na[V_2O_3(DL\text{-salser})_2]\cdot 5H_2O$, $[V_2O_3(S\text{-peida})_2]^-$, $[V_2O_3(NTA)_2]^{3-}$, $[V_2O_3(pmida)_2]^-$ and $[VO\{salen(5\text{-Me})_2\}_2]$ $[ClO_4]$. • $[PPh_4]_2[(VO)_2Cl_4(OR)_2]$ (R= CH_3 or CH_2CH_2Cl) and $[(VO)_2Cl_2$ MeC (CH_2OH) $(CH_2O)_2]$.
Nitrogen	• $[V(NMe_2)_4]$ • $V(NMe_2)_2(NHMe_2)_2(= NAr)]$ • $V(NMe_2)_2Cl_2(=NAr)]$ • $[VPX_2]$ (P = OEP, TPP, TMTP, TPTP; X =Cl or Br).
Others	• $[VX_4L_2]$ (X= halide; L = O-, S-, Se-, N-, P or As-donor ligand) • $[VCl_6]^{2-}$ • $[VOX_2L_2]$,$[VOF_4]^{2-}$ • $[VS(S_2CNR_2)_2]$,$[VX_2(S_2CNR_2)_2]$(X= halide; R = alkyl) • $[V(odt)_2]$, $[V(NCMe)_6][VCl_2(odt)]_2$, • $[VCl_2(NSSN)]$ (NSSN = 1,6-bis(*o*-pyridyl) -2,4, dithiahexane)

groups. The coordination chemistry of vanadium(IV) halides, oxovanadium (IV) compounds, and non-oxovanadium (IV) compounds have been reviewed and discussed. The mixed valence $[V^{IV}\text{–O–}V_V]^{3+}$ bridge is a common unit. It is found in complexes, viz. $Na[V_2O_3$ $(DL\text{-salser})_2]\cdot 5H_2O$, $[V_2O_3(S\text{-peida})_2]^-$, $[V_2O_3(NTA)_2]^{3-}$, $[V_2O_3(pmida)_2]^-$, and $[VO\{salen(5\text{-Me})_2\}_2][ClO_4]$. Some examples are listed in Table 2.8.

2.5.5 Coordination Complexes with Vanadium (V) [87]

The study of these complexes is very well documented and established. Their varying behavior in solvents, concentration, structures, spectroscopic studies, and many other conditions have been well studied. Some select complexes are listed in Table 2.9.

2.6 CONCLUDING REMARKS

Vanadium and its compounds have been known for ages. It is one of the trace metals found on earth. This trace metal has various applications, which have been documented by various scientific and non-scientific communities. Aligned with the aim of this book, this chapter is attributed to the basic chemistry of vanadium, which marks as an essential part for every learner working in this field. This chapter is compiled in a way where one can get information about vanadium from its occurrence to chemistry of its compound based on variable oxidation states.

TABLE 2.9
Vanadium (V) coordination complexes [87]

Ligand	Types of Complexes	Examples
Oxygen	• Neutral and cationic oxo-complexes containing VO^{3+} unit	• VOX_3 (X = F, Cl, Br), $[VO(NO_3)_3]$, • $[VOCl_2(OR)]$, $[VOCl(OR)_2]$, $[VO(OR)_3]$, $[VO(OC_6H_3Pr^i_2\text{-}2,6)_3]$, • $VO\{N(CH_2CH_2O)_3\}]$ • $[VO\{salen(5\text{-}X)\}]^+$ (X = OMe, Me, or Br). • $[VOClL_2]$(L = benzohydroxamate or *N*-phenylbenzohydroxamate) • $[(VOL)_2]$ (L = alkoxy substituents), $[\{VOL\}_2(\mu\text{-}F)]^+$,[(VOLL`)($\mu$-O)(VOL`L``)] (L Schiff base,or similar ligand; L` and L``= monodentate ligands)
	• Anionic oxo-complexes containing VO^{3+} unit[1]	• $[VOCl_4]^-$, $[VOCl_5]^{2-}$
	• Oxo-complexes containing the VO^{2+} unit.	• $[VO_2(NO_3)]$ • $[PPh_3Me][VCl_2O_2]$ • $[VO_2L]$ (L = Schiff base with NNO- or NO_2-atom ligation). • $[L_2V_2O_4(\mu\text{-}O)]$ (L = 1,4,7-trimethyl-1,4,7-triazacyclononane). • $[\{VO_2Ad\}_2]^{2-}$ (Ad = adenosine).
	• Peroxo -complexes	• $[VO_2(O_2)_2]^{3-}$ • $[V(O_2)]^{3+}$ • $KH_2[VO_2(O_2)_2]\cdot H_2O$, $(NH_4)_2H[VO_2(O_2)_2]\cdot nH_2O$. • $[VO(O_2)(L)]\_$ [L = tridentate, tetradentate or $(bidentate)_2$ ligand], $[VO(O_2)_2(L)]^{n-}$ [L = monodentate or 1/2(bidentate ligand)], $[V(O_2)_3(L)]^{2-}$, $[V(O_2)_4]^{3-}$.
Nitrogen		• [VN(acac)(1,4,7-triazacyclononane)]$^+$ • [V(NR)(L)] [R = alkyl or aryl; L =$N(CH_2CH_2S)_3^{3-}$ or $N(CH_2CH_2NR`)_3^{3-}$], • $[V\{N(CH_2CH_2S)_3\}(NNR_2)]$ (R_2 = CPh_2 or C_5H_{10}) • $[\{V[Me_3SiN(CH_2CH_2NSiMe_3)_2]\}_2(\mu\text{-}N)_2]$.
Others		• $[V(N_3S_2)L(L`)]$ (L` = bipy or phen; L = cat or dtbc) • $[V(dtbc)_2L`][SbF_6]$. • $[VS_2(S_2)SPh]^{2-}$ • $[NEt_4][VO(S_2)_2(bipy)]$. • $[V(=Y)\{N(CH_2CH_2NR`)_3\}]$ (Y=S or Se, R` = $SiMe_3$) • $[V(Se)\{Se(SiMe_3)_3\}N(SiMe_3)_2]$

REFERENCES

[1] Weeks, M. E. (1932). The discovery of the elements. VII. Columbium, tantalum, and vanadium. *Journal of Chemical Education*, *9*(5), 863. doi:10.1021/ed009p863

[2] Mukherjee, B., Patra, B., Mahapatra, S., Banerjee, P., Tiwari, A., & Chatterjee, M. (2004). Vanadium – an element of atypical biological significance. *Toxicology Letters*, *150*(2), 135–143. doi:10.1016/j.toxlet.2004.01.009

[3] Awan, R. S., Liu, C., Yang, S., Wu, Y., Zang, Q., Khan, A., & Li, G. (2021). The occurrence of vanadium in nature: Its biogeochemical cycling and relationship with organic matter—a case study of the Early Cambrian black rocks of the Niutitang Formation, western Hunan, China. *Acta Geochimica*, *40*(6), 973–997. doi:10.1007/s11631-021-00482-2

[4] Lide, D.R. (2008) *CRC Handbook of Chemistry and Physics*. CRC Press, Boca Raton, FL.

[5] Awan, R. S., Liu, C., Yang, S., Wu, Y., Zang, Q., Khan, A., & Li, G. (2021). The occurrence of vanadium in nature: its biogeochemical cycling and relationship with organic matter—a case study of the Early Cambrian black rocks of the Niutitang Formation, western Hunan, China. *Acta Geochimica*, *40*(6), 973–997. doi:10.1007/s11631-021-00482-2

[6] Schlesinger, W. H., Klein, E. M., & Vengosh, A. (2017). Global biogeochemical cycle of vanadium. *Proceedings of the National Academy of Sciences of the United States of America*, *114*(52), E11092–E11100. doi:10.1073/pnas.1715500114

[7] (a) Hope, B. K. (1994). A global biogeochemical budget for vanadium. *The Science of the Total Environment*, *141*(1–3), 1–10. doi:10.1016/0048-9697(94)90012-4. (b) Hope, B. K. (2008). A dynamic model for the global cycling of anthropogenic vanadium. *Global Biogeochemical Cycles*, *22*(4). doi:10.1029/2008gb003283

[8] Monakhov, I. N., Khromov, S. V., Chernousov, P. I., & Yusfin, Y. S. (2004). The flow of vanadium-bearing materials in industry. *Metallurgist (USSR)*, *48*(7/8), 381–385. doi:10.1023/b:mell.0000048420.68839.2a

[9] Watt, J. A. J., Burke, I. T., Edwards, R. A., Malcolm, H. M., Mayes, W. M., Olszewska, J. P., … Spears, B. M. (2018). Vanadium: A re-emerging environmental hazard. *Environmental Science & Technology*, *52*(21), 11973–11974. doi:10.1021/acs.est.8b05560

[10] Whitfield, M., & Turner, D. R. (1979). Water–rock partition coefficients and the composition of seawater and river water. *Nature*, *278*(5700), 132–137. doi:10.1038/278132a0.

[11] Gustafsson, J. P. (2019). Vanadium geochemistry in the biogeosphere –speciation, solid-solution interactions, and ecotoxicity. *Applied Geochemistry: Journal of the International Association of Geochemistry and Cosmochemistry*, *102*, 1–25. doi:10.1016/j.apgeochem.2018.12.027

[12] Wedepohl, K. H. (1971). Environmental influences on the chemical composition of shales and clays. *Physics and Chemistry of the Earth*, *8*, 307–333. doi:10.1016/0079-1946(71)90020-6

[13] Wedepohl, K.H. (1991). The composition of the upper earth's crust and the natural cycles of selected metals. Metal in natural raw materials, natural resources. *Metals and Their Compounds in the Environment. Occurrence, Analysis and Biological Relevance*, Wiley, Hoboken, NJ, pp. 3–17.

[14] Tribovillard, N., Algeo, T. J., Lyons, T., & Riboulleau, A. (2006). Trace metals as paleoredox and paleoproductivity proxies: An update. *Chemical Geology*, *232*(1–2), 12–32. doi:10.1016/j.chemgeo.2006.02.012

[15] Wehrli, B., & Stumm, W. (1989). Vanadyl in natural waters: Adsorption and hydrolysis promote oxygenation. *Geochimica et Cosmochimica Acta*, *53*(1), 69–77. doi:10.1016/0016-7037(89)90273-1

[16] Sutradhar, M., Da Silva, J. A. L., & Pombeiro, A. J. L. (2020). Introduction: vanadium, its compounds and applications. In *RSC Catalysis Series. Vanadium Catalysis*. Royal Society of Chemistry, pp. 1–11. doi:10.1039/9781839160882-00001

[17] Treibs, A. (1934). Chlorophyll- und Häminderivate in bituminösen Gesteinen, Erdölen, Erdwachsen und Asphalten. Ein Beitrag zur Entstehung des Erdöls. *Justus Liebigs Annalen Der Chemie*, *510*(1), 42–62. doi:10.1002/jlac.19345100103

[18] Treibs, A. (1936). Chlorophyll- und Häminderivate in organischen Mineralstoffen. *Angewandte Chemie (Weinheim an der Bergstrasse, Germany)*, *49*(38), 682–686. doi:10.1002/ange.19360493803

[19] Barwise, A. J. G., & Roberts, I. (1984). Diagenetic and catagenetic pathways for porphyrins in sediments. *Organic Geochemistry*, *6*, 167–176. doi:10.1016/0146-6380(84)90038-x

[20] Louda, J. W., Li, J., Liu, L., Winfree, M. N., & Baker, E. W. (1998). Chlorophyll—a degradation during cellular senescence and death. *Organic Geochemistry*, *29*(5–7), 1233–1251. doi:10.1016/s0146-6380(98)00186-7

[21] Lipiner, G., Willner, I., & Aizenshtat, Z. (1988). Correlation between geochemical environments and controlling factors in the metallation of porphyrins. *Organic Geochemistry*, *13*(4–6), 747–756. doi:10.1016/0146-6380(88)90097-6

[22] Premovic, P. (2000). Estimation of vanadyl porphyrin concentration in sedimentary kerogens and asphaltenes. *Fuel (London, England)*, *79*(7), 813–819. doi:10.1016/s0016-2361(99)00205-7

[23] Premovic, P. (2002). Vanadium of petroleum asphaltenes and source kerogens (La Luna Formation, Venezuela): isotopic study and origin. *Fuel (London, England)*, *81*(15), 2009–2016. doi:10.1016/s0016-2361(02)00104-7

[24] Filby, R. H. (1975). In: Yen, T.F. (ed)*The Nature of Metals in Petroleum*. Ann Arbor Science Publ., p. 31.

[25] Filby, R. H. (1994). Origin and nature of trace element species in crude oils, bitumens and kerogens: implications for correlation and other geochemical studies. *Geological Society, London, Special Publications*, *78*(1), 203–219.

[26] Fish, R. H., & Komlenic, J. J. (1984). Molecular characterization and profile identifications of vanadyl compounds in heavy crude petroleum by liquid chromatography/graphite furnace atomic absorption spectrometry. *Analytical Chemistry*, *56*(3), 510–517. doi:10.1021/ac00267a048

[27] Fish, R. H., Komlenic, J. J., & Wines, B. K. (1984). Characterization and comparison of vanadyl and nickel compounds in heavy crude petroleum and asphaltenes by reverse-phase and size-exclusion liquid chromatography/graphite furnace atomic absorption spectrometry. *Analytical Chemistry*, *56*(13), 2452–2460. doi:10.1021/ac00277a043

[28] Márquez, N., Ysambertt, F., & De La Cruz, C. (1999). Three analytical methods to isolate and characterize vanadium and nickel porphyrins from heavy crude oil. *Analytica Chimica Acta*, *395*(3), 343–349. doi:10.1016/s0003-2670(99)00304-9

[29] Varadaraj, R., & Brons, C. (2007). Molecular origins of heavy crude oil interfacial activity. Part 2: Fundamental interfacial properties of model naphthenic acids and naphthenic acids separated from heavy crude oils. *Energy & Fuels: An American Chemical Society Journal*, *21*(1), 199–204. doi:10.1021/ef060425s

[30] Lewan, M. D., & Maynard, J. B. (1982). Factors controlling enrichment of vanadium and nickel in the bitumen of organic sedimentary rocks. *Geochimica et Cosmochimica Acta*, *46*(12), 2547–2560. doi:10.1016/0016-7037(82)90377-5

[31] Yen, T. F. (1975). Chemical aspects of metals in native petroleum. *Trace Metals in Petroleum*. Ann Arbor Science Publishers, 221 pp.

[32] Barwise, A. J. G. (1990). Role of nickel and vanadium in petroleum classification. *Energy & Fuels: An American Chemical Society Journal*, *4*(6), 647–652. doi:10.1021/ef00024a005

[33] Comstock, M. J. (Ed.). (1987, July 6). Metal complexes in fossil fuels, copyright, ACS symposium series, foreword. *ACS Symposium Series*, pp. i–vi. doi:10.1021/bk-1987-0344.fw001

[34] Macara, I. G., McLeod, G. C., & Kustin, K. (1979). Tunichromes and metal ion accumulation in tunicate blood cells. *Comparative Biochemistry and Physiology. B, Comparative Biochemistry*, *63*(3), 299–302. doi:10.1016/0305-0491(79)90252-9

[35] Oltz, E. M., Bruening, R. C., Smith, M. J., Kustin, K., & Nakanishi, K. (1988). The Tunichromes. A class of reducing blood pigments from sea squirts: isolation, structures, and vanadium chemistry. *Journal of the American Chemical Society*, *110*(18), 6162–6172. doi:10.1021/ja00226a035

[36] Boucher, L. J., & Yen, T. F. (1968). Spectral properties of oxovanadium(IV) complexes. II. Bis trifluoro acetylacetone ethylenediimine. *Inorganic Chemistry*, *7*(12), 2665–2667. doi:10.1021/ic50070a049

[37] Ventura, G. T., Gall, L., Siebert, C., Prytulak, J., Szatmari, P., Hürlimann, M., & Halliday, A. N. (2015). The stable isotope composition of vanadium, nickel, and molybdenum in crude oils. *Applied Geochemistry: Journal of the International Association of Geochemistry and Cosmochemistry*, *59*, 104–117. doi:10.1016/j.apgeochem.2015.04.009

[38] Nielsen, S.G., Prytulak, J., & Halliday, A.N. (2011). Determination of high precision 51V/50V isotope ratios by MC-ICP-MS, Part 1: chemical separation of vanadium and mass spectrometric protocols. *Geostandards and Geoanalytical Research*, *35*, 293–306.

[39] Gao, P., He, Z., Li, S., Lash, G.G., Li. B., Huang, B., & Yan, D. (2018) Volcanic and hydrothermal activities recorded in phosphate nodules from the Lower Cambrian Niutitang Formation black shales in South China. Palaeogeography, *Palaeoclimatology*, Palaeoecology, *505*, 381–397.

[40] Wu, C., Tuo, J., Zhang, M., Sun, L., Qian, Y., & Liu, Y. (2016) Sedimentary and residual gas geochemical characteristics of the Lower Cambrian organic-rich shales in Southeastern Chongqing. *Marine and Petroleum Geology*, *75*, 140–150.

[41] Schuth, S., Horn, I., Brüske, A., Wolff, P. E., & Weyer, S. (2017). First vanadium isotope analyses of V-rich minerals by femtosecond laser ablation and solution-nebulization MC-ICP-MS. *Ore Geology Reviews*, *81*, 1271–1286. doi:10.1016/j.oregeorev.2016.09.028

[42] Wu, F., Qi, Y., Perfit, M. R., Gao, Y., Langmuir, C. H., Wanless, V. D., … Huang, F. (2018). Vanadium isotope compositions of mid-ocean ridge lavas and altered oceanic crust. *Earth and Planetary Science Letters*, *493*, 128–139. doi:10.1016/j.epsl.2018.04.009

[43] Manning, L. K., Frost, C. D., & Branthaver, J. F. (1991). A neodymium isotopic study of crude oils and source rocks: potential applications for petroleum exploration. *Chemical Geology*, *91*(2), 125–138. doi:10.1016/0009-2541(91)90086-7

[44] Selby, D., & Creaser, R. A. (2005). Direct radiometric dating of hydrocarbon deposits using rhenium-osmium isotopes. *Science (New York, N.Y.)*, *308*(5726), 1293–1295. doi:10.1126/science.1111081

[45] Aleshin, G., Altukhova, Z., Antipenko, V., Marchenko, S., & Kamyanov, V. (1984). Distribution of vanadium and vanadyl porphyrins in petroleum distillates of different chemical types. *Petroleum Chemistry U.S.S.R*, *24*(4), 191–195. doi:10.1016/0031-6458(84)90099-6

[46] Amorim, F. A. C., Welz, B., Costa, A. C. S., Lepri, F. G., Vale, M. G. R., & Ferreira, S. L. C. (2007). Determination of vanadium in petroleum and petroleum products using atomic spectrometric techniques. *Talanta*, *72*(2), 349–359. doi:10.1016/j.talanta.2006.12.01

[47] Lewan, M. D. (1984). Factors controlling the proportionality of vanadium to nickel in crude oils. *Geochimica et Cosmochimica Acta*, *48*(11), 2231–2238. doi:10.1016/0016-7037(84)90219-9

[48] Awan, R.S., Liu, C., Gong, H., Dun, C., Tong, C., & Chamssidini, L.G. (2020) Paleo-sedimentary environment in relation to enrichment of organic matter of Early Cambrian black rocks of Niutitang formation from Xiangxi area China. *Marine and Petroleum Geology*, *112*, 104057.

[49] Larsson, M.A. (2014) Vanadium in soils: Chemistry and cytotoxicity. Doctoral Thesis, Swedish University of Agricultural Sciences, Uppsala; ISBN (electronic version) 978-91-576-8153-9.

[50] Aydin, F., Saydut, A., Gunduz, B., Aydin, I., & Hamamci, C. (2012) Chemical speciation of vanadium in coal bottom ash. *Clean-Soil Air Water, 40*, 444–448.

[51] Nriagu, J.O. (1998) History, occurrence, and uses of vanadium. In: Nriagu, J.O., ed., *Vanadium in the Environment. Part 1: Chemistry and Biochemistry*, John Wiley & Sons, Inc., pp. 1–24.

[52] Gehring, A. U., Fry, I. V., Luster, J., & Sposito, G. (1993). The chemical form of Vanadium (IV) in kaolinite. *Clays and Clay Minerals, 41*(6), 662–667. doi:10.1346/ccmn.1993.0410604

[53] Gehring, A. U., Fry, I. V., Luster, J., & Sposito, G. (1994). Vanadium in sepiolite: A redox-indicator for an ancient closed brine System in the Madrid Basin, central Spain. *Geochimica et Cosmochimica Acta, 58*(16), 3345–3351. doi:10.1016/0016-7037(94)90090-6

[54] Bauer Boechat, C., Eon, J.-G., Malta Rossi, A., Andre´ de Castro Perez, C., & Aguiar da Silva San Gil, R. (2000). Structure of vanadate in calcium phosphate and vanadate apatite solid solutions. *Physical Chemistry Chemical Physics: PCCP, 2*(18), 4225–4230. doi:10.1039/b004339g

[55] Lu, X., Johnson, W. D., & Hook, J. (1998). Reaction of vanadate with aquatic humic substances: An ESR and 51V NMR study. *Environmental Science & Technology, 32*(15), 2257–2263. doi:10.1021/es970930r

[56] Venkataraman, B.V., & Sudha, S. (2005) Vanadium toxicity. Asian Journal *of* Experimental *Sciences, 19*, 127–134.

[57] Kabata-Pendias, A., & Pendias, H. (1979) Trace elements in the biological environment. In: *Warsaw Yard Geology*. CRC Press, Inc. ISBN 0-8493-6639-9

[58] Taner, M. F. (2002) Vanadium-geology, processing and applications. In: *Proceedings of International Symposium on Vanadium*. Canadian Institute of Mining, Metallurgy and Petroleum, p 265.

[59] Evans, H. T., & White, J. S. (1987). The colorful vanadium minerals: a brief review and a new classification. *The Mineralogical Record, 18*, 333–340.

[60] Kabata-Pendias, A., & Pendias, H. (1993) *Biochemistry of Trace Elements*. PWN, Warsaw.

[61] Dong, Y., Lin, H., Zhao, Y., & Gueret Yadiberet Menzembere, E. R. (2021). Remediation of vanadium-contaminated soils by the combination of natural clay mineral and humic acid. *Journal of Cleaner Production, 279*(123874), 123874. doi:10.1016/j.jclepro.2020.123874

[62] Zhang, B. C., Chen, K., Baddour, N., & Patnaik, P. C. (2019). Failure and life evaluation of EB-PVD thermal barrier coatings using temperature-process-dependent model parameters. *Corrosion Science, 156*, 1–9. doi:10.1016/j.corsci.2019.04.020

[63] Yan, B., Wang, D., Wu, L., & Dong, Y. (2018). A novel approach for pre-concentrating vanadium from stone coal ore. *Minerals Engineering, 125*, 231–238. doi:10.1016/j.mineng.2018.06.005

[64] Dong, Y.-B., Liu, Y., & Lin, H. (2018). Leaching behavior of V, Pb, Cd, Cr, and As from stone coal waste rock with different particle sizes. *International Journal of Minerals Metallurgy and Materials, 25*(8), 861–870. doi:10.1007/s12613-018-1635-2

[65] Shacklette, H. T., & Boerngen, J. G. (1984). *Element Concentrations in Soils and Other Surficial Materials of the Conterminous United States* (Vol. 1270). Washington, DC: US Government Printing Office.

[66] Cappuyns, V., & Slabbinck, E. (2012). Occurrence of vanadium in Belgian and European alluvial soils. *Applied and Environmental Soil Science, 2012*, 1–12. doi:10.1155/2012/979501

[67] Tribovillard, N., Algeo, T. J., Lyons, T., & Riboulleau, A. (2006). Trace metals as paleoredox and paleoproductivity proxies: An update. *Chemical Geology*, *232*(1–2), 12–32. doi:10.1016/j.chemgeo.2006.02.012

[68] Chermak, J. A., & Schreiber, M. E. (2014). Mineralogy and trace element geochemistry of gas shales in the United States: Environmental implications. *International Journal of Coal Geology*, *126*, 32–44. doi:10.1016/j.coal.2013.12.005

[69] Scott, C., Slack, J. F., & Kelley, K. D. (2017). The hyper-enrichment of V and Zn in black shales of the Late Devonian-Early Mississippian Bakken Formation (USA). *Chemical Geology*, *452*, 24–33. doi:10.1016/j.chemgeo.2017.01.026

[70] Imtiaz, M., Rizwan, M. S., Xiong, S., Li, H., Ashraf, M., Shahzad, S. M., ... Tu, S. (2015). Vanadium, recent advancements and research prospects: A review. *Environment International*, *80*, 79–88. doi:10.1016/j.envint.2015.03.018

[71] Ortiz-Bernad, I., Anderson, R. T., Vrionis, H. A., & Lovley, D. R. (2004). Vanadium respiration by *Geobacter metallireducens*: Novel strategy for in situ removal of vanadium from groundwater. *Applied and Environmental Microbiology*, *70*(5), 3091–3095.

[72] Rehder, D. (1991). The bioinorganic chemistry of vanadium. *Angewandte Chemie. International Edition in English*, *30*(2), 148–167. doi:10.1002/anie.199101481

[73] Wällstedt, T., Björkvald, L., & Gustafsson, J. P. (2010). Increasing concentrations of arsenic and vanadium in (southern) Swedish streams. *Applied Geochemistry: Journal of the International Association of Geochemistry and Cosmochemistry*, *25*(8), 1162–1175. doi:10.1016/j.apgeochem.2010.05.002

[74] Caccia, V. G., & Millero, F. J. (2003). The distribution and seasonal variation of dissolved trace metals in Florida bay and adjacent waters. *Aquatic Geochemistry*, *9*(2), 111–144. doi:10.1023/b:aqua.0000019486.07923.be

[75] Emerson, S. R., & Huested, S. S. (1991). Ocean anoxia and the concentrations of molybdenum and vanadium in seawater. *Marine Chemistry*, *34*(3–4), 177–196. doi:10.1016/0304-4203(91)90002-e

[76] Wang, D., & Sañudo Wilhelmy, S. A. (2009). Vanadium speciation and cycling in coastal waters. *Marine Chemistry*, *117*(1–4), 52–58. doi:10.1016/j.marchem.2009.06.001

[77] Hope, B. K. (2008). A dynamic model for the global cycling of anthropogenic vanadium. *Global Biogeochem. Cycles*, 22, article no. GB4021. doi:10.1029/2008GB003283

[78] Boström, K., & Fisher, D. E. (1971). Volcanogenic uranium, vanadium and iron in Indian Ocean sediments. *Earth and Planetary Science Letters*, *11*(1–5), 95–98. doi:10.1016/0012-821x(71)90148-8

[79] Wright, M. T., & Belitz, K. (2010). Factors controlling the regional distribution of vanadium in groundwater. *Ground Water*, *48*(4), 515–525. doi:10.1111/j.1745-6584.2009.00666.x

[80] Licheng, Z., & Kezhun, Z. (1992). Background values of trace elements in the source area of the Yangtze River. *The Science of the Total Environment*, *125*, 391–404. doi:10.1016/0048-9697(92)90403-f

[81] Sadiq, M. (1988). Thermodynamic solubility relationships of inorganic vanadium in the marine environment. *Marine Chemistry*, *23*(1–2), 87–96. doi:10.1016/0304-4203(88)90024-2

[82] Colina, M., Gardiner, P. H. E., Rivas, Z., & Troncone, F. (2005). Determination of vanadium species in sediment, mussel and fish muscle tissue samples by liquid chromatography–inductively coupled plasma-mass spectrometry. *Analytica Chimica Acta*, *538*(1–2), 107–115. doi:10.1016/j.aca.2005.02.044

[83] Wann CC, Jiang S-J (1997) Determination of vanadium species in water samples by liquid chromatography-inductively coupled plasma mass spectrometry. *Analytica Chimica Acta*, 357:211–218.

[84] Wanty, R. B., & Goldhaber, M. B. (1992). Thermodynamics and kinetics of reactions involving vanadium in natural systems: Accumulation of vanadium in sedimentary rocks. *Geochimica et Cosmochimica Acta*, *56*(4), 1471–1483. doi:10.1016/0016-7037(92)90217-7

[85] Kustin, K., Liu, S.-, Nicolini, C., & Toppen, D. (1974). Interaction of catechol and catechol derivatives with dioxovanadium:(V). i. Kinetics of complex formation in acidic media. *Journal of the American Chemical Society*, *96*(24), 7410–7415. doi:10.1021/ja00831a600

[86] Carpentier, W., Sandra, K., De Smet, I., Brigé, A., De Smet, L., & Van Beeumen, J. (2003). Microbial reduction and precipitation of vanadium by Shewanella oneidensis. *Applied and Environmental Microbiology*, *69*(6), 3636–3639. doi:10.1128/AEM.69.6.3636-3639.2003

[87] Richards, R. L. (2011). Vanadium: inorganic and coordination Chemistry. Based in part on the article 'Vanadium: Inorganic and Coordination Chemistry by Elizabeth M. Page & Sherilyn A.' In *Encyclopedia of Inorganic and Bioinorganic Chemistry*. John Wiley & Sons, Ltd. doi:10.1002/9781119951438.eibc0236

[88] Pourret, O., Dia, A. (2018). Vanadium. In: White, W.M. (eds), *Encyclopedia of Geochemistry. Encyclopedia of Earth Sciences Series*. Springer, Cham. https://doi.org/10.1007/978-3-319-39312-4_272.

[89] Esayed, A., & Northwood, D. (1992). Metal hydrides: A review of group V transition metals—niobium, vanadium and tantalum. *International Journal of Hydrogen Energy*, *17*(1), 41–52. doi:10.1016/0360-3199(92)90220-q

[90] Inorganic Syntheses Inc. (2007). *Inorganic Syntheses: 4* (J. C. Bailar Jr, Ed.). doi:10.1002/9780470132357

[91] Bennett, M. A., Huang, T.-N., Matheson, T. W., Smith, A. K., Ittel, S., & Nickerson, W. (2007). 16. (η6-Hexamethylbenzene)ruthenium complexes. in *Inorganic Syntheses. Inorganic Syntheses* (pp. 74–78). John Wiley & Sons. doi:10.1002/9780470132524.ch16

[92] Hitchcock, P. B., Hughes, D. L., Leigh, G. J., Sanders, J. R., De Souza, J., McGarry, C. J., & Larkworthy, L. F. (1994). Preparation of new vanadium(II) iodides and crystal structure of hexakis(acetonitrile)vanadium(II)(tetraiodide). *Journal of the Chemical Society, Dalton Transactions*, *24*, 3683. doi:10.1039/dt9940003683

[93] Juza, D., Giegling, D., & Schäfer, H. (1969). über die Vanadinjodide VJ2 und VJ3. *Zeitschrift fur anorganische und allgemeine Chemie*, *366*(3–4), 121–129. doi:10.1002/zaac.19693660303

[94] Groshens, T. J., & Klabunde, K. J. (1990). Molecular vapor synthesis: the use of titanium monoxide and vanadium monoxide vapors as reagents. *Inorganic Chemistry*, *29*(16), 2979–2982. doi:10.1021/ic00341a025

[95] King, R. B. (Ed.). (1994). *Encyclopedia of Inorganic Chemistry*. Chichester: John Wiley & Sons.

[96] Yang, D.-S., Baum, P., & Zewail, A. H. (2016). Ultrafast electron crystallography of the cooperative reaction path in vanadium dioxide. *Structural Dynamics (Melville, N.Y.)*, *3*(3), 034304. doi:10.1063/1.4953370

[97] Bauer, G., Güther, V., Hess, H., Otto, A., Roidl, O., Roller, H., & Sattelberger, S. (2000). Vanadium and vanadium compounds. In *Ullmann's Encyclopedia of Industrial Chemistry*. Wiley-VCH Verlag GmbH & Co. KGaA, Weinheim. doi:10.1002/14356007.a27_367

[98] King, R. B. (Ed.). (2005). *Encyclopedia of Inorganic Chemistry* (Vol. 10, p. 6696). New York: Wiley.

3 Chemistry of Chromium

Garima Pandey, Sangeeta Bajpai, Monika Kamboj and Bimal Krishan Banik

3.1 INTRODUCTION

Chromium has a glossy, fragile, hardened character. Its coloration is metallic-gray and it may acquire a high level of polish. Notably, chromium exhibits resistance to tarnishing when exposed to air; however, when subjected to heat, it oxidizes and generates the characteristic green chromic oxide [1]. In the presence of oxygen, chromium displays instability, promptly forming a thin oxide coating. This layer, impervious to O_2, swiftly develops, safeguarding the underneath metal from further oxidation. Chromium finds primary applications in various alloys, notably stainless steel, as well as the plating of chrome and metal ceramics. Historically, Cr coating was widely employed to provide steel with a reflecting, silvery mirror finish [2].In metallurgy, chromium serves to enhance corrosion resistance and bestow a glossy finish to metals. Its salts are utilized as dyes and paints, imparting an emerald green hue to glass and facilitating the production of synthetic rubies. Additionally, chromium acts as a catalyst in the process of dyeing and contributes to leather tanning. It is also utilized in the creation of molds utilized in the firing of bricks. CrO holds significance in the manufacturing of magnetic tape [3].

Table 3.1 illustrates the atomic and physical characteristics of chromium.

3.2 SOURCES

Chromium is naturally occurring in stones, living things, dirt, volcano dust, and gases. Cr typically found in two oxidation states in the environment: the trivalent form of chromium (Cr III), which exists naturally and serves as an essential nutrient, and a hexavalent form of chromium (Cr VI), which is combined with metallic chromium (Cr 0), primarily produced through processes in industry [4]. Chromium (III) plays an important role in appropriate glucose, protein, and lipid metabolism, making it an essential dietary ingredient. The body uses numerous methods to convert chromium (VI) to chromium (III), resulting in higher amounts of the latter. Trivalent chromium makes up the majority of airborne chromium emissions, which are generally in the form of tiny particles or aerosols. Key industrial operations responsible for releasing

DOI: 10.1201/9781032615110-4

TABLE 3.1
Atomic and physical characteristics of Cr

Symbol	Cr
At. No.	24
Molecular weight	51.98
Density	7.15 g/cm^3
State at 20°C	Solid
Electronic configuration	[Ar]3d^{4}4s^1
Electronegativity according to Pauling Scale	1.66
Van der Waals radius	189 pm
Atomic radius	140pm
Covalent radius	139 pm
Common oxidation states	+6, +3, +2, 6, 5, 4, 3, 2, 1, −1, −2, −4
The affinity for electrons	0.666 eV
Ionization energy (IE)	6.767 eV
IE (1st)	651.1 kJ/mol
IE (2nd)	1590.1 kJ/mol
IE (3rd)	2987 kJ/mol
Standard electrode potential	–0.71 V (Cr^{3+}/Cr)
Melting point	2180 K
Boiling point	2944 K
Discovered by	Vaughlin in 1797

chromium into the environment include ferrochrome manufacturing, ore refinement, chemicals and refractory processing, cement manufacturing, brakes of vehicle linings and catalytic conversions, leather tanning, and chrome paints [3].

3.3 ISOLATION

Chromium is extracted from its ores by reduction. An approach involves the utilization of coke within an electric arc furnace to manufacture ferrochromium alloy. Alternatively, pure chromium is obtained by fusing the ore with lime and soda ash, followed by leaching, filtration to remove aluminum impurities, acidification, and concentration to yield sodium dichromate. Chromium metal can then be recovered through electrolytic reduction or carbon reduction from sodium dichromate. Global chromite ore production is nearly 9.5×10^{-6} tons. Residues generated from the extraction process can undergo recycling, while spent ores necessitate eventual disposal. The complication emerges as these residues, frequently utilized for landfill purposes in residential regions, add to airborne dust via wind and traffic. Moreover, they contain chromium salts that can be leached by rain, leading to issues like acid rain. Additionally, residues may contaminate water systems when leachate enters aquifers or local water sources. Such movements can trigger changes in the oxidation state [4].

TABLE 3.2
Natural occurring isotopic composition of chromium

Isotopes	Atomic Weight	Isotopic Abundance	Nuclear Spin (I)	Half-life	Radioactive Decay
Chromium-50	49.946	4.345	0	$> 1.3 \times 10^{18}$ y	β+EC
Chromium-52	51.941	83.789	0	–	–
Chromium–53	52.941	9.501	3/2	–	–
Chromium–54	53.939	2.365	0	–	–

3.4 ISOTOPES

Chromium exhibits four naturally occurring isotopes: Cr-50, Cr-52, Cr-53, and Cr-54. Isotopes, differing by their mass numbers, are various forms of an element [5]. The mass number represents an amount of protons and neutrons within an atom's nucleus, enabling isotopic diversity within an element. In addition to these naturally occurring isotopes, seven radioactive isotopes of chromium have been identified. Radioactive isotopes decay and emit radiation, a phenomenon triggered by bombarding atoms with small particles. Consequently, these atoms become radioactive. Chromium compounds display a diverse range of colors, including purple, black, green, orange, and yellow. Among the radioactive chromium isotopes, chromium-51 finds application in medical research, particularly as a tracer in blood studies. Tracers refer to radioactive isotopes and their presence in a system may be readily identified. Injecting an isotopic substance into the system enables the tracking of its emitted radiation using detectors positioned around the system [6]. Chromium-51 is commonly employed in the examination of red blood cells, aiding in determining cell count and lifespan within an individual's body. It also facilitates the study of blood circulation in fetuses [7].

3.5 EXTRACTION

Cr can be produced using methods similar to those utilized for other metals. One approach involves heating Cr_2O_3 with either charcoal or aluminum.

In this process, charcoal, which is primarily composed of almost pure carbon, or Al, removes O_2 from Cr_2O_3, leading to the generation of pure chromium metal. This method shares similarities with the technique first developed by Vauquelin [8]. Another method entails passing an electric current through chromium compounds to obtain chromium metal. At times, chromite undergoes direct conversion into an alloy known as ferrochromium, or ferrochrome, which holds significant importance as a chromium alloy. Ferrochromium serves the purpose of enriching steel with chromium. During the initial stages of steel production, when the material is in a hot, liquid state, ferrochromium is introduced.

Inside the molten steel, chromium from ferrochromium dissolves and becomes incorporated into the steel. Upon solidification of the molten steel, the chromium becomes embedded within, resulting in the formation of Cr steel. The radioactive element Cr-51 is employed in investigating the flow of blood within a fetus (an unborn child) [9].

3.6 CHEMICAL REACTIONS OF CHROMIUM (STATE REACTIONS OF CHROMIUM)

3.6.1 Chromium Reaction with Air and Water

Chromium metal doesn't react with H_2O or O_2 at ambient temperature.

3.6.2 Reaction of Cr with the Halogens

At a temperature of 400°C and under pressures of 200–300 atm, chromium undergoes a direct reaction with fluorine (F_2) to produce chromium(VI) fluoride, CrF_6.

$$Cr_{(s)} + 3F_2\,(g) \longrightarrow CrF_6\,(s)\,(\text{Yellow})$$

Chromium(V) fluoride, known as CrF5, forms under milder conditions.

$$2Cr_{(s)} + 5F_2\,(g) \longrightarrow 2CrF_5\,(s)\,(\text{Red})$$

Under gentler conditions, chromium metal reacts with the halogens F_2, Cl_2, Br_2, and I_2 to yield the respective trihalides: Cr(III) fluoride (CrF_3), Cr(III) chloride ($CrCl_3$), Cr(III) bromide ($CrBr_3$), or Cr(III) iodide (CrI_3).

$$2Cr_{(s)} + 3F_2\,(g) \longrightarrow 2CrF_3\,(s)\,(\text{Green})$$

$$2Cr_{(s)} + 3Cl_2\,(g) \longrightarrow 2CrCl_3\,(s)\,(\text{red-violet})$$

$$2Cr_{(s)} + 3Br_2\,(g) \longrightarrow 2CrBr_3\,(s)\,(\text{very dark green})$$

$$2Cr_{(s)} + 3I_2\,(g) \longrightarrow 2CrI_3\,(s)\,(\text{very dark green})$$

3.6.3 Reaction of Cr with Acids

Chromium metal reacts with dilute hydrochloric acid to produce solutions containing the aquated Cr(II) ion, accompanied by the liberation of hydrogen gas, H_2. Typically, the Cr(II) species exists in the form of the complex ion $[Cr(OH_2)6]^{2+}$.

$$Cr_{(s)} + 2HCl(aq) \longrightarrow Cr^{2+}(aq) + 2Cl^-(aq)H_2(g)$$

$$Cr_{(s)} + 3H_2SO_4(aq) \longrightarrow Cr_2(SO_4)_3(aq) + 3SO_2(g) + 3H_2O(l)$$

3.6.4 Oxidation–Reduction Reactions

Chromium exhibits various oxidation states, such as +2, +3, and +6. In an acidic medium, the reaction between chromium(III) ions and hydrogen peroxide (H_2O_2) leads to the production of chromium(III) ions, water, and the simultaneous reduction of hydrogen peroxide to yield water and oxygen.

$$2Cr^{3+}_{(s)} + 3H_2O_2(aq) \longrightarrow 2Cr^{2+}(aq) + 3O_2(g) + 4H_2O(l)$$

3.6.5 Reaction of Chromium with Nitric Acid

When chromium metal undergoes a reaction with concentrated nitric acid, it produces chromium(III) nitrate ($Cr(NO_3)_3$), nitrogen dioxide gas (NO_2), and water.

$$2Cr_{(s)} + 6HNO_3(aq) \longrightarrow 2Cr(NO_3)_3(aq) + 3NO_2(g) + 4H_2O(l)$$

3.7 ORGANOCHROMIUM COMPOUNDS

The initial organochromium compound was documented by Franz Hein in 1919. When exposing C_6H_5MgBr to Cr(III)Cl_3, Hein got a unique compound, which he misidentified as Ph_5CrBr following hydrolysis. However, in 1957, H.H. Zeiss and coworkers replicated Hein's results and accurately determined the formation of a cationic bisarene chromium sandwich compound (ArH_2Cr^+)[10].

In 1956, Ernst Otto Fischer produced bis(benzene)chromium $Cr(\acute{\eta}_6\text{-}C_6H_6)_2$ by reacting chromium(III) chloride, benzene, and aluminum chloride during a similar timeframe. Fischer found chromocene, a similar chemical, a few decades earlier in 1953 [11,12].

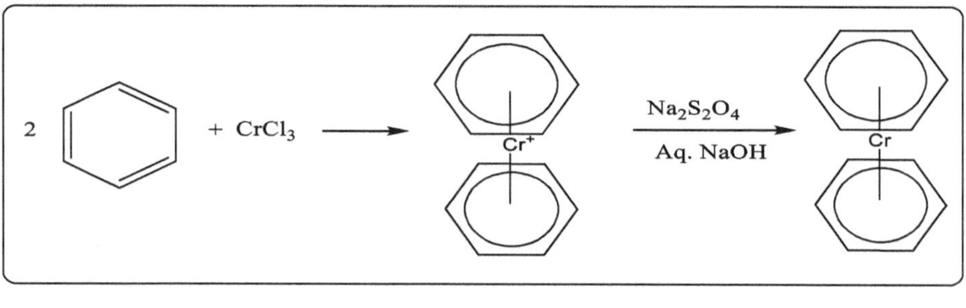

In 1963, G. Wilke and coworkers introduced tris-(η-allyl)Cr as one of the first Ziegler–Natta catalysts, yet its commercial success was limited [13]. Chromocene chemicals, which Union Carbide first used in ethylene polymerization in 1972, are still essential for the commercial production of high-density polyethylene [14].

These substances can be classified into the following general categories:

- Sandwich compounds include chromocene (Cp_2Cr) and derivatives of chromium bis(benzene) ($(ArH)_2Cr$). Half-sandwich molecules such as (η_6-$C_6H_5OMe)Cr(CO)_3$ are being studied more often.
- Carbenes of Cr metal(R_1)(R_2)C::CrL_n and carbynes of Cr metal (RC:::CrL_n)
- Cr(III) complexes $RCrL_5$ [15]
- Complexes of chromium carbonyl anion and cation (e.g., $Na_4Cr(CO)_4$) [16].

Organochromium compounds are utilized in ethylene polymerization and oligomerization processes. Chromium catalysts, including the Phillips catalyst and those by Union Carbide and DSM, are crucial in ethylene polymerization. They involve CrO(VI) impregnated on silica, activated in dry air. These catalysts, potentially reduced by ethylene, produce active species like Cr(II), though their exact mechanism remains unclear [17,18].

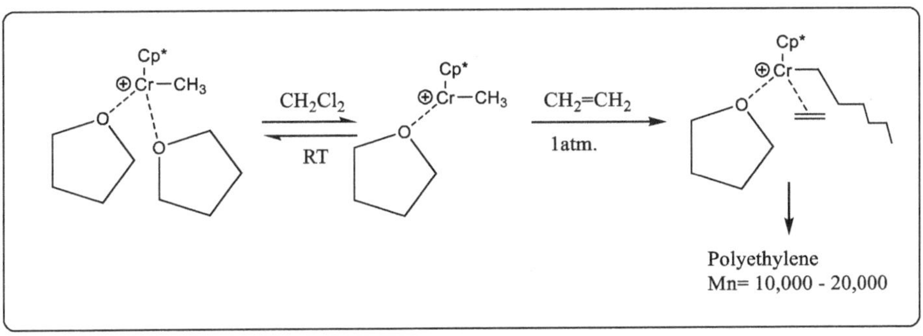

3.7.1 Preparation, Properties and Applications of Each Organochromium Compound

3.7.1.1 Bis(Arene)Chromium Π-Complex

3.7.1.1.1 Preparation of Bis(arene)Chromium π-Complex [19]

(i) Hein Grignard synthesis

The process entails the interaction of a chromium halide compound with a Grignard reagent. Typically, the reaction occurs in anhydrous and inert conditions to prevent side reactions.

$$CrCl_2 + 2RMgX \longrightarrow R_2Cr + 2MgXCl$$

R_2Cr= bis(arene)chromium π-complex

(ii) Fischer–Hafneralluminum synthesis

The Fischer–Hafneraluminum synthesis comprises the interaction of an alkyl halide with aluminum metal. Typically, it occurs under anhydrous and inert conditions to forestall undesired side reactions and to guarantee synthesis success.

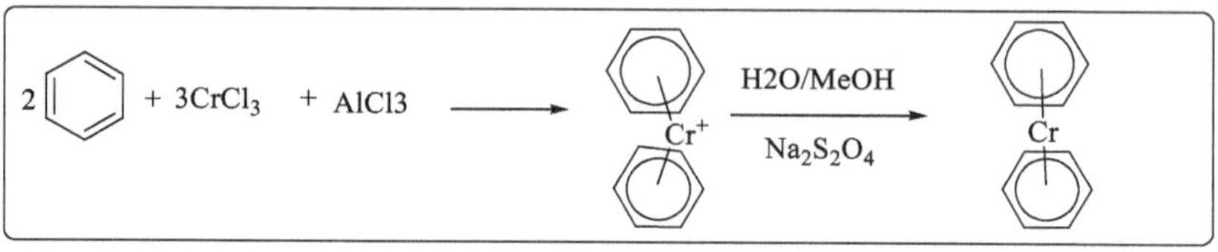

(iii) Herwig–Zeiss acetylation trimerization

Organochromium (II) and (III) compounds, bonded via sigma bonds, undergo reactions with substituted acetylenes, yielding bis-products.

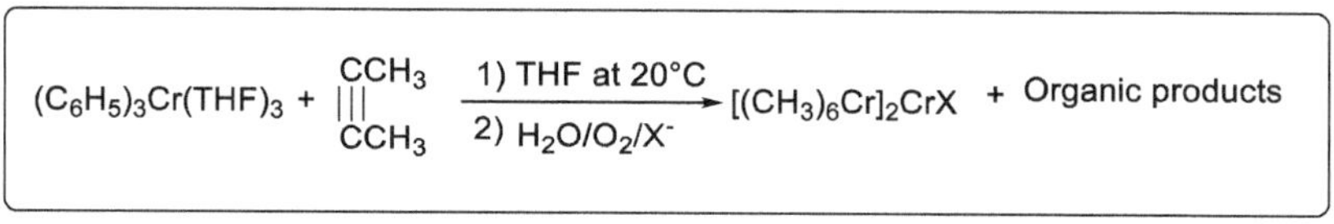

3.7.1.1.2 Chemical Properties of Bis (arene)Chromium π-Complex

Under an inert gas atmosphere, bis(benzene)chromium exhibits thermal stability. As expected, it shows diamagnetism, with a dipole moment of zero. Fischer and Weiss reported in 1956 that the crystal structure of bis(benzene)chromium is centrosymmetric and has cubic symmetry. $Cr(C_6H_6)_2$ with a molecular weight of 208.224 g/mol appears in brown-black crystals. Bis(benzene)chromium has a melting point of 284°Cand a boiling point of 160°C. It is insoluble in water [20].

3.7.1.1.3 Application of Bis (arene)Chromium π-Complex

Samuel et al. found that bis(benzene)Cr functions effectively as an organo-metallic radical scavenger. Unlike cobaltocene, which traps radicals (R) to form 19-valence e^- species(η_5-C_5H_5)(η_4-C_5H_5R)Co, bis(benzene)Cr interacts with radicals to form 17-valence e^- types (η_6-C_6H_6)(η_5-C_6H_6R)Cr (here R = H, D, isobutyronitrile) [21].

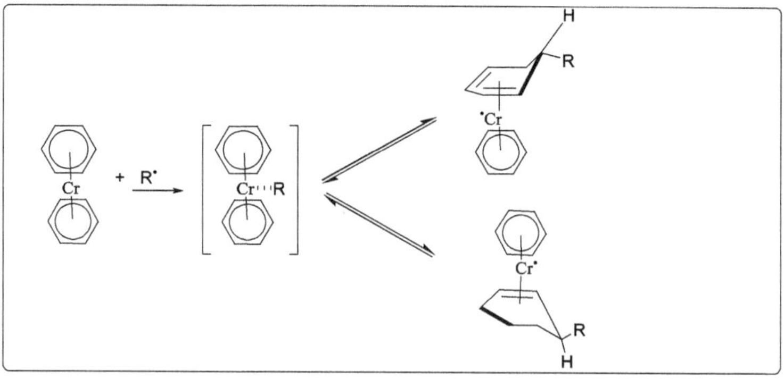

Bis(benzene)Cr was documented to facilitate the hydrosilation of alcohols and aldehydes.

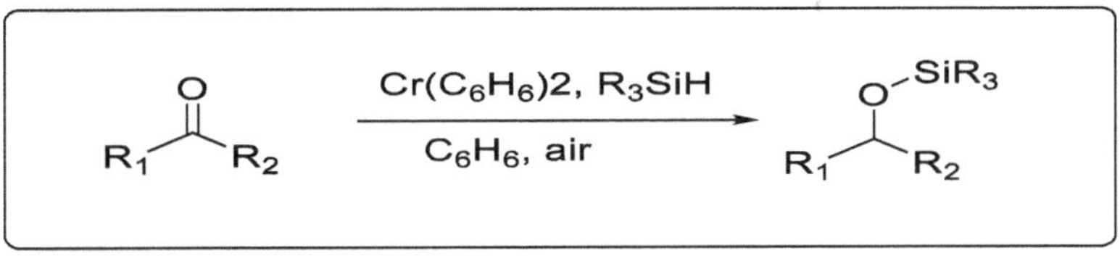

3.7.1.2 Carbonylchromium Π-Complex

3.7.1.2.1 Preparation of Carbonylchromium π-Complex [19]

(i) Synthesis from $Cr(CO)_6$

The aromatic compound when heated with under anaerobic condition, either alone or in an inert solvent, produces tricarbonyl compounds.

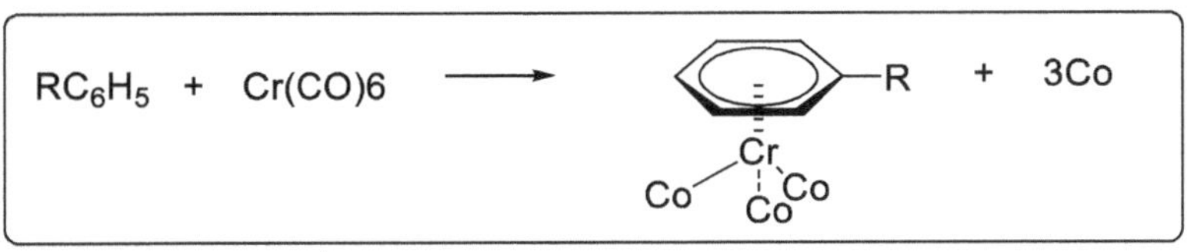

(ii) Ligand replacement reaction

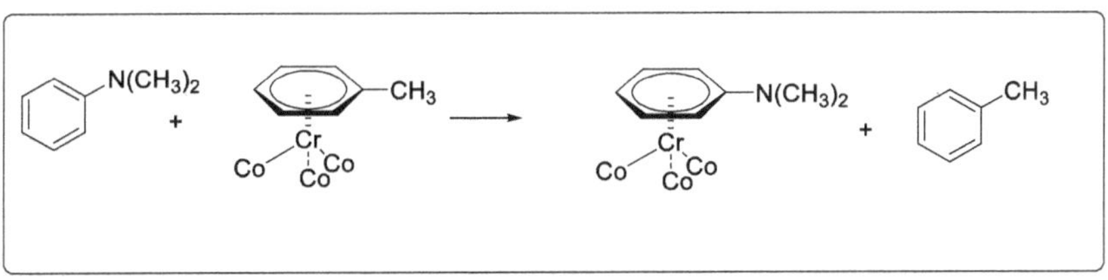

3.7.1.2.2 Chemical Properties of Carbonylchromium π-Complex

(Benzene)chromium tricarbonyl, with the formula $Cr(C_6H_6)(CO)_3$, is an organometallic compound. This solid, characterized by its yellow crystalline form, dissolves in standard nonpolar organic solvents. Its molecular structure is referred to as "piano stool" due to the planar orientation of the Ar and the existence of three cobalt ligands as "legs" along the Cr–bond axis [22]. It has a molar mass of 214.14 g/mol and is insoluble in water.

3.7.1.2.3 Application of Carbonylchromium π-Complex

Complexes such as $(Arene)Cr(CO)_3$ have undergone extensive investigations as reagents in chemical reactions. The aromatic ring of $[Cr(CO)_3(C_6H_6)]$ displays significantly higher electrophilicity compared to benzene, enabling it to engage in nucleophilic addition reactions [23].

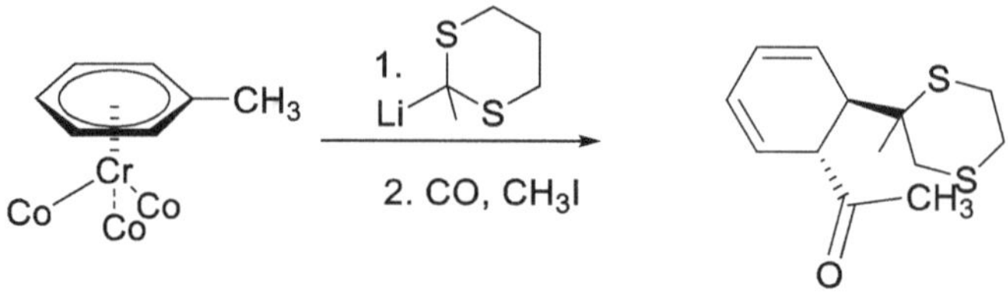

(Benzene)tricarbonylchromium is an effective catalyst for the hydrogenation process of 1,3-dienes, with the resultant alkene formed by 1,4-addition of H. Notably, the compound does not hydrogenate single double bonds.

3.8 COORDINATION CHEMISTRY OF CHROMIUM

The coordination chemistry of chromium is a distinct subfield within inorganic chemistry that revolves around investigating compounds containing chromium. This area places special emphasis on examining cases where chromium forms coordination complexes. The transition metal chromium has significantly contributed to the progress of inorganic chemistry. French scientist Vauquelin discovered the unusual $PbCrO_4$ in 1797. It was called for the different colors presented by its derivatives, with "chroma" (χρωμα) indicating color [24]. Cr is a transition metal located in Group 6 of the periodic table, belonging to the first row. It possesses an outer electronic configuration of $3d^5$ $4s^1$ and has the ability to form compounds in oxidation states ranging from II- to VI+ [25].Typically, only two Cr oxidation states, Cr^{3+} and Cr^{6+}, are considered physiologically and ecologically relevant, as they are stable in the presence of air and H_2O. Cr(III) complexes are stable in both kinetic and thermodynamic conditions. Conversely, while Cr(VI) complexes exhibit kinetic stability, they encounter challenges related to thermodynamic instability [26]. Various parameters are employed to evaluate the coordination in chromium, taking into account these distinct characteristics.

3.8.1 Oxidation States

- Transition metals have the capability to adopt multiple oxidation states, and chromium is frequently observed in oxidation states of +2, +3, and +6.
- Chromium can occupy oxidation states spanning from +2 to +6, each revealing specific coordination preferences. In the +2 state, chromium commonly forms complexes with ligands such as water, ammonia, or chloride, adopting either octahedral or tetrahedral geometries.
- Complexes of chromium (III), a prevalent oxidation state, frequently showcase octahedral coordination and vibrant colors attributed to electronic transitions involving 3d orbitals.
- Chromium in oxidation states IV and V is less prevalent but can create stable complexes with tetrahedral or square pyramidal geometries. Hexavalent

chromium (Cr(VI)) is notable in compounds like chromates and dichromates, showcasing potent oxidizing properties[24].

3.8.2 Coordination Number

- The coordination number represents the count of donor atoms from a ligand that surround the metal ion.
- Chromium often forms compounds with coordination numbers 4 and 6.
- Complexes of chromium (III) commonly display coordination arrangements in either octahedral or tetrahedral configurations. Examples include $[Cr(H_2O)_6]^{3+}$ and $[Cr(NH_3)_6]^{3+}$, $[CrCl_4]^-$

[i] $[Cr(H_2O)_6]^{3+}$

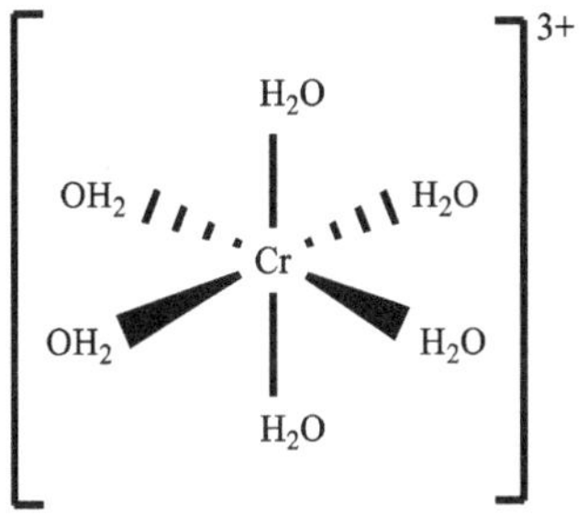

CN = 6, geometry = octahedral

[ii] $[CrCl_4]^-$

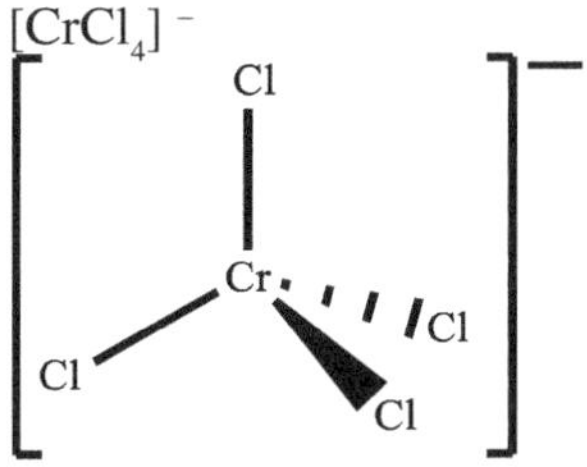

CN = 4, geometry = tetrahedral

- The colors displayed by these complexes result from electronic transitions.
- Chromium (II) complexes are infrequent and typically involve ligands such as CO or CN^-. Example: $CrCl_2(MeCN)_2$.

[i] $CrCl_2(MeCN)_2$

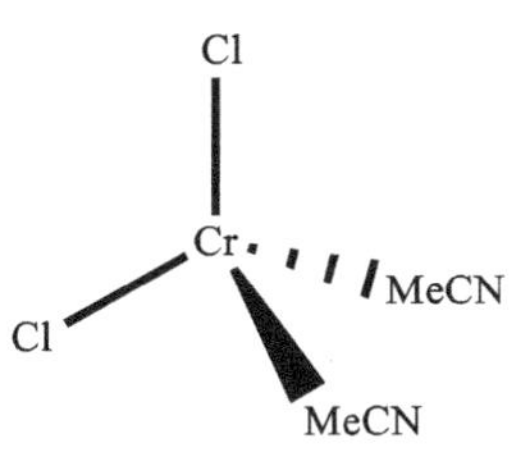

CN = 4, geometry = tetrahedral

- Chromium (VI) complexes, including chromates and dichromates, hold significance in oxidation–reduction reactions. Example: $[CrO_4]^{2-}$, CrO_2Cl_2.

  [i] $[CrO_4]^{2-}$

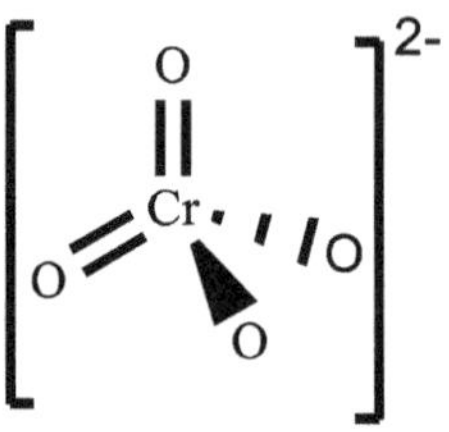

CN = 4, geometry = tetrahedral [24,27]

3.8.3 Electronic Configuration

- Chromium, a transition metal, possesses the electronic configuration [Ar] $3d^5\ 4s^1$.
- The capacity of chromium to lose electrons from both the 3d and 4s orbitals accounts for its ability to exhibit multiple oxidation states [28].

3.8.4 Ligands

- Chromium establishes coordination complexes through interactions with ligands.
- Ligands can be classified into two groups: monodentate, which donate a single electron pair, and polydentate, which donate multiple electron pairs.
- Chromium complexes commonly feature ligands such as water (H_2O), ammonia (NH_3), chloride ions (Cl^-), and a variety of organic ligands.

3.8.5 Bonding in Chromium Compounds

Bonding in coordination compounds can be examined through three perspectives:

- Valence bond theory
- Ligand field theory (a modification of molecular orbital (MO) theory)
- Crystal field theory (a concept grounded in purely electrostatic interactions).

3.8.5.1 Valence Bond Theory

- This hypothesis suggests that metal atoms or ions may use their (n-1)d, ns, np, or ns, np, nd orbitals to hybridize with ligands. Collection of analogous orbitals of various geometries, including square planar, octahedral, and tetrahedral, are produced by this hybridization process.
- The aforementioned hybridized orbitals can efficiently overlap with orbitals of ligands that can provide electron pairs to bond formation[29]. Table 3.3

shows the number of coordination mainly used in chromium and Table 3.4 shows some examples of chromium-coordinated compounds.

Example: Hexaamminechromium(III) complex

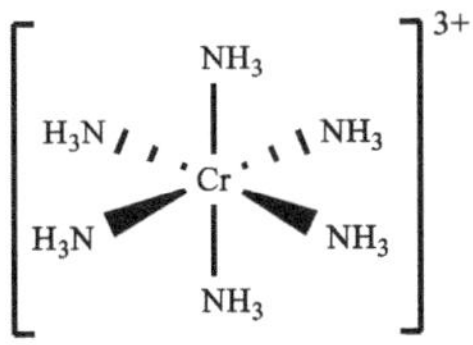

CN = 6, geometry = octahedral

The chromium atom (Z = 24) possesses the electronic configuration $3d^5\ 4s^1$ and exhibits a coordination number of six in this complex. In a +3 oxidation state, the chromium ion is formed by the loss of one 4s and two 3d-electrons. With the inner d-orbitals already vacant, the hybridization involves two vacant 3d, one 4s, and three 4p-orbitals, leading to the creation of six d^2sp^3 hybrid orbitals. These hybrid orbitals accommodate six pairs of e^-, one from each ammonia molecule (represented by xx), occupying the vacant hybrid orbitals. The molecule adopts an octahedral geometry. Paramagnetism is observed in the complex due to the existence of three unpaired electrons.

TABLE 3.3
Orbital count and hybridization patterns in chromium metal

Coordination Number	Hybridization Types	Distribution of Hybrid Orbitals in Space	Examples	Structures
6	d^2sp^3	Octahedral	$[Cr(NH_3)_6]^{3+}$	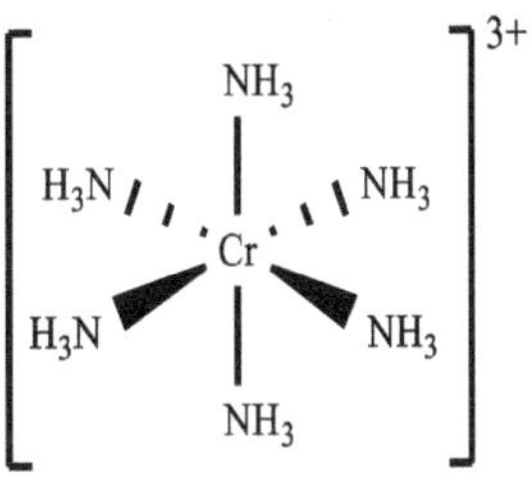
6	$sp^3\ d^2$	Octahedral	$[Cr(H_2O)]^{3+}$	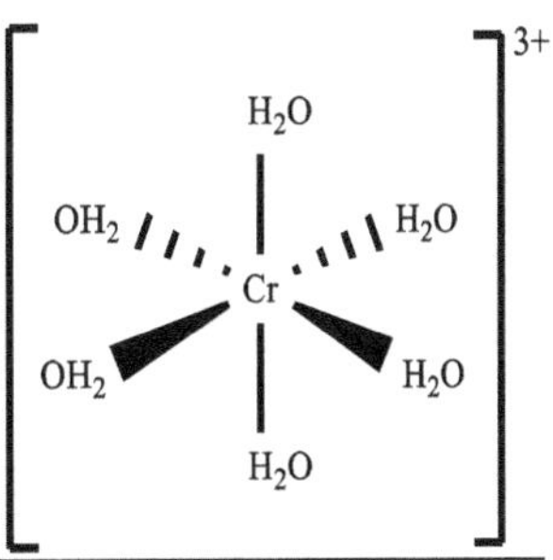

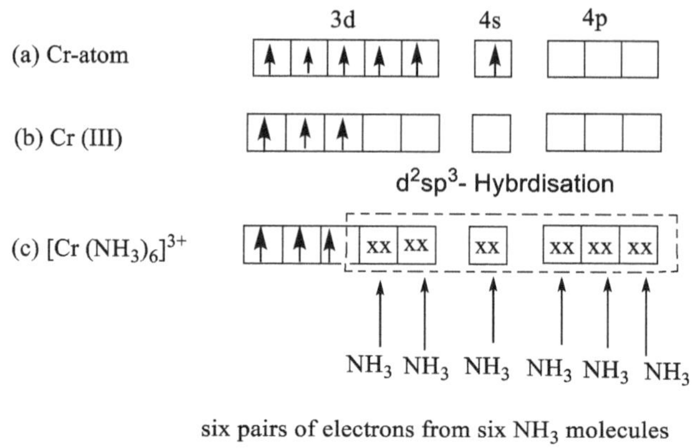

3.8.5.2 Crystal Field Theory

- The metal–ligand bond is understood by CFT, an electrostatic model, to be ionic and to result solely from electrostatic interactions between the ligand and the metal ions.
- For anions, ligands are viewed in CFT as point charges, while for neutral molecules, they are viewed as point dipoles.
- The 5d orbitals in a single gaseous metal atom or ion have the same energy levels, demonstrating their degeneracy. When a metal atom or ion is surrounded by a spherically symmetrical field of negative charges, the degeneracy remains preserved.
- It becomes asymmetrical, though, if the negative field inside a complex is produced by ligands, such as anions or the negative poles of dipolar molecules like ammonia and water. The separation of the d-orbitals is the result of this asymmetry lifting the degeneracy in the d-orbitals. This behavior is explained by a variety of crystal field theories, which interpret ligands as negative point charges and emphasize that the connection between the metal and ligands is only electrostatic [30].

What Happens When Ligands Approach A Metal

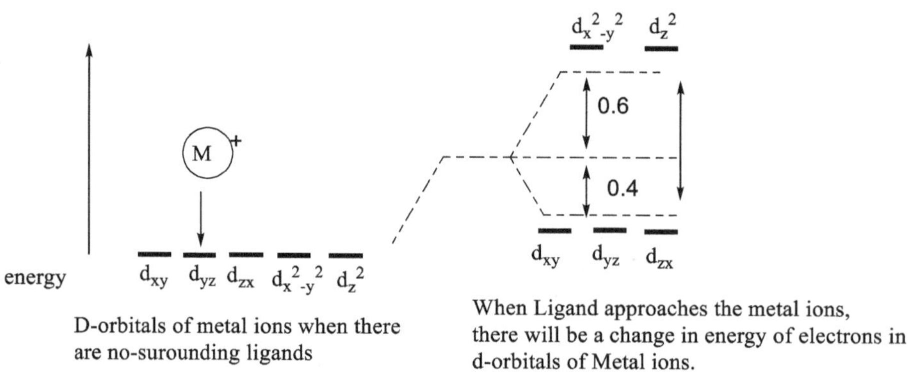

(a) CFS in octahedral coordination complexes

- Repulsion occurs between the electrons in the metal d-orbitals and the electrons of the ligands in the setting of an octahedral coordination entity, where the metal atom/ion is surrounded by six ligands. When the metal d-orbital is pointed toward the ligand as opposed to away from it, this repulsion is more noticeable (Figure 3.1).

Examples: $[Cr(NH_3)_6]^{3+}$, $[Cr(CN)_6]^{3-}$, $[CrF_6]^{2-}$ are examples of octahedral complexes and having coordination number 6. Since cyanide and ammonia are strong field ligands (from spectrochemical series), both are low-spin complexes. Conversely, F-forms high-spin compounds and is a weak field ligand.

(b) CFS in tetrahedral coordination compounds

- The d-orbital splitting experiences an inversion during the formation of a tetrahedral coordination entity, and it is less substantial than the splitting seen in an octahedral field.
- For identical metals, ligands, and metal–ligand distances, it can be demonstrated that Δt is equal to (4/9) Δ_0. Therefore, the energies responsible for orbital splitting are not considerable enough to enforce pairing, and as a result, configurations with low spin are infrequently observed (Figure 3.2) [31].

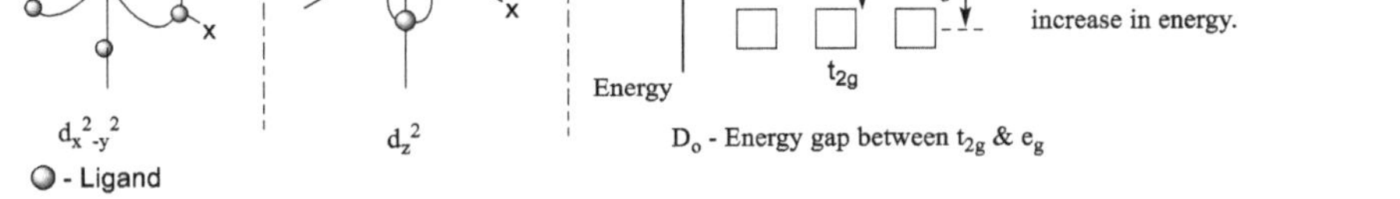

FIGURE 3.1 d-Orbital splitting in an octahedral crystal field.

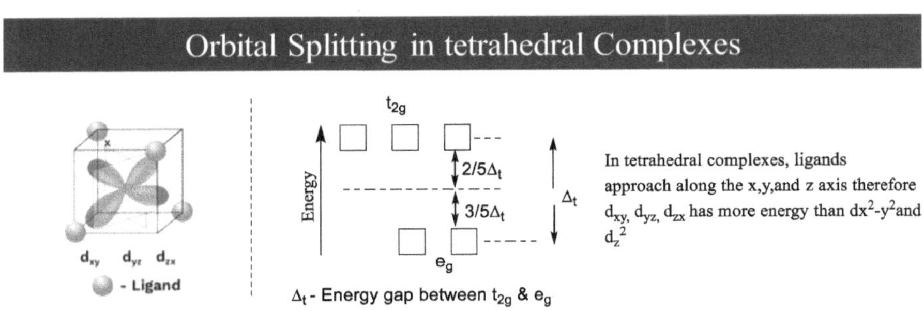

FIGURE 3.2 d-Orbital splitting in a tetrahedral crystal field.

Examples: $CrCl_2(MeCN)_2$

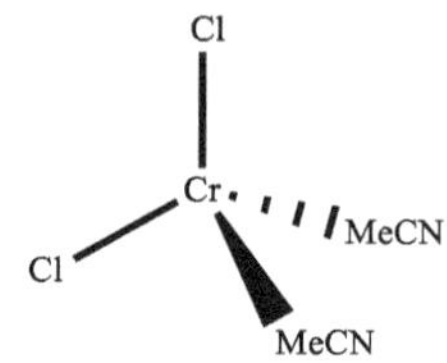

CN = 4, geometry= tetrahedral.

Spectrochemical series

$CO = CN^- > PPh_3 > NO^{2-} > NH_3 >$ Pyridine $> CH_3CH >$ NCG- $> H_2O = C_2O_4^{2-} >$ OH- $> NCO^- > F^- >$ Cl- $> SCN^- > S^{2-} > Br^- > I^- > O_2^{2-}$

Strength of ligand

Typically, ligands can be organized in a series based on increasing field strength, known as the spectrochemical series.

(i) When Δ_o is smaller than P, one of the eg orbitals is occupied by the fourth electron, giving rise to the configuration t_2g_3 eg_1. Weak field ligands are recognized as those that produce high spin complexes with $\Delta_o \leq P$.

(ii) When Δ_o exceeds P, the energetically favorable configuration involves the $4e^-$ occupying a t_{2g} orbital, resulting in $t_{2g}^{\,4}\, e_g^{\,0}$. Strong field ligands are ligands that produce low-spin complexes by inducing this phenomenon. A stronger ligand field results in a rise in Δo's magnitude.

Examples: $CrCl_2(MeCN)_{2,}[CrCl_4]^-$

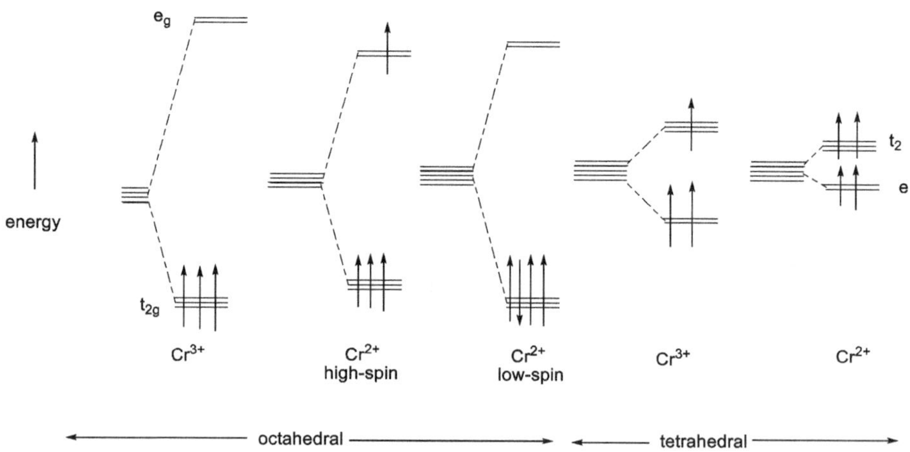

FIGURE 3.3 Electronic configurations of chromium cations in octahedral and tetrahedral coordination geometries [32].

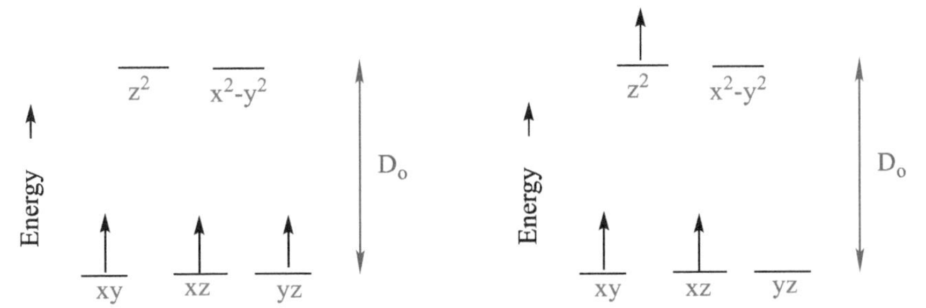

CrIII ground-state electron configuration in octahedral coordination.

CrIII excited-state electron configuration in octahedral coordination.

FIGURE 3.4 Cr^{III} ground-state electron and excited-state electronic configuration in octahedral coordination.

3.8.5.3 CFT of Chromium(III)

Two sets of five "conventional real (or directed)" 3d orbitals may be distinguished in the context of Cr(III), assuming perfect octahedral coordination: t_{2g} $(3_{dxy}, 3_{dxz}, 3_{dyz})$ at a lower energy level and e_g $(3_{dx}{}^{2}–y^{2}, 3_{dz}{}^{2})$ at a higher energy level. When d_1–d_3 systems are subjected to Hund's rule, it is anticipated that the three electrons would avoid pairing and inhabit the t_{2g} set with spin quantum number +½. The associated ground-state e- configuration in octahedral coordination is illustrated in Figure 3.4. Upon encountering a photon with energy Δ_o, Cr(III) has the capability to absorb the photon, causing the promotion of an electron from the lower-energy orbital set t_{2g} to the higher-energy set e_g. The e^-configuration in the resulting excited state is shown in Figure 3.4 [33].

3.8.5.4 Ligand Field Theory

LFT is a molecular orbital theory (MOT) utilized to explain the electronic structure and bonding in coordination compounds. LFT encompasses both σ (sigma) and π (pi) bonding interactions occurring between the metal center and ligands. Chromium metal typically exhibits a coordination number of 6, as exemplified by the complex hexaaquachromium(III) ion. The MO diagram (Figure 3.5) for hexaaquachromium(III) ion illustrates the interactions between the chromium (Cr) metal center and the six water (H_2O) ligands. With chromium in the +3 oxidation state, its electronic configuration is [Ar] $3d^3$.The donation of electron pairs from the six water ligands to the metal center results in the formation of both σ and π bonds. Specifically, σ bonding arises from the overlap of the metal 4s and 4p orbitals with ligand orbitals, while π bonding involves the interaction between the metal's 3d orbitals and the ligand orbitals [34].

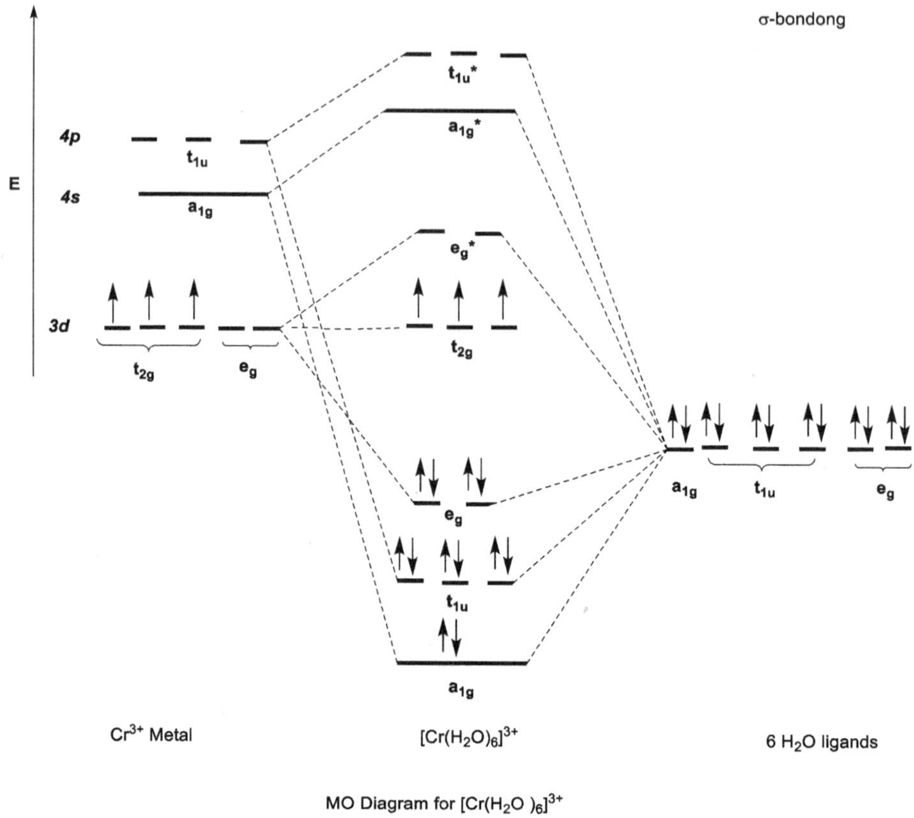

FIGURE 3.5 MOT diagram of complex hexaaquachromium(III).

3.8.6 Colors in Coordination Compound of Chromium

- Color of complex compound is due to d–d transition.
- Color of complex compound depends upon splitting energy.
- Complex become colorless in absence of ligand filled (no ligand, no splitting, no transition).
- Color of some compound is due to charge transfer (e-transition) from anion to metal ion.

Chromium shows different colors in different oxidation states because of d–d transition. In Figure 3.6, under strong octahedral field conditions, Cr(III) displays a violet hue, whereas in a weak octahedral field, it takes on a magenta coloration because of energy gap. Examples: $[CrF_6]^-$= green, hexaaquachromium(III) ion=violet, hexaamminechromium(III)= yellow, andhaving CN = 6[35].

3.9 HETEROCYCLIC CHEMISTRY OF CHROMIUM

Research in the heterocyclic coordinated chemistry of chromium focuses on examining chromium complexes within heterocyclic ligands. These ligands feature ring

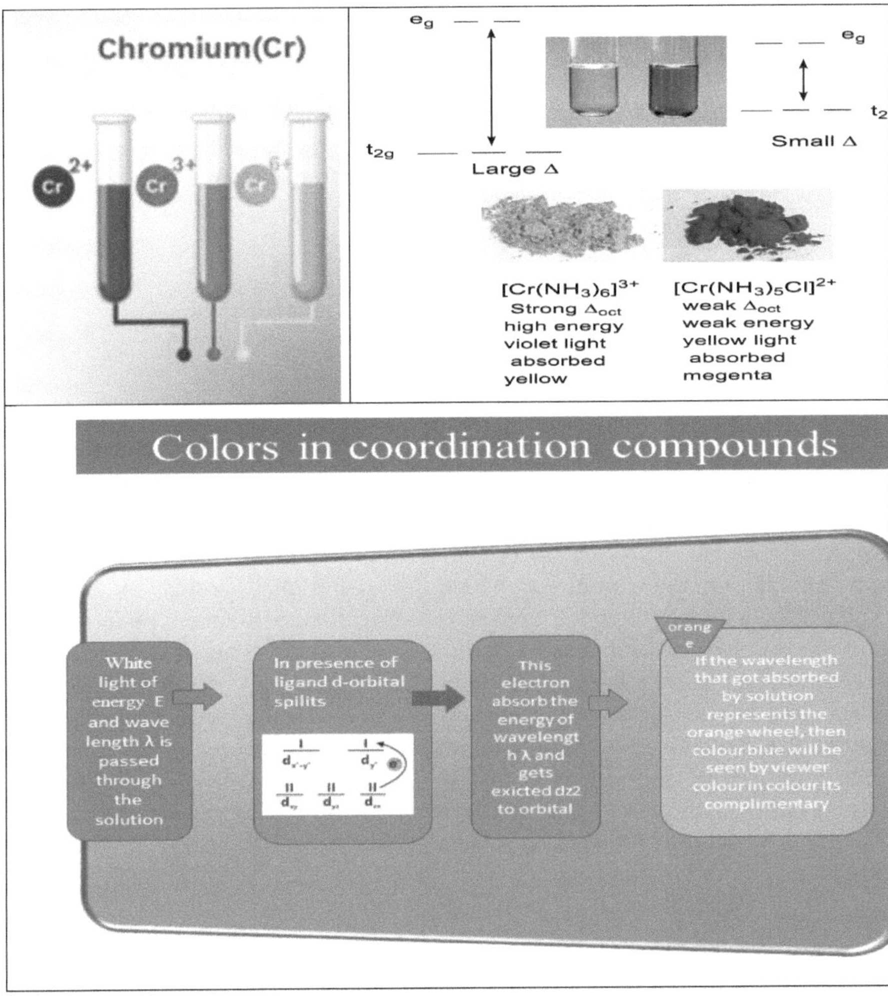

FIGURE 3.6 Color variations of Cr(III) complexes in strong and weak octahedral fields and across different oxidation states: Illustrating d–d transitions and ligand field effects.

structures that incorporate at least one non-carbon atom, such as nitrogen, oxygen, or sulfur, coordinating with the chromium center. Transition metal complexes can generate chelate complexes with heterocyclic ligands. They can participate in complex formation with nitrogen-donor ligands, oxygen-donor ligands, as well as monodentate, bidentate, tridentate, multidentate, and macrocyclic ligands. They exist in both homoleptic and heteroleptic forms. Examples tri-didentatehomoleptictris(1,10-phenanthroline)chromium(III) ion and heterolepticdichloridobis(1,10-phenanthroline) bis(bipyridine)chromium(III) ion or di-tridentate homoleptictris(2,2':6',2''-terpyridine)chromium(III) ion and heterolepticchlorido(2,2':6',2''-terpyridine)(4,4'-dimethyl-2,2':6',2''-terpyridine)chromium(III) ion. All these compounds exhibit a coordination number (CN) of six [36]. Typically, they have the capability to create complexes, with a specific focus on those involving α,α′-diimines like 2,2′-bipyridine (bpy), 1,10-phenanthroline (phen), or their substituted derivatives. Heterocyclic coordinated chromium compound plays a crucial role in biomedical field and having properties to form a DNA base.

Heterocyclic CrIII complex of isothipendyl is an antihistaminic drug [37].

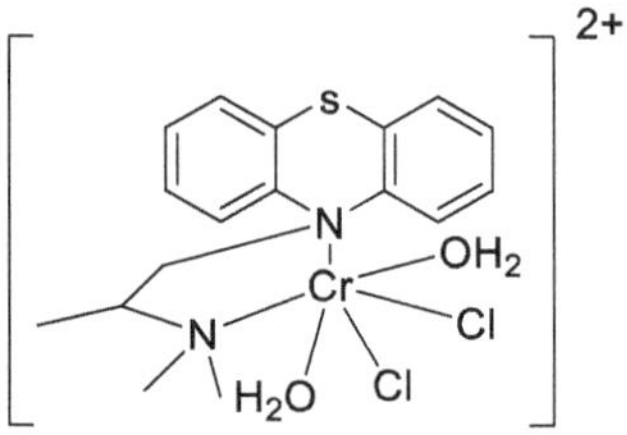

3.9.1 Homoleptic and Heteroleptic Coordinated Complex of Chromium

Heterocyclic ligands play a vital role in the formation of both homoleptic and heteroleptic chromium complexes within the realm of coordination chemistry. Homoleptic complexes exhibit a uniform coordination environment, with the chromium ion exclusively surrounded by ligands of the same type, such as a complex solely composed of pyridine ligands. In contrast, heteroleptic complexes showcase a diverse and intricate structure, incorporating various ligand types around the chromium center. This diversity allows for the inclusion of heterocyclic ligands like pyridine, imidazole, or pyrazole, alongside non-heterocyclic ligands. The nuanced interplay between chromium and heterocyclic ligands enriches our understanding of the expansive field of coordination chemistry [28].

Example: (i) $[Cr(Me_2bpy)_3]^{3+}$ (Homoleptic)

C.N=6 and geometry = octahedral

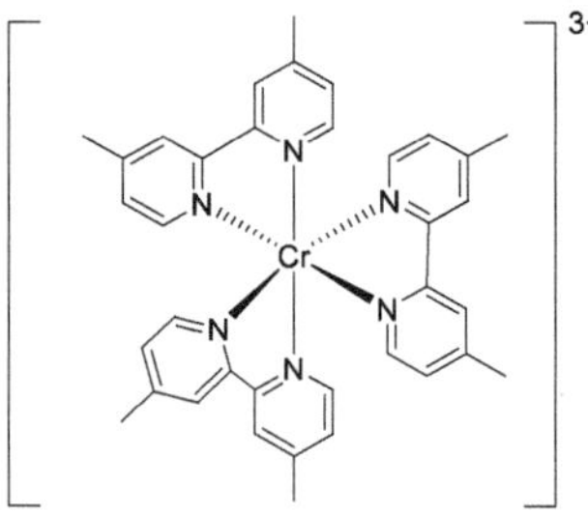

(ii) $[Cr(diimine)_2(DPPZ)]^{3+}$(Heterolaptic)

CN= 6

While diimine stands for the auxiliary ligands such as bpy (2,2’-bipyridine), phen (1,10-phenanthroline), DMP (5,6-dimethyl-1,10-phenanthroline), and

TMP (3,4,7,8-tetramethyl-1,10-phenanthroline), DPPZ refers to the intercalating dipyridophenazine ligand [38].

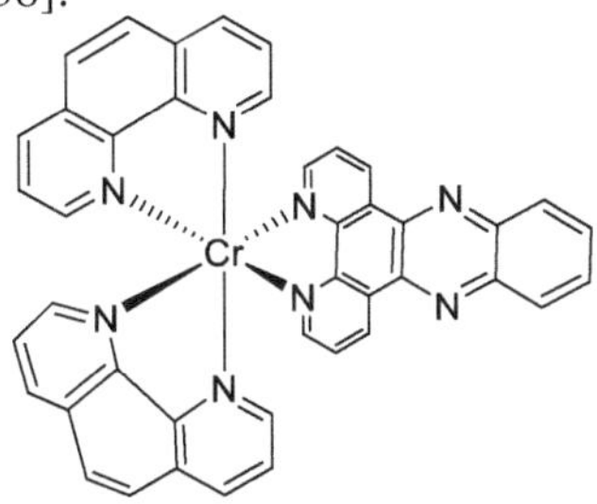

(iii) $[(phen)_2Cr(4\text{-}dmcbpy)](OTf)_3$ [39]
CN= 6

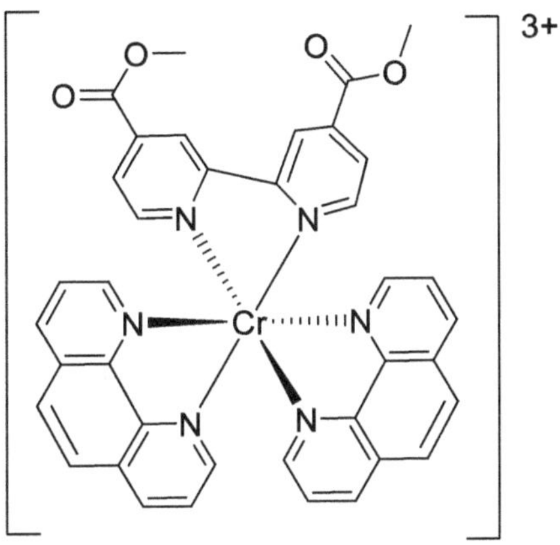

TABLE 3.4
Some examples of chromium-coordinated compounds

Oxidation State	Coordination Number	Distribution of Hybrid Orbitals in Space	Example Structures
Cr^{II}	4	Tetrahedral	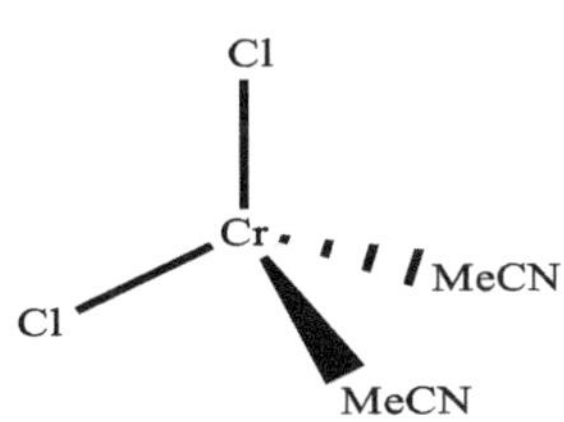 $CrCl_2(MeCN)_2$

(*continued*)

TABLE 3.4 (Continued)
Some examples of chromium-coordinated compounds

Oxidation State	Coordination Number	Distribution of Hybrid Orbitals in Space	Example Structures
	4	Square planner	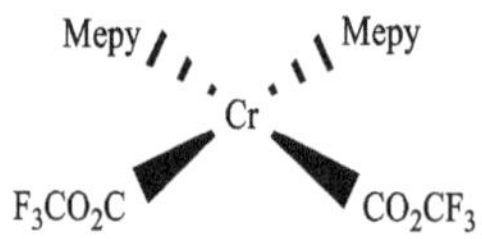 $Cr(O_2CCF_3)_2(Me_2py)2$
Cr^{III}	5	Trigonal bipyramidal	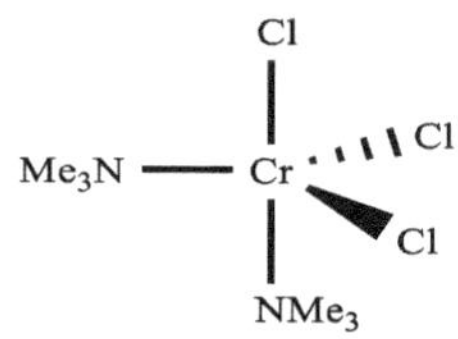 $CrCl_3(NMe_3)_2$
	3	Trigonal planar	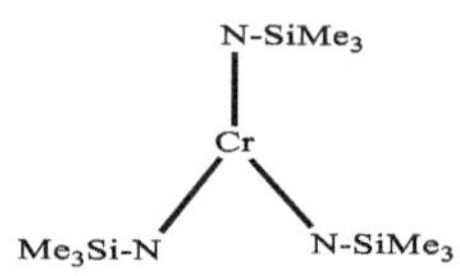 $Cr\{N(SiMe_3)_3\}$
	4	Tetrahedral	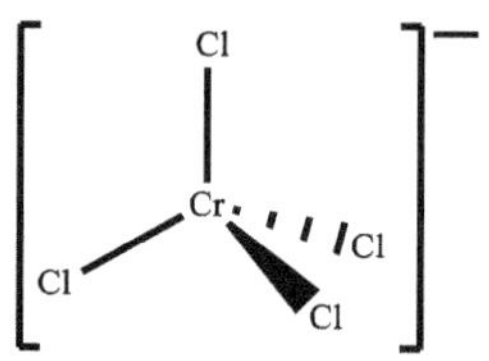 $[CrCl_4]^-$
	6	Octahedral	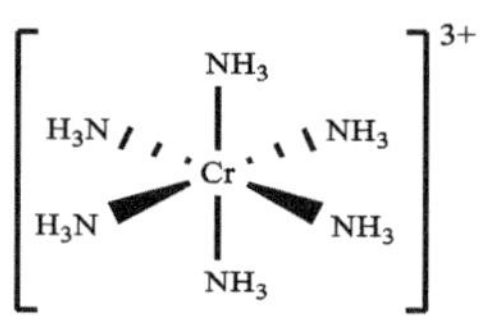 $[Cr(NH_3)_6]^{3+}$, $[Cr(CN)_6]^-$

TABLE 3.4 (Continued)
Some examples of chromium-coordinated compounds

Oxidation State	Coordination Number	Distribution of Hybrid Orbitals in Space	Example Structures
Cr^{IV}	6	Octahedral	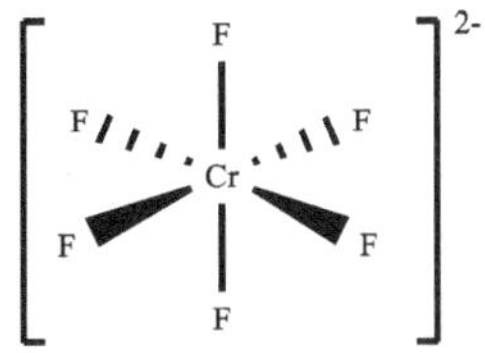 $[CrF_6]^{2-}$
	4	Tetrahedral	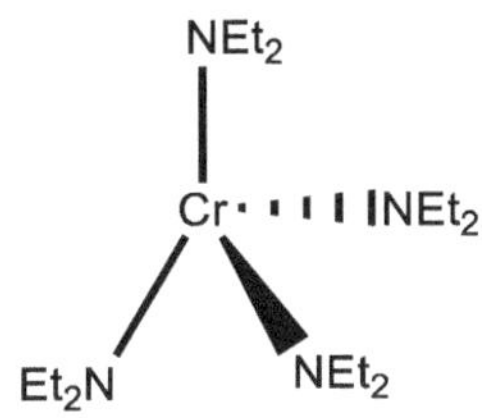 $Cr(NEt_2)_4$
Cr^{V}	4	Tetrahedral	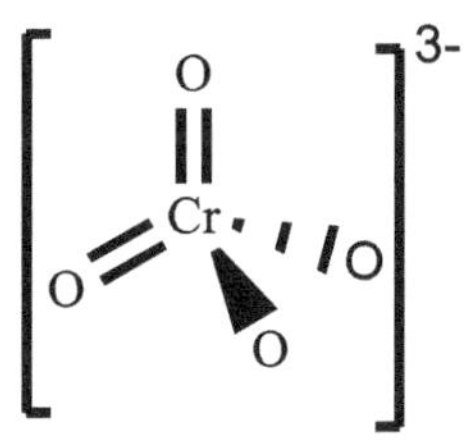 $[CrO_4]^{3-}$
	5	Square pyramidal	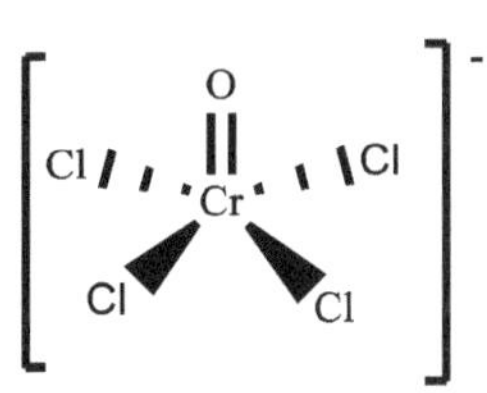 $[CrOCl_4]^{-}$

(*continued*)

TABLE 3.4 (Continued)
Some examples of chromium-coordinated compounds

Oxidation State	Coordination Number	Distribution of Hybrid Orbitals in Space	Example Structures
	6	Octahedral	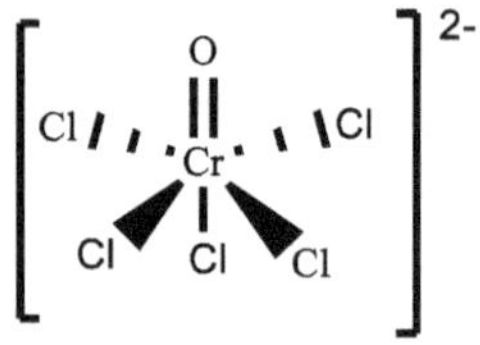 $[CrOCl_5]^{2-}$
Cr^{VI}	4	Tetrahedral	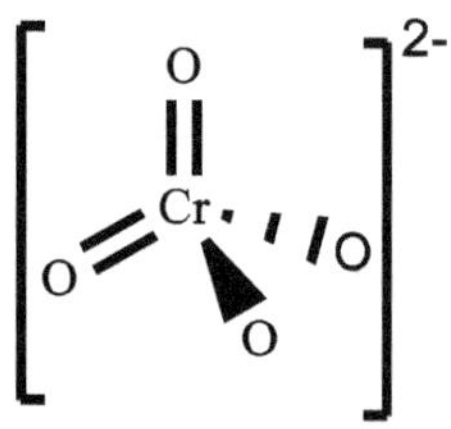 $[CrO_4]^{2-}$
Cr^{III}	6	Octahedral	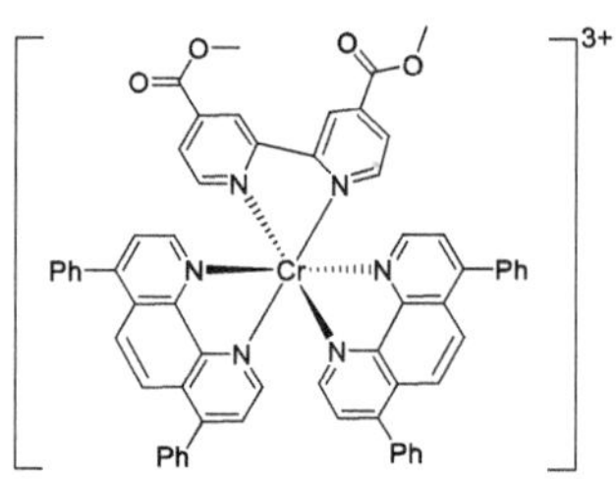 $[(Ph_2phen)_2Cr(4\text{-dmcbpy})](OTf)_3$

REFERENCES

[1] L. L. Majola, *Cr (VI) Occurrence in Groundwater, Health Impacts and Removal Using Chitosan*. University of Johannesburg (South Africa), 2019.

[2] J. Barnhart, "Occurrences, uses, and properties of chromium," *Regul. Toxicol. Pharmacol.*, vol. 26, no. 1, pp. S3–S7, 1997, doi: https://doi.org/10.1006/rtph.1997.1132.

[3] N. Omm-e-Hany, A. A. Asia, and H. Kanwal, "Determination of chromium in the tannery wastewater, Korangi, Karachi," *Int. J. Environ. Sci. Nat. Res.*, vol. 15, p. 555920, 2016.

[4] S. I. Shupack, "The chemistry of chromium and some resulting analytical problems," *Environ. Health Perspect.*, vol. 92, no. 1, pp. 7–11, 1991, doi: 10.1289/ehp.91927.
[5] L. Qin and X. Wang, "Chromium isotope geochemistry," *Non-Traditional Stable Isot.*, vol. 82, pp. 379–414, 2017, doi: 10.2138/rmg.2017.82.10.
[6] H. Sicius, "Copper group: Elements of the first subgroup," in *Handbook of the Chemical Elements*. Springer, 2024, pp. 795–836.
[7] L. Qin and X. Wang, "Chromium isotope geochemistry," *Rev. Mineral. Geochemistry*, vol. 82, no. 1, pp. 379–414, 2017.
[8] K. J. Sreeram and T. Ramasami, "Speciation and recovery of chromium from chromite ore processing residues," *J. Environ. Monit.*, vol. 3, no. 5, pp. 526–530, 2001.
[9] Q. Zhao et al., "A cleaner method for preparation of chromium oxide from chromite," *Process Saf. Environ. Prot.*, vol. 105, pp. 91–100, 2017.
[10] H. H. Zeiss and M. Tsutsui, "π-Complexes of the transition metals. I. Hein's polyaromatic chromium compounds 1, 2," *J. Am. Chem. Soc.*, vol. 79, no. 12, pp. 3062–3066, 1957.
[11] E. O. Fischer and W. Hafner, "Di-cyclopentadienyl-chrom," *Zeitschrift für Naturforsch. B*, vol. 8, no. 8, pp. 444–445, 1953.
[12] F. Hein, "Zur Frage der Struktur der Chrom-phenyl-Verbindungen. Bemerkungen zur Abhandlung von EO Fischer und D. Seus," *Chem. Ber.*, vol. 89, no. 8, pp. 1816–1821, 1956.
[13] G. Wilke et al., "Cyclooligomerisation von Butadien und Übergangsmetall-π-Komplexe," *Angew. Chemie*, vol. 75, no. 1, pp. 10–20, 1963.
[14] F. J. Karol, G. L. Karapinka, C. Wu, A. W. Dow, R. N. Johnson, and W. L. Carrick, "Chromocene catalysts for ethylene polymerization: Scope of the polymerization," *J. Polym. Sci. Part A-1 Polym. Chem.*, vol. 10, no. 9, pp. 2621–2637, 1972.
[15] J. H. Espenson, "Chemistry of organochromium (III) complexes," *Acc. Chem. Res.*, vol. 25, no. 5, pp. 222–227, 1992.
[16] W. A. Herrmann, *Synthetic Methods of Organometallic and Inorganic Chemistry*, vol. 8. Georg Thieme Verlag, 1996.
[17] K. H. Theopold, "Organochromium (III) chemistry: A neglected oxidation state," *Acc. Chem. Res.*, vol. 23, no. 8, pp. 263–270, 1990.
[18] K. S. Whiteley, T. G. Heggs, H. Koch, R. L. Mawer, and W. Immel, "Polyolefins," In *Ullmann's Encyclopedia of Industrial Chemistry*, 2000. https://doi.org/10.1002/14356007.a21_487
[19] R. Sneeden, *Organochromium Compounds*. Elsevier, 2012.
[20] N. G. Tsierkezos, "Investigation of electron-transfer kinetics for bis (benzene) chromium (1+/0) redox couple in acetonitrile/dichloromethane binary mixtures at 298.15 K," *Inorganica Chim. Acta*, vol. 360, no. 11, pp. 3626–3632, 2007.
[21] E. Samuel, D. Caurant, D. Gourier, C. Elschenbroich, and K. Agbaria, "Bis (benzene) chromium. A sandwich complex spin trap as revealed by ENDOR spectroscopy," *J. Am. Chem. Soc.*, vol. 120, no. 32, pp. 8088–8092, 1998.
[22] T. M. Gilbert, C. B. Bauer, and R. D. Rogers, "Structures of (η6-benzene dimethylacetal)-and (η6-benzene diethylacetal) chromium tricarbonyl: structural evidence for the near-electroneutrality of the dialkylacetal substituent," *J. Chem. Crystallogr.*, vol. 26, no. 5, pp. 355–360, 1996.
[23] E. P. Kündig, "Synthesis of transition metal η 6-arene complexes," *Transit. Met. Arene π-Complexes Org. Synth. Catal.*, pp. 3–20, 2004.
[24] K. H. Theopold, "Chromium: inorganic and coordination chemistry," In *Encyclopedia of Inorganic Chemistry* (Eds. R. B. King, R. H. Crabtree, C. M. Lukehart, D. A. Atwood and R. A. Scott), 2006, doi: 10.1002/0470862106.ia043.

[25] P. A. Lay and A. Levina, "Coordination chemistry of chromium," In *Reference Module in Chemistry, Molecular Sciences and Chemical Engineering*, (Ed. J. Reedijk), Elsevier Inc. doi: 10.1016/b978-0-12-409547-2.11126-6.

[26] J. B. Vincent, *Chromium: Properties and Determination*, 1st ed., no. III. Elsevier, 2015. doi: 10.1016/B978-0-12-384947-2.00161-6.

[27] J. Jiménez, B. Doistau, M. Poncet, and C. Piguet, "Heteroleptic trivalent chromium in coordination chemistry: Novel building blocks for addressing old challenges in multimetallic luminescent complexes," *Coordination Chemistry Reviews*, vol. 434, 2021, 213750, doi: 10.1016/j.ccr.2020.213750.

[28] P. A. Scattergood, "Recent advances in chromium coordination chemistry: luminescent materials and photocatalysis," In *Organometallic Chemistry*, Volume 43 (Eds. N. J. Patmore and P. I. P. Elliott), The Royal Society of Chemistry, 2020, pp. 1–34 2021. doi: 10.1039/9781788017077-00001.

[29] S. Shaik and P. C. Hiberty, "A chemist's guide to valence bond theory," *A Chemist's Guide to Valence Bond Theory*, pp. 1–316, John Wiley & Sons, Inc., 2007, doi: 10.1002/9780470192597.

[30] G. B. Kaufman, "Inorganic chemistry: principles of structure and reactivity, 4th ed. (Huheey, James E.; Keiter, Ellen A.; Keiter, Richard L.)," *J. Chem. Educ.*, vol. 70, no. 10, p. A279, Oct. 1993, doi: 10.1021/ed070pA279.1.

[31] F. M. Fernández, "Simple model for crystal field theory," no. March, 2016, [Online]. Available: http://arxiv.org/abs/1604.00340

[32] R. G. Burns, "Crystal field effects in chromium and its partitioning in the mantle," *Geochim. Cosmochim. Acta*, vol. 39, no. 6–7, pp. 857–864, 1975, doi: 10.1016/0016-7037(75)90031-9.

[33] H. J. Lunk, "Discovery, properties and applications of chromium and its compounds," *ChemTexts*, vol. 1, no. 1, 2015, doi: 10.1007/s40828-015-0007-z.

[34] J. S. Griffith, *The Theory of Transition-metal Ions*. Cambridge University Press, 1961.

[35] A. Bartecki and K. Kurzak, "Colour of 3d transition metal compounds. Part 1. Experimental quantitative studies and ligand-field based predictions for d l, d 4, d 6 and d 9 systems *Reviews in Inorganic Chemistry*, vol. 26, no. 4, 2006, pp. 389–442. https://doi.org/10.1515/REVIC.2006.26.4.389

[36] J. R. Jiménez, B. Doistau, M. Poncet, and C. Piguet, "Heteroleptic trivalent chromium in coordination chemistry: Novel building blocks for addressing old challenges in multimetallic luminescent complexes," *Coord. Chem. Rev.*, vol. 434, 2021, doi: 10.1016/j.ccr.2020.213750.

[37] S. Shanmukhappa and P. G. Ramappa, "Synthesis, spectral, magnetic, thermal and antibacterial properties of metal complexes of isothipendyl," *Synth. React. Inorg. Met. Chem.*, vol. 27, no. 7, 1997, pp. 967–983.

[38] M. S. Vandiver et al., "Effect of ancillary ligands on the DNA interaction of [Cr (diimine) 3] 3+ complexes containing the intercalating dipyridophenazine ligand," *Inorg. Chem.*, vol. 49, no. 3, pp. 839–848, 2010.

[39] A. M. McDaniel, H. W. Tseng, N. H. Damrauer, and M. P. Shores, "Synthesis and solution phase characterization of strongly photooxidizing heteroleptic Cr(III) tris-dipyridyl complexes," *Inorg. Chem.*, vol. 49, no. 17, pp. 7981–7991, 2010, doi: 10.1021/ic1009972.

Section II

Biological Properties of Se, V and Cr

4 Biological Properties of Selenium

Monika Kamboj, Sangeeta Bajpai, Garima Pandey and Bimal Krishan Banik

4.1 INTRODUCTION

Though selenium is regarded as toxic to humans, it has biological importance and plays an important role in human health, especially in the physiology of the brain. Klaus Schwartz and Calvin Foltz in 1957 established the role of Se as an essential element, responsible for preventing liver, muscle, and blood vessel lesions in rats and chickens [1,2]. It is required by all organisms to perform the proper functions of the body. It is well established that various neurodegenerative diseases may develop due to low levels of selenium in our bodies. Among the trace elements, selenium is the only element found in the enzyme's active site, in the form of amino acid selenocysteine. Selenoproteins are proteins with at least one selenocysteine residue that maintain cell redox balance and are responsible for many important physiological roles [3]. The biological effects of selenium are mostly due to the selenocysteine (one or many) residue in selenoprotein. Selenocysteine, an analog of cysteine, is the only amino acid with a Se atom and is the 21st amino acid [4a]. Various proteins contain selenocysteines and 30 such genes have been isolated [4b]. The essential source of Se is plants, both for humans and animals. Selenium enters the plants through the soil and then into the human and animal body when they consume it. Selenoprotein contains about 80 percent selenium [5]. The concentration of Se in our body depends on the amount of Se in various sources such as food, water, soil conditions, and plant crops. The level also shows variation among individuals. Selenium is an essential factor in several major metabolic routes such as the metabolism of the thyroid gland, immune system, defense, and reproductive system. Also, it is the cofactor of many enzymes like glutathione peroxidase, thioredoxin reductase, and deiodinase enzyme. Both glutathione peroxidase and thioredoxin reductase contain selenoprotein [6,7]. They catalyze biological processes and act as scavengers for free radicals and antioxidants. Approximately 25 selenoproteins have been identified in the human body [8a] and 30 in the mammalian system [8b]. High levels of selenium have adverse effects that may be associated with several diseases [8c].

4.2 SELENIUM METABOLISM

Selenium is an essential micronutrient. The World Health Organization (WHO) and Food and Agricultural Organization (FAO) recommended the daily Se intake levels in diet according to age [9] (Table 4.1). The daily intake of Se and its level found in

DOI: 10.1201/9781032615110-6

TABLE 4.1
Recommended daily Se intake in the diet

Age Group	Recommended Level (micrograms/day)
Infants (less than 1 year)	15–20
1- to 3-year-old children	20
4- to 8-year-old children	30
9- to 13-year-old children	40
Teens	55
Adults	55
Pregnant teens and women	60
Lactating teens and women	70

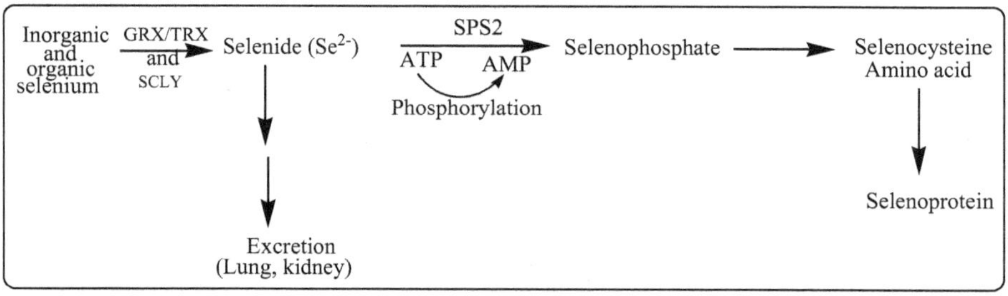

FIGURE 4.1 Metabolism of selenium.

humans vary from country to country [10–15]. The upper permissible level of Se is 400 micrograms/day. Exposure to high levels of Se may result in adverse effects. Se is essential for antioxidizing and immunity-boosting properties [15,16]. The metabolism of selenium occurs in a systemic procedure as shown in Figure 4.1.

The food or other supplementary diet is the source of selenium from where it is absorbed and then transformed into selenoprotein and finally excreted. The source of selenium from diet includes selenomethionine, selenocysteine (organic form), and selenite, selenate (inorganic form). The liver absorbs the Se and synthesizes selenoprotein P (SELENOP). Selenoprotein P transports selenium to various tissues and organs through blood circulation. Selenocysteine lyase enzyme metabolizes selenomethionine and selenocysteine to selenide. Glutathione and thioredoxin systems reduce the inorganic form of selenium to H_2Se. The hydrogen selenide is further reduced to selenophosphate by the enzyme selenophosphate synthetase 2. Selenide is transformed into the amino acid selenocysteine, which is incorporated into selenoprotein through translation as shown in Figure 4.1. Excretion of selenium takes place in the form of small metabolites formed by methylation through exhalation and urination [17].

4.3 BIOSYNTHESIS OF SELENOCYSTEINE AND SELENOPROTEINS

Selenoproteins are the group of proteins in which selenocysteine is incorporated. Selenocysteine, an analog of cysteine, in which Se replaces the S atom, is the only amino acid with a Se atom and is the 21st amino acid. The selenocysteine micronutrient acts as a biological catalyst in which Se is ionized at physiological pH [18].

Selenium is vital for the synthesis of selenoprotein, which is required in the regulation of thyroid hormones. It also acts as an antioxidant.

Glutathione peroxidases, thioredoxin reductase, Iodothyronine deiodinases, selenophosphate synthetase, and selenoprotein P are the major selenoproteins. Some of them are part of dietary supplements and multivitamins [19,20]. The biosynthesis of selenoprotein occurs by a complex process through several stages. The selenoprotein synthesis of selenoprotein differs in mechanism among prokaryotes, eukaryotes, and archaea [21]. UGA-selenocysteine codon, the specific tRNA(selenocysteinyl $tRNA^{(SeR)SeC}$), the selenocysteine insertion sequence SECIS element, and several protein factors are common features for all organisms.

Our body has no free residues of selenocysteine. Selenocysteine is synthesized on selenocysteinyl $tRNA^{(SeR)SeC}$, a specific tRNA. Selenocysteine is incorporated into amino acid through tRNA generating selenoprotein using UCA anticodon, complementary to UGA stop codon [22,23]. UGA codes for selenocysteine [24–26]. Synthesis begins with the acylation of $tRNA^{(SeR)SeC}$ with serine by seryl-tRNA synthetase in the presence of ATP. This results in the formation of Ser-$tRNA^{(SeR)Sec}$.

O-phosphoseryl-tRNA kinase (PSTK), selenophosphate synthetase 2 (SPS2), and Sec synthase (SepSerS) are the three enzymes that play a vital role in converting tRNA-bound serine residue into selenocysteine. The hydroxy group of serine in Ser-$tRNA^{(SeR)Sec}$ is further phosphorylated by PSTK and ATP to give *O*-phosphoseryl-$tRNA^{(SeR)Sec}$ [4a,27–29]. It is further substituted by a selenium atom to synthesize SeC-$tRNA^{(SeR)Sec}$ in the presence of selenophosphate synthetase 2 and Sec synthase (SepSerS). SeC-$tRNA^{(SeR)Sec}$ is then transferred to the A site of the ribosome. Selenide is phosphorylated to monoselenophosphate ($HSeH_2PO_3$) by selenophosphate synthetase 2 and adenosine triphosphate (ATP), which is further converted to selenocysteine amino acid. SPS2 contains its own SeC that contributes to the biosynthesis of selenoprotein. SeC-$tRNA^{(SeR)Sec}$ decodes codon UGA and selenocysteine is incorporated into amino acid arrangement, thereby forming selenoprotein. In the presence of selenocysteine insertion sequence (SECIS) and SECIS binding protein 2(SBP2), the UGA codon encodes a selenoprotein [30–33]. The biosynthesis of selenocysteine has been elucidated by Hatfield et al. [34]. The biosynthesis of selenocysteine and selenoproteins is depicted in Figure 4.2. The functions of various selenoproteins are depicted in Table 4.2 [35].

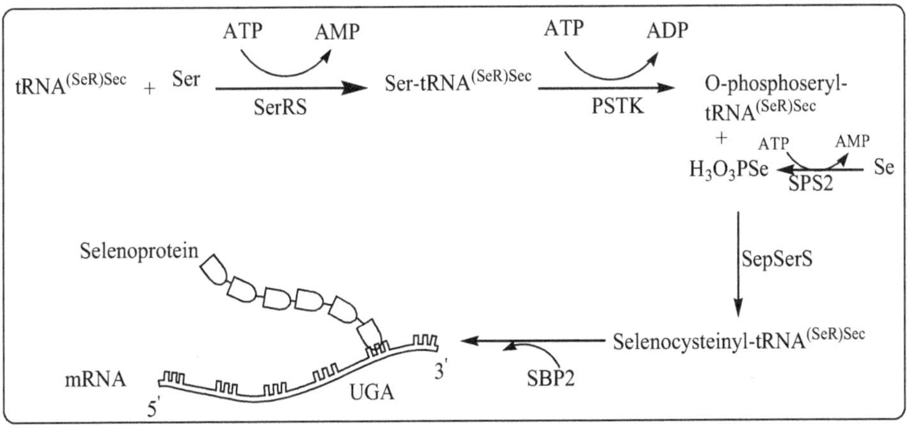

FIGURE 4.2 Pathway of biosynthesis of selenocysteine and selenoprotein.

TABLE 4.2
Types of selenoprotein in mammals and their functions

Class of Selenoprotein/ Selenoenzyme	Symbol (according to new selenoprotein gene nomenclature) [36]	Synonyms	Type	Location	Activity	Function	References
Glutathione peroxidase (GPX)	GPX1	Cytosolic glutathione peroxidase, GSHPX1	Mammalian antioxidant selenoenzymes	GPX1: lungs, kidneys, liver, and erythrocytes GPX2: Tissues of gastro intestine, lung, liver, and skin GPX3: plasma and extracellular fluid GPX4: mitochondria, cytoplasmic, and nucleus GPX6: embryonic tissues	Free radical scavenger (antioxidant)	Protect the cell damage by reducing • Hydrogen peroxide to water • lipid hydroperoxides to corresponding ethers	[7,37–41]
	GPX2	GSHPX-GI					
	GPX3	Plasma glutathione peroxidase					
	GPX4	Phospholipid hydroperoxide glutathione peroxidase, PHGPX					
	GPX6						
Iodothyronine deiodinase (DIOs)	DIO1	D1	Membrane protein selenoenzymes	D1O1: plasma membrane proteins present in the thyroid, liver, kidneys, and pituitary gland, D1O2: endoplasmic reticulum (ER) membrane in thyroid, central nervous system, and skeletal muscle	Thyroid hormone metabolism	• Convert thyroxine into triiodothyronine (active form) • Participate in oxidation-reduction enzymatic reaction • •	[42–46]
	DIO2	D2					
	DIO3	D3					

				DI03: plasma membrane proteins, present in the embryonic and neonatal tissues, uterus, and central nervous system.			
Thioredoxin reductase (TrxR)	*TXNRD1* *TXNRD2* *TXNRD3*	TR1, TRXR1 TRXR2, TR3, mitochondrial thioredoxin reductase TGR, TRXR3, TR2	selenoenzymes	TrxRl:cytoplasm TrxR2: mitochondria, TrxR3: testes. cell	Antioxidants	Reduces oxidized thioredoxin that provides a reducing equivalent to the disulfide bond of thioredoxin reductase, ribonucleotide reductase, thioredoxin peroxidase, and some transcription factors acting as a cell growth factor in DNA synthesis	[7, 47–50]
Selenophosphate synthetase 2	*SEPHS2*	SPS2	selenoenzymes	-	selenoprotein synthesis regulatory enzyme in mammals, utilizes Selenium	catalyzes the synthesis of selenophosphate (Figure 4.1) in the presence of ATP, selenide is phospahorylated to monoselenophosphate ($HSeH_2PO_3$), thereby providing active selenium donor	[51–57]

(*continued*)

TABLE 4.2 (Continued)
Types of selenoprotein in mammals and their functions

Class of Selenoprotein/ Selenoenzyme	Symbol (according to new selenoprotein gene nomenclature) [36]	Synonyms	Type	Location	Activity	Function	References
Methionine sulfoxide reductase Bl (Selenoprotein R)	*MSRB1*	SELR, SELX, SEPX1	selenoenzymes	Cytosol and nucleus	Antioxidant	Reduces methionine sulfoxide to methionine and maintains intracellular redox balance. Repair, regulate, and protect damaged proteins. It interacts with many proteins and is responsible for neurodegeneration thereby protecting oxidative damage to the brain	[46, 58–60]
Selenoprotein F (SELENOF)	SELENOF	Selenoprotein 15, SEP15	Selenoprotein	endoplasmic reticulum (ER) organelle	Oxidoreductase	Responsible for protein folding and the secretion process in the endoplasmic reticulum	[61–65]
Selenoprotein H (SELENOH)	SELENOH	SELH, C11 orf31	DNA binding-Selenoprotein	nucleolus	Oxidoreductase	protect neurons against UVB-induced damage by inhibiting apoptotic cell death pathways, promote mitochondrial	[66–76]

						biogenesis and mitochondrial function, and suppress cellular senescence through genome maintenance, redox regulation, and act as a transcription factor	
Selenoprotein I	SELENOI	SELI, EPT1	Selenoprotein	It is present in the endoplasmic reticulum and located in the brain, placenta, liver, pancreas, and small intestine, heart, skeletal muscle, lung, kidney, etc.	ethanolamine phospho transferase and antioxidant property	Essential for neuronal growth and phospholipid signaling pathways.	[77–80]
Selenoprotein K	SELENOK	SELK	Selenoprotein	transmembrane selenoprotein, localized in the endoplasmic reticulum (ER)	antioxidant and peroxidase activity	oxidation resistance, calcium (Ca^{2+}) flux regulation, and the ER-associated degradation (ERAD) pathway in immune cells. protection of cells from ER stress-induced apoptosis defense, calcium regulation and the ER-associated protein degradation pathway. Cellular redox balance	[81–88]

(continued)

TABLE 4.2 (Continued)
Types of selenoprotein in mammals and their functions

Class of Selenoprotein/ Selenoenzyme	Symbol (according to new selenoprotein gene nomenclature) [36]	Synonyms	Type	Location	Activity	Function	References
Selenoprotein M	SELENOM	SELM, SEPM	Selenoprotein	perinuclear region of the brain	Oxidoreductase	It mediates the exchange of thiol-disulfide and promotes intracellular calcium homeostasis. Also protects against oxidative stress and has neuroprotective properties. Also responsible for protein folding	[8, 89–93]
Selenoprotein N	SELENON	SEPN1, SELN	Selenoprotein (type II transmembrane protein)	transmembrane selenoprotein, localized in the endoplasmic reticulum (ER)	Antioxidant and peroxidase activity	Responsible for refilling calcium stores in the ER. It senses the fluctuations in calcium level and responds to depletion in calcium level. It acts as a reductase to refill calcium stores. The essential role is proper muscle development and protein folding	[94–100]

Selenoprotein O	SELENOO	SELO	Selenoprotein	Located in the Mitochondria and expressed in the brain, heart, liver, kidneys, lungs, and stomach	Redox activity	It catalyzes AMPylation (addition of AMP from ATP to a hydroxyl side chain of protein during post-translational modification) and protects against oxidative stress. Play an essential role in Chondrogenesis	[101–104]
Selenoprotein P	SELENOP	SEPP1, SeP, SELP, SEPP	Selenoprotein	Selenoprotein P is a very abundant glycoprotein in plasma and is formed by two domains: the C-terminal very rich in selenocysteine (10 residues) and the N-terminal, larger than the C-terminal. It is also present in the liver, brain, and testes	Antioxidant	helps to eliminate peroxynitrite formed in the reaction of superoxide and nitric oxide. Also store and transport selenium	[105–107]
Selenoprotein S	SELENOS	SELS, SEPS1, VIMP	transmembrane Selenoprotein	Located in ER. Present in tissues of the liver, kidneys, adipose tissue, skeletal muscle, pancreatic islets, and blood vessels	Reductase	It is associated with inflammation processes and endoplasmic reticulum (ER)--associated degradation by removing misfolded protein. Also, protects against oxidative and endoplasmic stress	[108–114]

(continued)

TABLE 4.2 (Continued)
Types of selenoprotein in mammals and their functions

Class of Selenoprotein/ Selenoenzyme	Symbol (according to new selenoprotein gene nomenclature) [36]	Synonyms	Type	Location	Activity	Function	References
Selenoprotein T	SELENOT	SELT	Selenoprotein	It is located in the Golgi apparatus and ER and is present in the plasma membrane. abundant in endocrine organs such as the pituitary, pancreas, thyroid, and testis.	Oxidoreductase	Protect the neuron cell against oxidative stress. Play an essential role in hormone production	[115–122]
Selenoprotein V	SELENOV	SELV	Selenoprotein	It is located in seminiferous tubules of the testes	Redox activity	The main function is the redox-related process	[123–126]
Selenoprotein W	SELENOW	SELW, SEPW1	Selenoprotein	Cytoplasm	Antioxidant	Involved in the development of muscular degeneration. Plays an important role in myopathies of selenium deficiency	[127–133]

4.4 CONCLUSION

Selenium is an essential micronutrient, required by our body to synthesize a biological catalyst selenocysteine, which is further incorporated into selenoprotein. Selenium has a crucial role in the synthesis of selenoprotein, which regulates various enzymatic activities and regulates the metabolism of thyroid hormones. The role of selenium in maintaining the various physiological functions of human beings is carried out by selenoprotein containing amino acids, selenocysteine. Selenium is absorbed in the form of selenoprotein. So, selenium performs its biological function through selenoprotein. Selenocysteine, an analog of cysteine, is the core biological form of selenoprotein that is found in different parts of the cell and is responsible for various functions of the body. There is a narrow range of safety for this micronutrient so it is important to maintain a proper level of Se. Deficiency and excess of this micronutrient may cause adverse effects on the health of humans. This makes selenium essential for the well-being of humans.

REFERENCES

1. Kieliszek, M. Selenium—Fascinating microelement: Properties and sources in food. *Molecules* **2019**, *24* (7), 1298. DOI: 10.3390/molecules24071298
2. Schwarz, K.; Foltz, C. M. Selenium as an integral part of factor 3 against dietary necrotic liver degeneration. *J. Am. Chem. Soc.* **1957**, *79* (12), 3292–3293. DOI: 10.1021/ja01569a087
3. Kuršvietienė, L.; Mongirdienė, A.; Bernatonienė, J.; Šulinskienė, J.; Stanevičienė, I. Selenium anticancer properties and impact on cellular redox status. *Antioxidants (Basel)* **2020**, *9*(1), 80. DOI: 10.3390/antiox9010080
4. (a) Schmidt, R.L.; Simonović, M. Synthesis and decoding of selenocysteine and human health. *Croat Med J.* 2012, *53*(6):535–50. doi: 10.3325/cmj.2012.53.535; (b) Sarode, V.; Kothapalli, L.; Thomas, A.; Sarode, V.; Bhole, R. Organoselenium compounds as potential therapeutic agents for treating brain disorders, *Curr. Ind. Sci.*, **2023**, *1*, e271022210428.
5. Fan, A. M.; Book, S. A.; Neutra, R. R.; Epstein, D. M. Selenium and human health implications in California's San Joaquin Valley. *J. Toxicol. Environ. Health* **1988**, *23*(4), 539–559. DOI: 10.1080/15287398809531135
6. Liu, Z. Antioxidant activity of the thioredoxin system. *Biophys. Rep.* **2023**, *9*(1), 26–32. DOI: 10.52601/bpr.2023.230002
7. Lewin, M. H.; Hume, R.; Howie, A. F.; Richard, K.; Arthur, J. R.; Nicol, F.; Walker, S. W.; Beckett, G. J. Thioredoxin reductase and cytoplasmic glutathione peroxidase activity in human foetal and neonatal liver. *Biochim. Biophys. Acta* **2001**, *1526*(3), 237–241. DOI: 10.1016/s0304-4165(01)00133-7
8. (a) Nunes, L. G. A.; Cain, A.; Comyns, C.; Hoffmann, P. R.; Krahn, N. Deciphering the role of selenoprotein M. *Antioxidants (Basel)* 2023, *12*(11), 1906. DOI: 10.3390/antiox12111906. (b) Perrone, D.; Monteiro, M.; Nunes, J. C. The chemistry of selenium in Selenium: chemistry, analysis, function and effects, ed. V. R. Preedy, *The Royal Society of Chemistry*, 2015, pp. 3–15. (c) Genchi, G.; Lauria, G.; Catalano, A.; Sinicropi, M.S.; Carocci, A. Biological activity of selenium and its impact on human health. *Int. J. Mol. Sci.* **2023**, *24*, 2633. https://doi.org/10.3390/ijms24032633es

9. Institute of Medicine (US) Panel on Dietary Antioxidants and Related Compounds. *Dietary Reference Intakes for Vitamin C, Vitamin E, Selenium, and Carotenoids.* Washington, DC: National Academies Press (US); **2000**. PMID: 25077263.
10. Wu, J.; Xu, G. L. Plasma selenium content, platelet glutathione peroxidase and superoxide dismutase activity of residents in Kashin-Beck disease affected area in China. *J. Trace Elem. Electrolytes Health Dis.* **1987**, *1* (1), 39–43.
11. Alfthan, G.; Eurola, M.; Ekholm, P.; Venäläinen, E.-R.; Root, T.; Korkalainen, K.; Hartikainen, H.; Salminen, P.; Hietaniemi, V.; Aspila, P.; Aro, A.; Selenium Working Group. Effects of nationwide addition of selenium to fertilizers on foods, and animal and human health in Finland: From deficiency to optimal selenium status of the population. *J. Trace Elem. Med. Biol.*, **2015**, *31*, 142–147. DOI: https://doi.org/10.1016/j.jtemb.2014.04.009
12. Aaseth, J.; Alexander, J.; Bjørklund, G.; Hestad, K.; Dusek, P.; Roos, P. M.; Alehagen, U. Treatment strategies in Alzheimer's disease: A review with focus on selenium supplementation. *Biometals* **2016**, *29*(5), 827–839. DOI: 10.1007/s10534-016-9959-8 (Epub August 16, 2016).
13. Stoffaneller, R.; Morse, N. L. A review of dietary selenium intake and selenium status in Europe and the Middle East. *Nutrients* **2015**, *7*(3), 1494–1537. DOI: 10.3390/nu7031494
14. Wang, Y.; Rijntjes, E.; Wu, Q.; Lv, H.; Gao, C.; Shi, B.; Schomburg, L. Selenium deficiency is linearly associated with hypoglycemia in healthy adults. *Redox Biol.*, **2020**, *37*, 101709. DOI: https://doi.org/10.1016/j.redox.2020.101709
15. Sun, Y.; Wang, Z.; Gong, P.; Yao, W.; Ba, Q.; Wang, H., Review on the health-promoting effect of adequate selenium status. *Front. Nutr.* **2023**, *10*, 1136458. DOI: 10.3389/fnut.2023.1136458
16. National Center for Biotechnology Information. PubChem Pathway Summary for Pathway WP28, Selenium Metabolism and Selenoproteins, Source: WikiPathways. https://pubchem.ncbi.nlm.nih.gov/pathway/WikiPathways:WP28. (2024).
17. Hassoun, B. S.; Palmer, I. S.; Dwivedi, C. Selenium detoxification by methylation. *Res. Commun. Mol. Pathol. Pharmacol.* **1995**, *90*, 133–142.
18. Arthur, J. R.; Brown, K. M.; Fairweather-Tait, S. J.; Crews, H.M., Dietary selenium: why do we need it and how much is enough? *Nutr. Food Sci.* **(1997)**, *97*(6), 225–228. DOI: https://doi.org/10.1108/00346659710180343
19. Mugesh, G.; du Mont, W. W.; Sies, H. Chemistry of biologically important synthetic organoselenium compounds. *Chem. Rev.* **2001**, *101* (7), 2125–2179. DOI: 10.1021/cr000426w
20. Steinbrenner, H.; Speckmann, B.; Klotz, L.-O. Selenoproteins: Antioxidant selenoenzymes and beyond. *Arch. Biochem. Biophys.* **2016**, 595, 113–119. DOI: https://doi.org/10.1016/j.abb.2015.06.024
21. (a) Papp, L. L. V.; Lu, J .; Holmgren, A.; Khanna, K. K. From selenium to selenoproteins: synthesis, identity, and their role in human health. *Antioxid. Redox Signal.* **2007**, 9 (7), 775–806. DOI: 10.1089/ars.2007.1528. (b) Carlson, B. A.; Xu, X. M.; Kryukov, G. V.; Rao, M.; Berry, M. J.; Gladyshev, V. N.; Hatfield, D. L. Identification and characterization of phosphoseryl tRNA[ser]sec kinase. *Proc. Natl. Acad. Sci. U. S. A.* **2004**, *101*(35), 12848–12853. DOI: 10.1073/pnas.0402636101, Google Scholar.
22. Minich, W. B. Selenium metabolism and biosynthesis of selenoproteins in the human body. *Biochemistry (Mosc)* **2022**, *87* (Suppl.1), S168–S102. DOI: CrossRef10.1134/S0006297922140139

23. Flohé, L. The labour pains of biochemical selenology: the history of selenoprotein biosynthesis. *Biochim. Biophys. Acta* **2009**, *1790* (11), 1389–1403. DOI: 10.1016/j.bbagen.2009.03.031
24. Squires, J. E.; Berry, M. J. Eukaryotic selenoprotein synthesis: mechanistic insight incorporating new factors and new functions for old factors. *IUBMB Life* **2008**, *60* (4), 232–235. DOI: 10.1002/iub.38
25. Bulteau, A. L.; Chavatte, L. Update on selenoprotein biosynthesis. *Antioxid. Redox Signal.* 2015, *23*(10), 775–794. DOI: CrossRef10.1089/ars.2015.6391
26. Vindry, C.; Ohlmann, T.; Chavatte, L. Translation regulation of mammalian selenoproteins. *Biochim. Biophys. Acta Gen. Subj.* 2018, *1862*(11), 2480–2492. DOI: 10.1016/j.bbagen.2018.05.010
27. Puppala, A.K.; Castillo Suchkou, J.; French, R. R. L.; Kiernan, K. A.; Simonović, M. Structural basis for the tRNA-dependent activation of the terminal complex of selenocysteine synthesis in humans. *Nucleic Acids Res.*, 2023, *51*(8), 4012–4026. DOI: https://doi.org/10.1093/nar/gkad182
28. Yuan, J.; Palioura, S.; Salazar, J. C.; Su, D.; O'Donoghue, P.; Hohn, M. J.; Cardoso, A. M.; Whitman, W. B.; Söll, D. RNA-dependent conversion of phosphoserine forms selenocysteine in eukaryotes and archaea. *Proc. Natl. Acad. Sci. U. S. A.*, 2006, *103* (50), 18923–18927. DOI: 10.1073/pnas.0609703104
29. Xu, X. M.; Carlson, B. A.; Mix, H.; Zhang, Y.; Saira, K.; Glass, R. S.; Berry, M. J.; Gladyshev, V. N.; Hatfield, D. L. Biosynthesis of Selenocysteine on Its tRNA in Eukaryotes. *PLoS Biol.* 2007, *5* (1), e4. DOI: https://doi.org/10.1371/journal.pbio.0050004
30. Lu, J.; Holmgren, A. Selenoproteins. *J. Biol. Chem.* 2009, *284* (2), 723–727. DOI: 10.1074/jbc.R800045200
31. Seeher, S.; Mahdi, Y.; Schweizer, U. Post-transcriptional control of selenoprotein biosynthesis. *Curr. Protein Pept. Sci.* 2012, *13*(4), 337–346. DOI: 10.2174/138920312801619448
32. Hatfield, D. L.; Tsuji, P. A.; Carlson, B. A.; Gladyshev, V. N. Selenium and selenium cysteine: Roles in cancer, health, and development. *Trends Biochem. Sci.* 2014, *39*(3), 112–120. DOI: CrossRef10.1016/j.tibs.2013.12.007
33. Bubenik, J. L.; Miniard, A. C.; Driscoll, D. M. Characterization of the UGA-recoding and SECIS-binding activities of SECIS-binding protein 2. *RNA Biol.* 2014, *11*(11), 1402–1413. DOI: CrossRef10.1080/15476286.2014.996472
34. Turanov, A. A.; Xu, X. M.; Carlson, B. A.; Yoo, M. H.; Gladyshev, V. N.; Hatfield, D. L. Biosynthesis of selenocysteine, the 21st amino acid in the genetic code, and a novel pathway for cysteine biosynthesis. *Adv. Nutr.* 2011, *2* (2), 122–128. DOI: 10.3945/an.110.000265.
35. (a) Bellinger, F. P.; Raman, A. V.; Reeves, M. A.; Berry, M. J. Regulation and function of selenoproteins in human disease. *Biochem. J.* 2009, *422* (1), 11–22. DOI: 10.1042/BJ20090219. (b) Zhang, Y.; Roh, Y.J.; Han, S.-J.; Park, I.; Lee, H.; Zhang, Y.; Roh, Y. J.; Han, S. J.; Park, I.; Lee, H. M.; Ok, Y. S.; Lee, B. C.; Lee, S. R. Role of selenoproteins in redox regulation of signaling and the antioxidant system: A review. *Antioxidants (Basel).* (2020), *9*(5), 383. DOI: 10.3390/antiox9050383. Common to all (c) Kryukov, G. V.; Castellano, S.; Novoselov, S. V.; Lobanov, A. V.; Zehtab, O.; Guigó, R.; Gladyshev, V. N. Characterization of mammalian selenoproteomes. *Science* 2003, *300*(5624), 1439–1443. DOI: 10.1126/science.1083516.
36. Gladyshev, V. N.; Arnér, E. S.; Berry, M. J.; Brigelius-Flohé, R.; Bruford, E. A.; Burk, R. F.; Carlson, B. A.; Castellano, S.; Chavatte, L.; Conrad, M.; Copeland, P. R.;

Diamond, A. M.; Driscoll, D. M.; Ferreiro, A.; Flohé, L.; Green, F. R.; Guigó, R.; Handy, D. E.; Hatfield, D. L.; Hesketh, J.; Hoffmann, P. R.; Holmgren, A.; Hondal, R. J.; Howard, M. T.; Huang, K.; Kim, H.-Y.; Kim, IckI. Y.; Köhrle, J.; Krol, A.; Kryukov, G. V.; Lee, B. J.; Lee, B. C.; Lei, Xin GenX. G.; Liu, Q.; Lescure, A.; Lobanov, A. V.; Loscalzo, J.; Maiorino, M.; Mariotti, M.; Sandeep Prabhu, K.; Rayman, M. P.; Rozovsky, S.; Salinas, G.; Schmidt, E. E.; Schomburg, L.; Schweizer, U.; Simonović, M.; Sunde, R. A.; Tsuji, P. A.; Tweedie, S.; Ursini, F.; Whanger, P. D.; Zhang, Y. Selenoprotein gene nomenclature. *J. Biol. Chem.*, **2016**, 291 (46), 24036–24040, ISSN 0021-9258. DOI: https://doi.org/10.1074/jbc.M116.756155

37. Zhao, L.; Zong, W.; Zhang, H.; Liu, R. Kidney toxicity and response of selenium containing protein-glutathione peroxidase (Gpx3) to CdTe QDs on different levels. *Toxicol. Sci.* **2019**, *168* (1), 201–208. DOI: 10.1093/toxsci/kfy297
38. Pei, J.; Pan, X.; Wei, G.; Hua, Y. Research progress of glutathione peroxidase family (GPX) in redoxidation. *Front. Pharmacol.* **2023**, *14*, 1147414. DOI: 10.3389/fphar.2023.1147414
39. Barbosa, N. V.; Nogueira, C. W.; Nogara, P. A.; de Bem, A. F.; Aschner, M.; Rocha, J. B. T. Organoselenium compounds as mimics of selenoproteins and thiol modifier agents. *Metallomics* **2017**, *9* (12), 1703–1734. DOI: 10.1039/c7mt00083a
40. Brigelius-Flohé, R.; Maiorino, M. Glutathione peroxidases. *Biochim. Biophys. Acta* **2013**, *51830* (5), 3289–3303. DOI: 10.1016/j.bbagen.2012.11.020
41. Handy, D. E.; Loscalzo, J. The role of glutathione peroxidase-1 in health and disease. *Free Radic. Biol. Med.* **2022**, *188*, 146–161. doiDOI: 10.1016/j.freeradbiomed.2022.06.004
42. Ren, X. M.; Zhang, L.; Xin, B.; Liu, Q. L.; Qian, W. W.; Zhang, R. Q. Biological functions of selenoprotein iodothyronine deiodinase and its expression in osteoarthritis. *Chin. Med. Sci. J.* **2022**, *37* (2), 142–150. DOI: 10.24920/003991.
43. Antoniak, H. L.; Babych, N. O.; Solohub, L. I.; Snityns'kyĭ, V. V. Rol' ĭodotyronin-deĭodynaz u Mekhanizmakh Diï Tireoïdnykh Hormoniv u Klitynakh Tvaryn Ta Liudyny [Role of iodothyronine-deiodinase in thyroid hormone mechanisms in animal and human cells]. *Ukr. Biokhim. Zh.* **1999**, *74* (1):5–18. Ukrainian., *2002* Jan-Feb
44. Paragliola, R. M.; Corsello, A.; Concolino, P.; Ianni, F.; Papi, G.; Pontecorvi, A.; Corsello, S. M. Iodothyronine deiodinases and reduced sensitivity to thyroid hormones. *Front. Biosci. (Landmark Ed.)* **2020**, *25* (2), 201–228. DOI: 10.2741/4803
45. Xu, X. M.; Carlson, B. A.; Irons, R.; Mix, H.; Zhong, N.; Gladyshev, V. N.; Hatfield, D. L. Selenophosphate synthetase 2 is essential for selenoprotein biosynthesis. *Biochem. J.* **2007**, *404* (1), 115–120. DOI: 10.1042/BJ20070165
46. Kim, H. Y. The methionine sulfoxide reduction system: Selenium utilization and methionine sulfoxide reductase enzymes and their functions. *Antioxid. Redox Signal.* **2013**, *19* (9), 958–969. DOI: 10.1089/ars.2012.5081.
47. Mustacich, D.; Powis, G. Thioredoxin reductase. *Biochem. J.* **2000**, *346*(1) (Pt 1), 1–8. DOI: 10.1042/bj3460001
48. Salmain, M.; Gaschard, M.; Baroud, M.; Lepeltier, E.; Jaouen, G.; Passirani, C.; Vessières, A. Thioredoxin reductase and organometallic complexes: A pivotal system to tackle multidrug resistant tumors? *Cancers (Basel)* **2023**, *15*(18), 4448. DOI: 10.3390/cancers15184448
49. Söderberg, A.; Sahaf, B.; Rosén, A. Thioredoxin reductase, a redox-active selenoprotein, is secreted by normal and neoplastic cells: presence in human plasma. *Cancer Res.* **2000**, *60*(8), 2281–2289.

50. Greetham, D.; Grant, C.M. Antioxidant activity of the yeast mitochondrial one-Cys peroxiredoxin is dependent on thioredoxin reductase and glutathione in vivo. *Mol Cell Biol.* **2009**, *29*(11):3229–3240. 10.1128/MCB.01918-08
51. Xu, X. M.; Carlson, B. A.; Irons, R.; Mix, H.; Zhong, N.; Gladyshev, V. N.; Hatfield, D. L. Selenophosphate synthetase 2 is essential for selenoprotein biosynthesis. *Biochem. J.* **2007**, *404*(1), 115–120. DOI: 10.1042/BJ20070165
52. Guimarães, M. J.; Peterson, D.; Vicari, A.; Cocks, B. G.; Copeland, N. G.; Gilbert, D. J.; Jenkins, N. A.; Ferrick, D. A.; Kastelein, R. A.; Bazan, J. F.; Zlotnik, A. Identification of a novel selD homolog from eukaryotes, bacteria, and archaea: Is there an autoregulatory mechanism in selenocysteine metabolism? *Proc. Natl. Acad. Sci. U. S. A.* **1996**, *93* (26), 15086–15091. 10.1073/pnas.93.26.15086
53. Kim, I. Y.; Guimarães, M. J.; Zlotnik, A.; Bazan, J. F.; Stadtman, T. C. Fetal mouse selenophosphate synthetase 2 (SPS2): Characterization of the cysteine mutant form overproduced in a baculovirus-insect cell system. *Proc. Natl. Acad. Sci. U.S.A.* **1997**, *94*(2), 418–421. 10.1073/pnas.94.2.418
54. Nunziata, C.; Polo, A.; Sorice, A.; Capone, F.; Accardo, M.; Guerriero, E.; Marino, F. Z.; Orditura, M.; Budillon, A.; Costantini, S. Structural analysis of human SEPHS2 protein, a selenocysteine machinery component, over-expressed in triple negative breast cancer. *Sci. Rep.* **2019,** *9*(1), 16131. DOI: 10.1038/s41598-019-52718-0
55. Itoh, Y.; Sekine, S.; Matsumoto, E.; Akasaka, R.; Takemoto, C.; Shirouzu, M.; Yokoyama, S. Structure of selenophosphate synthetase essential for selenium incorporation into proteins and RNAs. *J. Mol. Biol.* **2009**, *385*(5), 1456–1469. DOI: 10.1016/j.jmb.2008.08.042.
56. Oudouhou, F.; Casu, B.; Dopgwa Puemi, A. S.; Sygusch, J.; Baron, C. Analysis of novel interactions between components of the selenocysteine biosynthesis pathway, SEPHS1, SEPHS2, SEPSECS, and SECp43. *Biochemistry* **2017**, *56*(17), 2261–2270. DOI: 10.1021/acs.biochem.6b01116.
57. Manta, B.; Makarova, N. E.; Mariotti, M. The selenophosphate synthetase family: A review. *Free Radic. Biol. Med.*, **2022**, *192*, 63–76. DOI: https://doi.org/10.1016/j.freeradbiomed.2022.09.007
58. Kryukov, G. V.; Kumar, R. A.; Koc, A.; Sun, Z.; Gladyshev, V. N. Selenoprotein R is a zinc-containing stereo-specific methionine sulfoxide reductase. *Proc. Natl. Acad. Sci. U. S. A.* **2002**, *99*(7), 4245–4250. DOI: 10.1073/pnas.072603099
59. Shi, T.; Yang, Y.; Liu, Q.; Li, N.; Selenoprotein, R. A unique methionine sulfoxide reductase. *Prog. Chem.* **2018**, *30* (10), 1496–1502. DOI: 10.7536/PC180625
60. Chen, P.; Wang, C.; Ma, X.; Zhang, Y.; Liu, Q.; Qiu, S.; Liu, Q.; Tian, J.; Ni, J. Direct interaction of selenoprotein R with clusterin and its possible role in Alzheimer's disease. *PLOS ONE* **(2013)**, *8* (6), e66384. DOI: https://doi.org/10.1371/journal.pone.0066384
61. Tian, J.; Liu, J.; Li, J.; Zheng, J.; Chen, L.; Wang, Y.; Liu, Q.; Ni, J. The interaction of selenoprotein F (SELENOF) with retinol dehydrogenase 11 (RDH11) implied a role of SELENOF in vitamin A metabolism. *Nutr. Metab. (Lond)* **2018**, *15*, 7. DOI: 10.1186/s12986-017-0235-x
62. Li, M.; Zhang, Y.; Zhou, J.; Liu, H.; Selenoprotein, F. Selenoprotein F knockout caused glucose metabolism disorder in young mice by disrupting redox homeostasis. *Antioxidants (Basel)* **2022**, *11* (11), 2105. DOI: https://doi.org/10.3390/antiox11112105
63. Liao, P. S.; Liu, H. M.; He, C. M. Chemical synthesis of human selenoprotein F and elucidation of its thiol-disulfide oxidoreductase activity. *Chem. Sci.* **2022**, *13* (21), 6322–6327. DOI: 10.1039/d2sc00492e.

64. Ren, B. Y.; Liu, M.; Ni, J. Z.; Tian, J. Role of selenoprotein F in protein folding and secretion: Potential involvement in human disease. *Nutrients* **2018**, *10*(11), 1619. DOI: 10.3390/nu10111619
65. Zheng, X. X.; Ren, B. Y.; Li, X. M.; Yan, H. H.; Xie, Q. G.; Liu, H. M.; Zhou, J.; Tian, J.; Huang, K. X. Selenoprotein F knockout leads to glucose and lipid metabolism disorders in mice. *J. Biol. Inorg. Chem.* **2020**, *25*(7), 1009–1022. DOI: 10.1007/s00775-020-01821-z
66. (a) Bertz, M.; Kühn, K.; Koeberle, S. C.; Müller, M. F.; Hoelzer, D.; Thies, K.; Deubel, S.; Thierbach, R.; Kipp, A. P. Selenoprotein H controls cell cycle progression and proliferation of human colorectal cancer cells. *Free Radic. Biol. Med.* **2018**, *127* , 98–107. DOI: 10.1016/j.freeradbiomed.2018.01.010. (b) Cao, L.; Pechan, T.; Lee, S.; Cheng, W. H. Identification of selenoprotein H isoforms and impact of selenoprotein H overexpression on protein but not mRNA levels of 2 other selenoproteins in 293T cells. *J. Nutr.* **2021**, *151*(11), 3329–3338. DOI: 10.1093/jn/nxab290. Erratum in: *J. Nutr.* 2021, *151*(11), 3597. DOI: 10.1093/jn/nxab290. (c) Barage, S. H.; Deobagkar, D. D.; Baladhye, V. B. Characterization of structural and functional role of selenocysteine in selenoprotein H and its impact on DNA binding. *Amino Acids* **2018**, *50*(5), 593–607. DOI: 10.1007/s00726-018-2543-5.
67. Novoselov, S. V.; Kryukov, G. V.; Xu, X. M.; Carlson, B. A.; Hatfield, D. L.; Gladyshev, V. N. Selenoprotein H is a nucleolar thioredoxin-like protein with a unique expression pattern. *J. Biol. Chem.* **2007**, *282*(16), 11960–11968. DOI: 10.1074/jbc.M701605200
68. Ben Jilani, K. E.; Panee, J.; He, Q.; Berry, M. J.; Li, P. A. Overexpression of selenoprotein H reduces ht22 neuronal cell death after UVB irradiation by preventing superoxide formation. *Int. J. Biol. Sci.* **2007**, *3*(4), 198–204. DOI: 10.7150/ijbs.3.198
69. Mendelev, N.; Witherspoon, S.; Li, P. A. Overexpression of human selenoprotein H in neuronal cells ameliorates ultraviolet irradiation-induced damage by modulating cell signaling pathways. *Exp. Neurol.* **2009**, *220*(2), 328–334. DOI: 10.1016/j.expneurol.2009.09.008
70. Mehta, S. L.; Mendelev, N.; Kumari, S.; Andy Li, P. AndyA. Overexpression of human selenoprotein H in neuronal cells enhances mitochondrial biogenesis and function through activation of protein kinase A, protein kinase B, and cyclic adenosine monophosphate response element-binding protein pathway. *Int. J. Biochem. Cell Biol.* **2013**, *45* (3), 604–611. DOI: 10.1016/j.biocel.2012.11.022
71. Wu, R. T.; Cao, L.; Chen, B. P.; Cheng, W. H. Selenoprotein H suppresses cellular senescence through genome maintenance and redox regulation. *J. Biol. Chem.* **2014**, *289* (49), 34378–34388. DOI: 10.1074/jbc.M114.611970
72. Mendelev, N.; Witherspoon, S.; Li, P. A. Overexpression of human selenoprotein H in neuronal cells ameliorates ultraviolet irradiation-induced damage by modulating cell signaling pathways. *Exp. Neurol.* **2009**, *220* (2), 328–334. (PMC free article). DOI: 10.1016/j.expneurol.2009.09.008
73. Cox, A. G.; Tsomides, A.; Kim, A. J.; Saunders, D.; Hwang, K. L.; Evason, K. J.; Heidel, J.; Brown, K. K.; Yuan, M.; Lien, E. C.,; Lee, B. C.,; Nissim, S.,; Dickinson, B.,; Chhangawala, S.,; Chang, C. J.,; Asara, J. M.,; Houvras, Y.,; Gladyshev, V. N.; Goessling, W. Selenoprotein H is an essential regulator of redox homeostasis that cooperates with p53 in development and tumorigenesis. *Proc. Natl. Acad. Sci. U. S. A.* **2016**, *113* (38), E5562–E5571. DOI: 10.1073/pnas.1600204113
74. Novoselov, S. V.; Kryukov, G. V.; Xu, X. M.; Carlson, B. A.; Hatfield, D. L.; Gladyshev, V. N. Selenoprotein H is a nucleolar thioredoxin-like protein with a unique expression pattern. *J. Biol. Chem.* **2007**, *282*(16), 11960–11968. DOI: 10.1074/jbc.M701605200

75. Panee, J.; Stoytcheva, Z. R.; Liu, W.; Berry, M. J. Selenoprotein H is a redox-sensing high mobility group family DNA-binding protein that up-regulates genes involved in glutathione synthesis and phase II detoxification. *J. Biol. Chem.* **2007**, *282*(33), 23759–23765. DOI: 10.1074/jbc.M702267200
76. Stoytcheva, Z. R.; Vladimirov, V.; Douet, V.; Stoychev, I.; Berry, M. J. Metal transcription factor-1 regulation via MRes in the transcribed regions of selenoprotein H and other metal-responsive genes. *Biochim. Biophys. Acta* **2010**, *1800*(3), 416–424. DOI: 10.1016/j.bbagen.2009.11.003
77. Li, F.; Shi, Z.; Cheng, M.; Zhou, Z.; Chu, M.; Sun, L.; Zhou, J. C. Biology and roles in diseases of selenoprotein I characterized by ethanolamine phosphotransferase activity and antioxidant potential. *J. Nutr.*, **2023**, *153*(11), 3164–3172. DOI: https://doi.org/10.1016/j.tjnut.2023.03.023
78. Nunes, L. G. A.; Pitts, M. W.; Hoffmann, P. R.; Selenoprotein, I. Selenoprotein I (selenoi) as a critical enzyme in the central nervous system. *Arch. Biochem. Biophys.*, **2022**, *729*, 109376, ISSN 0003-9861. DOI: https://doi.org/10.1016/j.abb.2022.109376
79. Ma, C.; Martinez-Rodriguez, V.; Hoffmann, P. R. Roles for selenoprotein I and ethanolamine phospholipid synthesis in T cell activation. *Int. J. Mol. Sci.* **2021**, *22* (20), 11174. DOI: 10.3390/ijms222011174
80. Ma, C.; Hoffmann, F. W.; Nunes, L. G.; Urena, F.; Andrukhiv, A.; Gerschenson, M.; Pitts, M. W.; Hoffmann, P. R. Selenoprotein I deficiency in T cells promotes differentiation into tolerant phenotypes while decreasing th17 pathology. *J. Leukoc. Biol.* **2022**, *112* (6), 1387–1397. DOI: https://doi.org/10.1002/JLB.1A0122-080R
81. Liu, J.; Zhang, Z.; Rozovsky, S. Selenoprotein K form an intermolecular diselenide bond with unusually high redox potential. *FEBS Lett.* **2014**, *588* (18), 3311–3321. DOI: 10.1016/j.febslet.2014.07.037
82. Lu, C.; Qiu, F.; Zhou, H.; Peng, Y.; Hao, W.; Xu, J.; Yuan, J.; Wang, S.; Qiang, B.; Xu, C.; Peng, X. Identification and characterization of selenoprotein K: an antioxidant in cardiomyocytes. *FEBS Lett.* **2006,** *580* (22), 5189–5197. DOI: 10.1016/j.febslet.2006.08.065
83. Du, Shaoqing.; Zhou, Jun.; Jia, Yi.; Kaixun Huang, S. K. Is a novel ER stress-regulated protein and protects hep G2 cells from ER stress agent-induced apoptosis. *Arch. Biochem. Biophys.*, **2010**, *502*(2), 137–143, DOI: https://doi.org/10.1016/j.abb.2010.08.001
84. Jia, S. Z.; Xu, X-W. W.; Zhang, Z-H.H.; Chen, C.; Chen, Y.-B.; Huang, S-.L.; Liu, Q.; Hoffmann, P. R.; Song, G-L.L. Selenoprotein K deficiency-induced apoptosis: A role for calpain and the ERS pathway. *Redox Biol.*, **2021**, *47*, 102154, DOI: https://doi.org/10.1016/j.redox.2021.102154
85. Shchedrina, V. A.; Everley, R. A.; Zhang, Y.; Gygi, S. P.; Hatfield, D. L.; Gladyshev, V. N.; Selenoprotein, K. Multiprotein binds selenoprotein K binds multiprotein complexes and is involved in the regulation of endoplasmic reticulum homeostasis. *J. Biol. Chem.*, **2011**, *286*(50), 42937–42948. DOI: 10.1074/jbc.M111.310920
86. Fredericks, G. J.; Hoffmann, P. R. Selenoprotein K and protein palmitoylation. *Antioxid. Redox Signal.* 2015, *23* (10), 854–862. DOI: 10.1089/ars.2015.6375
87. Verma, S.; Hoffmann, F. W.; Kumar, M.; Huang, Z.; Roe, K.; Nguyen-Wu, E.; Hashimoto, A. S.; Hoffmann, P. R. Selenoprotein K knockout mice exhibit deficient calcium flux in immune cells and impaired immune responses. *J. Immunol.* **2011**, 186 (4), 2127–2137. DOI: 10.4049/jimmunol.1002878
88. Wang C.; Li, R.; Huang, Y.;Wang, M.; Yang, F.; Huang, D.; Wu, C.; Li, Y.; Tang, Y.; Zhang, R.; Cheng, J. Selenoprotein K modulate intracellular free Ca^{2+} by regulating

expression of calcium homoeostasis endoplasmic reticulum protein. *Biochem Biophys Res Commun.* **2017 Mar 18**, *484* (4), 734–739. doi: 10.1016/j.bbrc.2017.01.117. Epub 2017 Jan 24. PMID: 28130108.

89. Xu, Z. N.; Lin, Z. Q.; Zhao, G. S.; Guo, Y. B. Biogeochemical behavior of selenium in soil-air-water environment and its effects on human health. *Int. J. Environ. Sci. Technol.* **2024**, *21* (1), 1159–1180. DOI: https://doi.org/10.1007/s13762-023-05169-0
90. Gong, T.; Berry, M. J.; Pitts, M. W. Selenoprotein M: Structure, expression and functional relevance. In *Selenium*. Hatfield, D., Schweizer, U., Tsuji, P., Gladyshev, V., Eds.; Springer, **2016**. DOI: https://doi.org/10.1007/978-3-319-41283-2_21
91. Varlamova, E. G.; Goltyaev, M. V.; Fesenko, E. E. Protein partners of selenoprotein SELM and the role of selenium compounds in regulation of its expression in human cancer cells. *Dokl. Biochem. Biophys.* **2019**, *488* (1), 300–303. DOI: 10.1134/S1607672919050065
92. Jiang, H.; Shi, Q. Q.; Ge, L. Y.; Zhuang, Q. F.; Xue, D.; Xu, H. Y.; He, X. Z. Selenoprotein M stimulates the proliferative and metastatic capacities of renal cell carcinoma through activating the PI3K/AKT/mTOR pathway. *Cancer Med.* **2019**, *8* (10), 4836–4844. DOI: 10.1002/cam4.2403
93. Dogaru, C. B.; Muscurel, C.; Duță, C.; Stoian, I. 'Alphabet' selenoproteins: Their characteristics and physiological roles. *Int. J. Mol. Sci.* **2023**, *24*(21), 15992. DOI: 10.3390/ijms242115992
94. Reeves, M. A.; Bellinger, F. P.; Berry, M. J. The neuroprotective functions of selenoprotein M and its role in cytosolic calcium regulation. *Antioxid. Redox Signal.* **2010**, *12* (7), 809–818. DOI: 10.1089/ars.2009.2883
95. Castets, P.; Lescure, A.; Guicheney, P.; Allamand, V. Selenoprotein N in skeletal muscle: from diseases to function. *J. Mol. Med. (Berl.)* **2012**, *90* (10), 1095–1107. DOI: 10.1007/s00109-012-0896-x.
96. Castets, P.; Maugenre, S.; Gartioux, C.; Rederstorff, M.; Krol, A.; Lescure, A.; Tajbakhsh, S.; Allamand, V.; Guicheney, P. Selenoprotein N is dynamically expressed during mouse development and detected early in muscle precursors. *BMC Dev. Biol.* **2009**, *9*, 46. DOI: 10.1186/1471-213X-9-46
97. Santesmasses, D.; Gladyshev, V. N. Pathogenic variants in selenoproteins and selenocysteine biosynthesis machinery. *Int. J. Mol. Sci.* **2021**, *22* (21), 11593. DOI: 10.3390/ijms222111593
98. Petit, N.; Lescure, A.; Rederstorff, M.; Krol, A.; Moghadaszadeh, B.; Wewer, U. M.; Guicheney, P. Selenoprotein N: An endoplasmic reticulum glycoprotein with an early developmental expression pattern. *Hum. Mol. Genet.*, **2003**, *12*(9), 1045–1053. DOI: https://doi.org/10.1093/hmg/ddg115
99. Hamilton, S.; Terentyev, D. ER stress and calcium-dependent arrhythmias. *Front. Physiol.* **2022**, *13*, 1041940. DOI: https://doi.org/10.3389/fphys.2022.1041940
100. Chernorudskiy, A.; Varone, E.; Colombo, S. F.; Fumagalli, S.; Cagnotto, A.; Cattaneo, A.; Briens, M.; Baltzinger, M.; Kuhn, L.; Bachi, A.; Berardi, A.; Salmona, M.; Musco, G.; Borgese, N.; Lescure, A.; Zito, E. Selenoprotein N is an endoplasmic reticulum calcium sensor that links luminal calcium levels to a redox activity. *Proc. Natl. Acad. Sci. U. S. A. Proceedings of the National Academy of Sciences Volume* **2020**, *117* (35, September), 21288–21298. DOI: 10.1073/pnas.2003847117
101. Petit, N.; Lescure, A.; Rederstorff, M.; Krol, A.; Moghadaszadeh, B.; Wewer, U. M.; Guicheney, P. Selenoprotein N: An endoplasmic reticulum glycoprotein with an early developmental expression pattern. *Hum. Mol. Genet.* **2003**, *12* (9), 1045–1053. DOI: 10.1093/hmg/ddg115

102. Mukherjee, M.; Sreelatha, A. Identification of selenoprotein O substrates using a biotinylated ATP analog. *Methods Enzymol.* **2022**, *662*, 275–296. DOI: 10.1016/bs.mie.2021.10.001. (Epub November 12, 2021).
103. Han, S. J.; Lee, B. C.; Yim, S. H.; Gladyshev, V. N.; Lee, S. R. Characterization of mammalian selenoprotein O: A redox-active mitochondrial protein. *PLOS ONE* **2014**, *9* (4), e95518. DOI: 10.1371/journal.pone.0095518
104. Casey, A. K.; Orth, K. Enzymes involved in AMPylation and deAMPylation. *Chem. Rev.* **2018**, *118* (3), 1199–1215. DOI: 10.1021/acs.chemrev.7b00145
105. Yan, J.; Fei, Y.; Han, Y.; Lu, S. Selenoprotein O deficiencies suppress chondrogenic differentiation of ATDC5 cells. *Cell Biol. Int.* **2016**, 40 (10), 1033–1040. 10.1002/cbin.10644DOI: 10.1002/cbin.10644
106. (a) Burk, R. F.; Hill, K. E. Selenoprotein P-expression, functions, and roles in mammals. *Biochim. Biophys. Acta* 2009, *11790* (11), 1441–1447. DOI: 10.1016/j.bbagen.2009.03.026. (Epub April 1, 2009). (b) Burk, R. F.; Hill, K. E. Selenoprotein P: an extracellular protein with unique physical characteristics and a role in selenium homeostasis. *Annu. Rev. Nutr.* **2005**, *25*, 215–235. DOI: 10.1146/annurev.nutr.24.012003.132120
107. Zheng, X.; Toyama, T.; Siu, S.; Kaneko, T.; Sugiura, H.; Yamashita, S.; Shimoda, Y.; Kanamori, M.; Arisawa, K.; Endo, H.; Saito, Y. Selenoprotein P expression in glioblastoma as a regulator of ferroptosis sensitivity: Preservation of GPX4 via the cycling-selenium storage. *Sci. Rep.* **2024**, *14* (1), 682. DOI: 10.1038/s41598-024-51259-5
108. Ye, X.; Toyama, T.; Taguchi, K.; Arisawa, K.; Kaneko, T.; Tsutsumi, R.; Yamamoto, M.; Saito, Y. Sulforaphane decreases serum selenoprotein P levels through enhancement of lysosomal degradation independent of nrf2. *commun. Biol.* **2023**, *6* (1), 1060. DOI: 10.1038/s42003-023-05449-y
109. Ghelichkhani, F.; Gonzalez, F. A.; Kapitonova, M. A.; Rozovsky, S. Selenoprotein S interacts with the replication and transcription complex of SARS-CoV-2 by binding nsp7. *J. Mol. Biol.* **2023**, *435* (8), 168008. DOI: 10.1016/j.jmb.2023.168008
110. Ghelichkhani, F.; Gonzalez, F. A.; Kapitonova, M. A.; Schaefer-Ramadan, S.; Liu, J.; Cheng, R.; Rozovsky, S. Selenoprotein S: A versatile disordered protein. *Arch. Biochem. Biophys.* 2022, *731*, 109427. DOI: 10.1016/j.abb.2022.109427
111. Bar-Nun, S. The role of p97/cdc48p in endoplasmic reticulum-associated degradation: from the immune system to yeast. *Curr. Top. Microbiol. Immunol.* **2005**, *300*, 95–125. DOI: 10.1007/3-540-28007-3_5
112. Yu, S. S.; Du, J. L. Selenoprotein S: A therapeutic target for diabetes and macroangiopathy? *Cardiovasc. Diabetol.* **2017**, *16* (1), 101. DOI: 10.1186/s12933-017-0585-8
113. Turanov, A. A.; Shchedrina, V. A.; Everley, R. A.; Lobanov, A. V.; Yim, Sun HeeS. H.; Marino, S. M.; Gygi, S. P.; Hatfield, D. L.; Gladyshev, V. N. Selenoprotein S is involved in maintenance and transport of multiprotein complexes. *Biochem. J.* **2014**, *462* (3), 555–565. DOI: 10.1042/BJ20140076
114. Zhong, Y.; Yu, S.; Yu, H.; Yao, J.; Men, L.; Li, Y.; Wang, Q.; Du, J. Selenoprotein S attenuates endothelial dysfunction in a diabetic vascular chip. *Exp. Gerontol.*, **2020**, *137*, 110963. DOI: https://doi.org/10.1016/j.exger.2020.110963
115. Wang, Z.; Li, Y.; Yao, J.; Yu, S.; Yu, H.; Men, L.; Du, J. Selenoprotein S attenuates high glucose and/or ox-LDL-induced endothelium injury by regulating akt/MTOR signaling and autophagy. *Int. J. Biochem. Cell Biol.*, **2021**, *141*, 106111. DOI: https://doi.org/10.1016/j.biocel.2021.106111

116. Anouar, Y.; Lihrmann, I.; Falluel-Morel, A.; Boukhzar, L. Selenoprotein T is a key player in ER proteostasis, endocrine homeostasis and neuroprotection. *Free Radic. Biol. Med.* **2018**, *127*, 145–152. DOI: 10.1016/j.freeradbiomed.2018.05.076
117. Pothion, H.; Jehan, C.; Tostivint, H.; Cartier, D.; Bucharles, C.; Falluel-Morel, A.; Boukhzar, L.; Anouar, Y.; Lihrmann, I.. Selenoprotein T: An essential oxidoreductase serving as a guardian of endoplasmic reticulum homeostasis. *Antioxid. Redox Signal.* **2020**, *33*(17), 1257–1275. DOI: 10.1089/ars.2019.7931
118. Li, K.; Feng, T.; Liu, L.; Liu, H.; Huang, K.; Zhou, J. Hepatic proteomic analysis of selenoprotein T knockout mice by TMT: Implications for the role of selenoprotein T in glucose and lipid metabolism. *Int. J. Mol. Sci.* **2021**, *22* (16), 8515. DOI: https://doi.org/10.3390/ijms22168515
119. Moustafa, M. E.; Antar, H. A. A bioinformatics approach to characterize mammalian selenoprotein T. *Biochem. Genet.* **2012**, *50* (9–10), 736–747. DOI: 10.1007/s10528-012-9516-2
120. SenGupta, A.; Carlson, B. A.; Labunskyy, V. M.; Gladyshev, V. N.; Hatfield, D. L. Selenoprotein T deficiency alters cell adhesion and elevates selenoprotein W expression in murine fibroblast cells. *Biochem. Cell Biol.* **2009**, *87* (6), 953–961. DOI: 10.1139/o09-064
121. Grumolato, L.; Ghzili, H.; Montero-Hadjadje, M.; Gasman, S.; Lesage, J.; Tanguy, Y.; Galas, L.; Ait-Ali, D.; Leprince, J.; Guérineau, N. C.; Elkahloun, A. G.; Fournier, A.; Vieau, D.; Vaudry, H.; Anouar, Y. Selenoprotein T is a PACAP-regulated gene involved in intracellular Ca^{2+} mobilization and neuroendocrine secretion. *FASEB J.* **2008**, *22* (6), 1756–1768. DOI: 10.1096/fj.06-075820
122. Boukhzar, L.; Hamieh, A.; Cartier, D.; Tanguy, Y.; Alsharif, I.; Castex, M.; Arabo, A.; El Hajji, S. E.; Bonnet, J. J.; Errami, M.; Falluel-Morel, A.; Chagraoui, A.; Lihrmann, I.; Anouar, Y. Selenoprotein T exerts an essential oxidoreductase activity that protects dopaminergic neurons in mouse models of Parkinson's disease. *Antioxid. Redox Signal.* **2016**, *24* (11), 557–574. DOI: 10.1089/ars.2015.6478
123. Zhang, X.; Xiong, W.; Chen, L. L.; Huang, J. Q.; Lei, X. G. Selenoprotein V protects against endoplasmic reticulum stress and oxidative injury induced by pro-oxidants. *Free Radic. Biol. Med.* **2020**, *160*, 670–679. DOI: 10.1016/j.freeradbiomed.2020.08.011.
124. Zhang, X.; Xiong, W.; Huang, J.; Lei, Xin GenX. G. Selenoprotein V protects against oxidative stress, endoplasmic reticulum stress, and apoptosis. *Curr. Dev. Nutr.*, **2020**, *4*(Suppl. 2), nzaa067_85 DOI: https://doi.org/10.1093/cdn/nzaa067_085
125. Moustafa, M. E.-S.; El-Srouji, R. Characterization Selenoprotein v. FASEB. *The. FASEB J.* **2010**, *24*(S1), 916.12–916.12. DOI: https://doi.org/10.1096/fasebj.24.1_supplement.916.12
126. Dessen, P.; *SELENOV (selenoprotein V).* Atlas Genet Cytogenet Oncol Haematol. 2019-04-01.
127. Whanger, P. D. Selenoprotein W: A review. *Cell. Mol. Life Sci.* **2000**, *57* (13–14), 1846–1852. DOI: 10.1007/PL00000666
128. Kim, H.; Lee, K.; Kim, J. M.; Kim, M. Y.; Kim, J. R.; Lee, H. W.; Chung, Y. W.; Shin, H. I.; Kim, T.; Park, E. S.; Rho, J.; Lee, S. H.; Kim, N.; Lee, S. Y.; Choi, Y.; Jeong, D. Selenoprotein W ensures physiological bone remodeling by preventing hyperactivity of osteoclasts. *Nat. Commun.* **2021**, *12* (1), 2258. DOI: https://doi.org/10.1038/s41467-021-22565-7

129. Nettleford, S. K.; Liao, C.; Short, S. P.; Rossi, R. M.; Singh, V.; Prabhu, K. S.; Selenoprotein, W. Selenoprotein W ameliorates experimental colitis and promotes intestinal epithelial repair. *Antioxidants (Basel)* **2023**, *12* (4), 850. DOI: https://doi.org/10.3390/antiox12040850
130. Vendeland, S. C.; Beilstein, M. A.; Chen, C. L.; Jensen, O. N.; Barofsky, E.; Whanger, P. D. Purification and properties of selenoprotein W from rat muscle. *J. Biol. Chem.* **1993**, *268*(23), 17103–17107. DOI: 10.1016/S0021-9258(19)85307-3
131. Beilstein, M. A.; Vendeland, S. C.; Barofsky, E.; Jensen, O. N.; Whanger, P. D. Selenoprotein W of rat muscle binds glutathione and an unknown small molecular weight moiety. *J. Inorg. Biochem.* **1996**, *61* (2), 117–124. DOI: 10.1016/0162-0134(95)00045-3
132. Alkan, Z.; Duong, F. L.; Hawkes, W. C. Selenoprotein W controls epidermal growth factor receptor surface expression, activation and degradation via receptor ubiquitination. *Biochim. Biophys. Acta* **2015**, *1853* (5), 1087–1095. DOI: 10.1016/j.bbamcr.2015.02.016
133. Jeon, Y. H.; Park, Y. H.; Lee, J. H.; Hong, J. H.; Kim, I. Y. Selenoprotein W enhances skeletal muscle differentiation by inhibiting TAZ binding to 14-3-3 protein. *Biochim. Biophys. Acta* **2014**, *1843* (7), 1356–1364. 10.1016/j.bbamcr.2014.04.002

5 Biological Properties of Vanadium

Sangeeta Bajpai, Garima Pandey, Monika Kamboj and Bimal Krishan Banik

5.1 INTRODUCTION

Minerals are the essential nutrients for living organisms. They are present in small amounts in almost all parts of body tissues and fluids. They do not produce energy but could be essential nutrient markers for boosting metabolic activities such as immunity, growth and development, and reproduction. They have been broadly classified into (i) macro or major elements, (ii) micro or trace elements and recently (iii) ultra trace or possibly essential elements which includes aluminum, arsenic, cobalt, chromium, fluorine, molybdenum, nickel, silicon, tin, and vanadium [1,2]. A very small dose, around 1 ug/g, of these ultra trace elements plays a significant role in metabolism. Vanadium is a naturally occurring ultra trace metal exhibiting unique property of having variable oxidation states from –III to +V where (+IV) oxidation state is found to be most stable [3,4]. Having unique property of variable oxidation states, vanadium has attracted attention from the scientific community and shows various applications in the industrial front. Reports on its chemical behavior in the field of (i) developing electrochemical storage systems (vanadium redox flow batteries), (ii) reducing agents, (iii) photography, (iv) drying agents for paints and varnishes, (v) production of pesticides, (vi) production of black dyes, ink, and pigments in ceramic, printing, and textile industries are available [5–8]. Apart from these applications, the biological role of vanadium is also not untouched.

Although vanadium is relatively less abundant on the Earth (0.013% w/w), but still, it is an essential micronutrient for certain organisms, playing critical roles in various biological processes. It is broadly distributed among all living organisms. It is absorbed by the plants from the environment, and it travels along the food chain into the bodies of living organisms. In human beings, reports reveal [5] the presence of vanadium content mainly in liver, kidney, and bones; in blood plasma, it is around 200 nM and in tissues it is reported to be around 0.3 mg/kg. In vertebrates, vanadium content enters through food and respiration via digestive and respiratory tracts respectively [9,10]. Various marine species are reported to have this ultra trace metal as an essential element [11–13]. Vanadium content is found to be more than 40 mg/g in various food sources, viz. black pepper, dill seed, mushrooms, parsley, shellfishes etc. where it is present in the form of vanadium pentoxide, vanadyl sulfate, sodium

 DOI: 10.1201/9781032615110-7

metavanadate and sodium orthovanadate [13–15]. Vanadium concentration (~1 μg/L) is also reported in potable water and thus its intake by this source depends upon the daily ingested volume [5]. Hence, the consumption of vanadium contents by human beings corresponds to 10–30 μg/day. Its accumulation in human bodies does not create hazardous conditions, because most of the dietary vanadium content is usually excreted in the feces [5, 16–18].

Vanadium ions possess many structural roles as reflected by its structural and electronic analogy with phosphorous (Reactions 5.1–5.3) [19–29]. These ions act as an enzyme cofactor and is found in sea squirts, and probably in mammals. The action and function of vanadium in biological systems is based on its coordination behavior and redox chemistry along with structural and functional aspects of biological systems [29]:

$$H_3VO_4 + H_2O \rightleftharpoons H_2VO_4 + H_3O^{+} pKa\ 3.5 \tag{5.1}$$

$$H_2VO_4 + H_2O \rightleftharpoons HVO_4^{2-} + H_3O^{+} pKa\ 7.8 \tag{5.2}$$

$$HVO_4^{2-} + OH \rightleftharpoons VO_4^{3-} + H_2O\ pKa\ 12.7 \tag{5.3}$$

The role of vanadium ions in biology can be as counterions for protein, nucleic acids and in several biological organelles. The structural role often displays the care of several biological structures. Omnipresent vanadium in the Earth's crust shows cumulative and diverse functions in organisms such as algae, fungi, bacteria, plants, and animals. Various vanadium-dependent enzymes viz. Halo peroxidases in algae and fungi, nitrogenases in nitrogen-fixing bacteria are reported for their correct functioning [30].

The discovery of the first vanadium-containing enzymes "haloperoxidases" is very well documented and formulates the active sites of these enzymes along with its structural and functional relation with phosphatases [29]. The inhibition of phosphatases by vanadium compounds is well established. Vanadium nitrogenases commonly contain a molybdenum–iron cofactor. Some bacteria produce additional nitrogenase enzyme that are genetically distinct and instead contain V–Fe or Fe–Fe central metal coordination systems. Multiple groups of diazotrophic microorganisms have been identified to contain vanadium nitrogenases synthesized under molybdenum-controlled conditions [29, 31].

Computational modeling (both structural and functional) for biological activities has long been of interest to chemists as they are most informative and helpful in achieving designed targets in this area. Structural studies, based on oxidation states, are defining its effect in biological systems in many ways. During the last few decades, the false picture of vanadium being toxic and carcinogenic has been eventually ratified to an essential trace transition element with various medicinal applications [29]. Herein, we have tried to compile biological properties of vanadium including its biological history, presence in living organisms, aqueous, redox, and bio-coordination chemistry, enzymatic behavior and metabolic actions.

5.2 BIOLOGICAL HISTORY OF VANADIUM

The significance of Na-K-ATPase enzyme in maintaining homeostasis and role of endogenous modulators of this enzyme in various physiological and pathophysiological states is known since 1965 [32,33]. The research findings of vanadium, as a natural constituent of most animal tissues and as a potent inhibitor of Na-K-ATPase, promoted research to establish its biological properties. Perhaps, Gaymee and Larmuth were the first to report the biological properties of vanadium in frog hearts in the year 1876 [34]. The pharmacological effects of vanadium showed similarity with that of digitalis [35]. Interest in vanadium was triggered by an important discovery made by Charney et al. [36] who extracted commercial ATP from horse skeletal muscle. This finding was confirmed by other group, where they reported the presence of 'Na-K -ATPase modifier' from commercial ATP. Later, Josephson and Cantley et al. successfully isolated this "modifier" from striped muscle and identified it as vanadate and demonstrated it as a potent inhibitor of the sodium pump. [37,38].

5.3 VANADIUM IN LIVING ORGANISMS AND ITS DISTRIBUTION

The selection of a metal "as a constituent essential for life process" during evolution depends on major three factors: (i) availability, particularly in the aqua sphere (known to be the cradle of all terrestrial lives); (ii) showing negligible toxic behavior at physiological levels; (iii) possessing compatible reactivity with biological processes such as Lewis's acidity and redox conversions. Zn^{2+}- and Fe^{2+}, Fe^{3+} systems are the examples of latter and former factors, respectively. Omnipresent vanadium constitutes 0.013 percent of the earth crust and is as abundant as zinc. Accumulated vanadium in ores as vanadinite $Pb_5(VO_4)_3Cl$ or VS_4 (patronite) is rare. Vanadium in sea water is present in high concentration in an average concentration of 30 Nm as vanadate ions in its contact ionic pair form $Na^+H_2VO_4^-$ [39,40].

Like other trace transition elements, vanadium possesses various oxidation states. Vanadium compounds, in their most common valence states V (III) to V(V), exhibit biological activities. In living systems, vanadyl ions (IV, VO^{2+}, V^{4+}) and vanadate ions (V, $VO_4^{\ 3-}$, V^{5+}) are the most prevalent forms and play important biological roles. Vanadium with lower oxidation states is found only in sea creatures, which are capable of reducing V^{5+} to V^{3+} within their bodies [41].

Vanadium switches between its oxidation states IV and V easily. As per reaction I, the redox potential (–0.341) at pH 7 is within the range where oxidation of vanadyl ions to vanadate ions and takes place under aerobic conditions and reduction of vanadate ions to vanadyl ions (VO^{2+}) is done by cellular components such as NADH, phenolic compounds and cysteine containing peptides and proteins. The acid base pair species $H_2VO_4^- \rightleftharpoons HVO_4^{2-} + H^+$ ($pK_a = 8.1$) plays the key role in physiological aerobic conditions.

$$H_2VO_4^- + 4H^+O^{-1} \rightleftharpoons VO^{2+} + 3H_2O \text{ (redox potential at pH 7 = - 0.341) ..... (Reaction I)}$$

Cationic vanadium (V) species VO_3^+ and VO_2^+ show stability around pH =7 only when coordinated with strong ligands, which prohibits the formation of hydroxides. Another oxidation species of vanadium VO^{2+} (predominant species under anaerobic physiological conditions) exists in the form of acid base pair shown in reaction II:

$$[VO\,(H_2O)_5]^{2+} \rightleftharpoons [VO(OH)(H_2O)_4]^+ + H^+ \text{ (at lower pH)} \quad \ldots\ldots\ \textbf{(Reaction II)}$$

Oxovanadium cations are also strong Lewis acids. Thus, following two conditions support vanadium to be a potential biometal:

(i) Showing redox activity within physiological frames of biological systems
(ii) Susceptibility toward nucleophilic species.

Vanadium, in its biologically active compounds, shows either property.

The entry of vanadates in living systems takes place through anionic channels or via phosphate transportation systems due to its structural resemblance with phosphorous systems, whereas entry of vanadyl cations adopts passive transport or diffusion methodology [42–44]. Due to structural and electronic analogy between vanadate and phosphate, vanadate can inhibit various phosphate driven physiological activities. It shuffles both the energy and matter cycles by replacing phosphate ions. Reports reveal the possible inhibition activity of enzymes by vanadium.

It can inhibit the activity of sodium–potassium adenosine triphosphatase (Na+ /K+-ATPase), plasma membrane hydrogen-translocation ATPase, phosphoryl transfer enzymes, phosphotyrosine, protein tyrosine kinase and phosphatases [19,29,37,38,45,46].

5.3.1 In Plants

Plants act as major vanadium flux from earth crust to organisms. Very low concentrations of vanadium uptake (<2 mg/kg) by the plants enhance synthesis of chlorophyll and utilization of potassium uptake from the soil. An enhanced rate of nitrogen fixation process is also reported. In plants, vanadium concentrations vary with their leaf stalk ratio. Its concentration is more in leaves than in seeds and fruits. It is present as vanadyl (VO^{2+}) in other naturally occurring food stuffs. Vanadium concentration ranges from 0.05 to 2 mg/kg in mushroom, parsley, dill seed, black pepper, cereals, and fruits [47].

In most plants, vanadium concentration is found in the order of stem < leaves < roots, and at subcellular level, its concentration is in the order of cell wall < soluble fraction < nucleus < mitochondria < chloroplast [41,48–50]. Vanadium enhances chlorophyll contents in leaves. The possible reason is the elevated uptake of iron (fe) & magnesium (mg) (essential elements for chlorophyll biosynthesis) [41,51]. But then, reduced carotenoid content in plants is also reported by researchers, which might be due to the destruction of these elements by oxidative stress and permeability difference of cellular membranes [52–56]. Reduced water uptake by vanadium-treated plants have been reported due to reduced plant physiological factors. These plant physiological factors (photosynthetic anabolism, stomatal conductance, and transpiration) provide the potential force for water lifting via roots to leaves followed by transport of minerals and nutrients to the aerial parts of the plants [49,52,55,57,58]. The effect of vanadium on plant physiology is shown in Table 5.1.

TABLE 5.1
Vanadium effect on plant physiology

Species	Parameters	Effect	Vanadium Level	Vanadium Salt
Bunny cactus	Cladodes water content	Reduction	9–14.5 mg/L	Vanadate V(V)
Bush bean	Leaf Mn	Elevation	2 mg/L	NH_4VO_3
Bush bean	Leaf Fe	Significant (p<0.05) reduction	2 mg/L	NH_4VO_3
Bush bean	Root Fe, Mn, Cu	Elevation	50.9 mg/L	NH_4VO_3; reduced in aerial tissue
Cabbage	Fe, Mn	Reduction	>0. 4mg/L	VCl_3
Chickpea	Seedlings H_2O_2, MDA	Elevation	25 mg/L	NH_4VO_3
Chickpea	P uptake	Elevation	<30 mg/L	NH_4VO_3
Corn	Leaf plasma membrane ATPase	Reduction	25.5 mg/L	vanadate
Corn	K+ transport	Significant (p<0.05) reduction	25.5 mg/L	vanadate V(V)
Corn	Root ATPase activity	Reduction	~1 mg/L	vanadate V(V)
Corn	Root plasma membrane ATPase	Reduction	0.31 mg/L	$NaVO_3$
Corn seedling	Phyto chelatins, glutathione, Thiols	Reduction	5 mg/L	NH_4VO_3
Dog tail grass	Seedling POD,CAT	Elevation	< 16.35 mg/L	mainly vanadate V(V)
Dog tail grass	SOD	Elevation	< 40.6 mg/L	mainly vanadate V(V)
Dog tail grass	MDA	Elevation	>16.4mg/L	mainly vanadate V(V)
Dog tail grass	Leaf P, Fe	Significant (p<0.05) elevation	>47.4 mg/L	mainly vanadate V(V)
Horseradish	POD	Reduction	2.2 mg/L	vanadate V(V)
Lettuce	Leaf Ca, Fe	Elevation	1 mg/kg	NH_4VO_3
Lettuce	Root K, Mg	Reduction	1 mg/kg	NH_4VO_3
Onion	Root cell mitotic index	Reduction	25-100 mg/kg	V_2O_5
Pepper	Root P, Ca, Mg, Cu, B	Elevation	0.76 mg/L	NH_4VO_3

TABLE 5.1 (Continued)
Vanadium effect on plant physiology

Species	Parameters	Effect	Vanadium Level	Vanadium Salt
Pepper	Stem N, P, K, Ca, Mg, Cu, Mn, B	Elevation	0.25 mg/L	NH_4VO_3
Pepper	Leaf Mg, Mn	Reduction	0.76 mg/L	NH_4VO_3
Pepper	Leaf amino acids, sugars	Elevation	0.25 mg/L	NH_4VO_3
Pennyroyal	Leaf Ca, Mg, K, Mn	Significant (p<0.05) reduction	40 mg/L	NH_4VO_3
Rape	O_2, MDA, cell death	Elevation	17.8 mg/L	$NaVO_3$
Rape	SOD	Reduction	17.8 mg/L	$NaVO_3$
Red beet	Root plasma membrane ATPase	Reduction	0.76 mg/L	$NaVO_3$
Red clover	Transport of isoflavones, secondary metabolites across plasma membrane	Significant (p<0.05) reduction	0.51 mg/L	NH_4VO_3
Rice	Root auxin, ABA, JA, GSH transport, ABC	Elevation	51 mg/L	$NaVO_4$
Rice	DCT, DMT, ZIP	Reduction	51 mg/L	$NaVO_4$
Soybean	Root Ca	Significant (p<0.05) elevation	6 mg/L	$VOSO_4$
Soybean	Aerial parts Ca; root K, Mg	Reduction	6 mg/L	$VOSO_4$
Soybean	Nitrate N	Elevation	10 mg/L	$VOCl_2$
Sugar beet	Sucrose content	Elevation	509 mg/L	$VOSO_4$
Sunflower	Total proteins	Significant (p<0.05) reduction	7.5–15mg/L	NH_4VO_3
Sunflower	Anthocyanin, flavonoids, tannins , lignin	Elevation	7.5 mg/L	NH_4VO_3
Sunflower	Soluble sugar	Significant (p<0.05) elevation	7.5–15mg/L	NH_4VO_3
Tomato	Leaf Fe	Significant (p<0.05) elevation	0.25 mg/L	NH_4VO_3
Watermelon	Chlorophyll content	Elevation	50 mg/L	NH_4VO_3
Watermelon	Seedlings respiration	Reduction	50 mg/L	NH_4VO_3

The accumulation of vanadium in various parts of plants and species have been discussed by Jiang et al. [59,60]. Vanadium accumulated plant species possess good accumulation and exclusion mechanism by which toxic levels are maintained. The uptake of toxic elements is thus prohibited, beyond limit. The maximum vanadium accumulation has been reported in roots. Plants like Chinese green mustard, chickpea, and bunny cactus were reported to have highest vanadium accumulation levels with the approximate accumulation around 10,334, 8500, and 8640 mg/kg, respectively. High concentrations of vanadium in roots and aerial parts of vegetables and plants such as cabbage, corn, tomato, lettuce, rice, and green bean have also been reported even in the presence of relatively low vanadium uptake. In contrast, plants like Chinese cabbage, rape, box thorn, and rape take up low vanadium in the presence of rich vanadium concentrations in the growth media. The accumulation of vanadium in rape is found to be 0.56 mg/kg in roots and 0.24 mg/kg in its aerial parts, thereby exhibiting efficient exclusion and elimination strategy. These types of plants show high potential for cultivation in vanadium-contaminated substrata [41].

The plant cell wall is the first protective screen for against the uptake of toxic effects of heavy metals [48,50]. It is rich in proteins, polysaccharides (cellulose, hemicellulose and lignin) and amidogen ($-NH_2$), carboxyl (–COOH) and oxhydryl (–OH) ligands. It is the main region for vanadium accumulation at subcellular levels. The mechanism of vanadium accumulation involves chelation and complexation process where vanadium compounds undergo complexation with polar ligands present in cell wall of the roots, thereby reducing the mobility of toxic elements. The low levels of vanadium accumulation in aerial parts might be due to the formation of stable calcium compounds in root tissues, and sections in vacuoles of the roots and stems which prohibit the free translocation of vanadium ions [41].

5.3.2 In Animals and Humans

Vanadium is a transition metal that is an endogenous product of most mammalian tissues. Low doses of vanadium have a positive effect at the cellular and subcellular levels. Although the importance of vanadium in humans and animals is unclear, but its low concentration is necessary for bacteria, fungi, and animals and its deficiency causes many side effects. Studies reported various functional behavior of vanadium compounds in human diseases. Vanadium compounds regulate iodine metabolism and thyroid function and are necessary for the development of the musculoskeletal system as vanadium salts promote calcium deposition in bones. Further studies also reported that its deficiency can lead to reduced growth, poor quality of bones and teeth, and decreased reproductive ability. Like vanadate, decavanadate ions affect mitochondrial functions due to its interaction with proteins such as myosin, actin, and Ca^{2+}-ATPase apart from further inducing changes in various cellular antioxidant markers [47].

Vanadium is essential for many species of green algae (*Scenedesmus obliquus, Chlorella pyrenidasa*), yellow-green algae (*Bumileriquisis filiformia*), and brown algae (*Fucus algae*). Vanadium haloperoxidase enzymes (chloro-, bromo- and iodoperoxidase) have also been isolated and characterized from brown seaweed [61] and *Curvularia inaequalis* fungi [33,41]. Among haloperoxidase enzymes, thyroid peroxidases play an important role in higher animals. Vanadium haloperoxidases have been extracted from various species of Laminariaceae family [43]. Vanadium chloroperoxidases have been detected in soil organisms only, whereas bromoperoxidase and iodoperoxidase enzymes of vanadium are more common in the marine environment. The antimicrobial effect of vanadium chloroperoxidase enzymes was reported by Rehder [44]. It is potentially used in the treatment of dental biofilms produced by *Streptococcus mutans* and *Enterococcus faecalis* like bacteria. Recently, *Pseudomonas isachenkovii* [45] bacteria have been found active in producing molybdenum-free nitrate reductase enzyme.

Among animals, vanadium has been found to be an essential element in rats and chicks [47]. A very low concentration of vanadium (<0.1 to 2 ppm) is required for growth in rats [48]. It is thought to be important in mice [49] because less than 100 ppb vanadium in the mouse diet reduces body growth [50]. In addition to growth retardation, vanadium deficiency is also associated with reproductive failure, altered erythrocyte formation, iron metabolism, and changes in blood lipid levels in mice [51].

In human beings, the permissible level of vanadium intake is less than 1.8 mg per day [53]. However, the average diet can provide 6–20 μg of vanadium/day [54]. It has pure medicinal properties and is recommended by many national institutes of health as a dietary supplement for diabetics, bodybuilders, and athletes at a dose of about 60 mg/day [55]. Vanadium is also present in many human multivitamins and dietary supplements at a level of about 0.025 mg/day [56]. Nowadays, it has become an ingredient in many drugs and other dietary supplements to improve the immune system and treat diabetes due to its ability to mimic insulin in animal models. Vanadium deficiency does not affect hemoglobin [58] but a vanadium-deficient diet increases blood concentration of triglycerides, creatinine, lipoprotein levels, and glutamyl transferase and citric acid enzyme activity without changing amino acid metabolism [59]. No toxicity has been observed after long-term supplementation of vanadium compounds into the blood [60].

5.4 BIOLOGICAL CHEMISTRY OF VANADIUM

Vanadium has at least two unique properties: (i) analogy with phosphates where tetrahedral anion vanadate (V) shows similarity with phosphate anion and (ii) vanadium can expand its sphere beyond tetrahedral coordination and interchange between the +III, +IV, and +V oxidation states in physiological media easily. The similarity of vanadate and phosphate may explain the anti-inflammatory effects of vanadium with parasites such as maltolate and picolinic acid, as well as the mediation of vanadium in cardiovascular and neuronal defects. Other medical applications involving many combinations of vanadium, such as the treatment of parasitic tropical diseases, may also result from the special properties of the ligand spheres [10].

The biological chemistry of vanadium is based on its four major properties:

i. Aqueous chemistry of vanadium
ii. Coordination behavior of vanadium

iii. Vanadate–phosphate analogy
iv. Enzymatic action

These governing three factors have been discussed earlier by different research groups [28,29]. Here, we have tried to compile all the above major properties of vanadium, which will be beneficial to all learning class and will come up with new prospects in this field.

5.4.1 Aqueous Chemistry of Vanadium

In aqueous medium, vanadate and vanadyl oxo ions both undergo numerous hydrolytic, acid/base, self-condensation, and redox reactions, which distinguish the aqueous and non-aqueous chemistry of vanadium. These reactions are very sensitive to (i) pH of the solution and (ii) presence of potential ligands. The coordination behavior of vanadium and ligand systems results in several complexes with different coordination geometries [28].

Furthermore, some complications also arise because of the facile interconversion between oxidation states of vanadium. Vanadate, under some conditions, is an oxidizing agent and can be readily reduced to either V(IV) or V(III) along with the production of an oxidized product. In the solutions having neutral/basic pH range, air oxidation will quickly regenerate vanadate unless the V(IV) is tightly complexed with good stabilizing ligands. For instance, free aqueous V(IV) in the form of vanadyl is stable to air oxidation only under acidic conditions. The appropriate use of reducing compounds can maintain the V(IV) oxidation state and extend the stability of such systems to include physiological conditions.

The chemistry of V(III), V(IV), and V(V) species differ widely in aqueous systems because of facile interconversion among oxidation states of vanadium. This is readily exemplified by the fact that, in aqueous solution, both V(III) and V(IV) species are cationic while V(V) species are anionic.

The study of aqueous chemistry of V(III) compounds is not easy as the characterization of the solvated V(III) species and often requires a combination of potentiometric and electrochemical studies, the interpretation of which is often problematic [62]. The stability of this oxidation state limits over its potential and pH range. Like V(IV) and V(V) species, V(III) also undergoes hydrolytic reactions forming both monomeric and higher oligomeric cationic species. The deprotonation of the monomeric species $[V\ (H_2O)_6]^{3+}$ and $[V(OH)(H_2O)_5]^{2+}$ (Figure 5.1) occurs readily since their pK_a values are about 2.6 and 4.2, respectively. Thus the major and stable monomeric species of V(III), in the neutral pH range, was found to be $[V(OH)_2(H2O)_4]^+$. Dimeric and trimeric species are also stable and form in solutions with millimolar vanadium concentrations.

The study of aqueous chemistry of V(IV) species reveals the formation of vanadyl cation $(VO\ (H_2O)_5{}^{2+}$ (Figure 5.1) by dissolution of $VOSO_4$ in acidic aqueous solutions commonly abbreviated as VO^{2+} [63]. This species is air-stable in acidic solutions [64,65]. Several oligomeric and polymeric spin-paired species are generated, on raising the pH depending upon the physiological range comes complicated. A detailed discussion on the speciation and aqueous equilibria had been described [63].

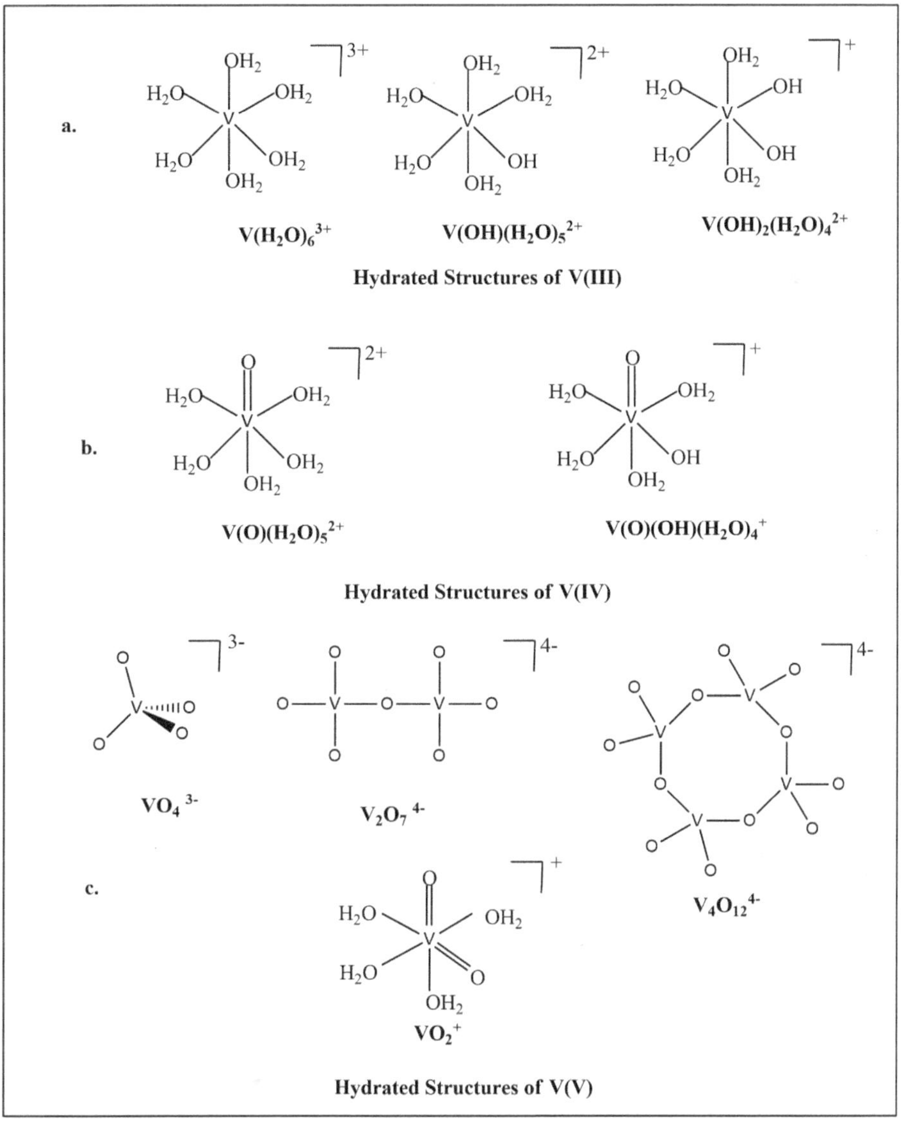

FIGURE 5.1 Hydrated compounds of vanadium in aqueous medium.

Aqueous chemistry of V(V) species depicts that at concentrations below ~1 mM and within the typical physiological pH range, vanadate predominantly exists as the monomeric $H_2VO_4^-$ anion [20,28]. This fact is confirmed with when stock solutions are prepared using solids like $NaVO_3$, NH_4VO_3, Na_3VO_4, and V_2O_5. Parameters such as stoichiometries, charges, formation constants, pK_a values, and ^{51}V NMR studies of the principal V(V) species present in aqueous solution have been discussed by Crans et al. [28]. The various protonation states can be expressed by the general formula $H_nVO_4^{(3-n)-}$, with n representing integer values from 0 to 3.

Monomeric vanadate easily undergoes condensation reactions with a variety of nucleophilic ligands, including other derivatives of vanadate. In solutions where vanadium(V) concentrations reach millimolar levels or higher, the predominant species present are the different oligomers (Figure 5.1). Decavanadate, represented as $V_{10}O_{28}^{6-}$ or V_{10}, emerges as the predominant oligomer in acidic medium, having pH values between 3 and 6. This complex anion is notable for being the only major colored vanadium(V) oxoanion, producing striking yellow-orange solutions. As the pH increases from 6 to 10, the main oligomeric species shift to include colorless dimer (V_2), cyclic tetramers (V_4), and cyclic pentamers (V_5) [66,67]. Above pH 10, the anionic monomer becomes the dominant form, while below pH 3, the cationic monomer prevails. Additionally, limited concentrations of other oligomers such as linear V_3, V_4, V_5, and V_6 are also generated across various pH ranges [66,67]. Colorless oxovanadium(V) ions interconvert in the aqueous medium. Both in vitro or in vivo systems, stock solution of oxovanadium (V) ions may contain a variety of oxoions that quickly generate $H_2VO_4^-$ upon dilution.

In the past, the condensation reactions responsible for the formation of V_2, V_4, and V_5 species were often overlooked in biochemical investigations, even when using vanadate concentrations of millimolar or greater, as noted in Tracey and Crans' work, "Vanadium Compounds," published in the ACS Symposium Series by the American Chemical Society in 1998 [28]. It may come as a surprise that the effects of these oxoanions vary, as enzymes are capable of distinguishing among them due to their sufficiently long lifetimes. This variation poses a significant challenge when interpreting outcomes from vanadate mixtures. For instance, V_2 serves as a strong inhibitor of phosphoglycerate mutase, whereas V_4 affects various enzymes associated with the pentose phosphate pathway and glucose metabolism—some of which are also inhibited by V_2. To accurately interpret the biochemical data generated from these mixtures, it is essential to analyze each oxoanion species individually.

The structural characterization of how these enzymes interact with oxometalates remains incomplete, although initial investigations in this domain have been documented [28]. While oxovanadium anions engage with enzymes that bind phosphates and polyphosphates, similar affinities for these anions can also be found in other enzymes that do not typically showcase such interactions. This implies that it would be overly simplistic to assume that the interactions between vanadium(V) ions and proteins are solely determined by the interactions observed with the related phosphates [28].

5.4.2 Coordination Behavior of Vanadium

The study of vanadium's interactions with various ligands, especially within biochemical contexts, is gaining increased attention due to the metal's possible influence on biological functions. Vanadium can exist in several oxidation states, particularly +4 [vanadium (IV)] and +5 [vanadium(V)], affecting its reactivity and coordination characteristics. In biological settings, vanadium forms complexes with ligands that feature functional groups like hydroxylates, carboxylates, phenolates, imidazoles, and thiolates. These functional groups, found in amino acids and other biomolecules, facilitate effective coordination with vanadium ions. The coordination chemistry of vanadium is crucial for mimicking the behavior of certain metalloenzymes, influencing enzymatic reactions and affecting protein structure and function.

The biochemical relevance of vanadium is evident in various scenarios: first, it may participate in enzyme catalysis, acting as a substitute for other metal ions in metalloenzymes and potentially affecting their activity or stability. Second, it can be incorporated into transport proteins, thereby influencing the binding and transport of vital substrates or cofactors across cellular membranes. Additionally, in the cytoplasm, vanadium may interact with multiple biomolecules, affecting cellular signaling pathways and metabolic processes. The arrangement of vanadium-binding residues plays a significant role; the geometry and orientation of ligands can enhance or impede coordination, thereby influencing overall chemical reactivity. A deeper understanding of these molecular interactions is crucial for unraveling the role of vanadium in biological systems, exploring its potential therapeutic applications (such as in treating diabetes), and developing vanadium-based compounds as biochemical tools or medications. In summary, vanadium's chemistry with a wide range of ligands underscores its versatility and importance in biological processes, highlighting the need for further exploration of its biochemical roles and mechanisms [28].

5.4.2.1 Aqueous Interaction of V(IV) in Aqueous Solutions with Ligands

With inorganic ligands

Crans et al. reviewed coordination behavior of vanadium in aqueous medium [28]. Hydrated vanadyl, commonly referred to as VO_2^+, readily interacts with ligands containing oxygen, nitrogen, and sulfur. For instance, the vanadyl cation is known to combine with various ligands, including carbonates, phosphonates, and phosphates, as well as other inorganic species (ligands). Additionally, it forms complexes with organic ligands such as pyridine, imidazole, and other amine bases, along with deprotonated sulfides. Notably, nucleotides and nucleic acids serve as significant ligands, where the phosphate group plays a crucial role in coordinating with vanadium. By electron-nuclear double resonance (ENDOR) spectroscopy, the structural details of complexation of vanadyl cation with nucleotides and other ligands have been extensively studied. Furthermore, vanadyl cation can easily form complexes with halides in aqueous medium [68]. The stability and capability of these adducts to engage in redox or hydrolytic reactions influence the characteristics of the resulting complexes. Importantly, certain ligands that successfully create complexes with vanadium(V) in solution may not yield stable complexes with vanadium (IV). A noteworthy example is hydrogen peroxide, which fails to form an isolatable vanadium (IV) complex [28].

The range of coordination geometries seen in vanadium (IV) complexes with oxygen and nitrogen donor functionalities reveal interesting interactions:

(i) The vanadyl cation exhibits a strong affinity for carboxylate functionalities as compared to those formed with aryloxide or alkoxide functionalities.
(ii) Neutral groups, like the carbonyl, can also act as effective ligands for the vanadyl cation.

(iii) Ligands such as aromatic amines, including pyridine and imidazole, tend to form more stable complexes with vanadyl than their alkyl amine counterparts.
(iv) Biogenic ligands, which often possess multiple functional groups that interact favorably with the vanadyl cation, provide numerous opportunities for potential reactions.

Ligands of physiological importance frequently contain several oxygen functionalities, enabling them to form robust complexes with vanadyl cation. Hydroxycarboxylic acids (citric, tartaric acid, glyceric acid, and lactic acids) act as important metabolites, some of which are directly linked with glucose metabolism. Nucleotides and nucleosides both form complexes with vanadyl cation, with more to lesser extent. The strong affinity of vanadyl for catechol has been well documented, contributing to its colorful chemistry across different oxidation states. Moreover, the vanadyl cation readily reacts with peptides, resulting in a mixture of complexes that have primarily been characterized in terms of stoichiometry and stability. Recent spectroscopic studies suggest that the complexation process is mainly driven by strong coordination with carboxylate groups, with only weak interactions observed with amine nitrogens. Additionally, sulfur-containing ligands, particularly those that include carboxylate groups like cysteine and glutathione, form strong complexes near the pK_a value of the thiol group. Given the high affinity of various peptides for the vanadyl cation, it is expected that robust complexes would also be formed with proteins, including serum albumin.

With organic ligands

Various stable and less stable complexes of vanadium with nitrogen and oxygen donor ligands are documented. Vanadyl cations in its +4 oxidation states form coordination complexes with N- and O-donor ligands. The coordination geometries of these complexes are depicted in Figure 5.2. Some key points are as follows:

1. Ligands with single oxygen donor site, viz. carboxylates and carbonyls form stable complexes with vanadyl cation than aryl oxides and alkoxides.
2. Ligands with nitrogen donor sites such as aromatic amines, including pyridine and imidazole, form stronger complexes with vanadyl cation than alkyl amine bases [69].
3. Ligands with more than one oxygen donor sites, viz. hydroxycarboxylic acid (citric acid, tartaric acid, glyceric acid, lactic acid), phosphocarboxylates, and catechols, form stable complexes with the vanadyl cation.
4. Hydroxycarboxylates, viz. citric acid, tartaric acid, glyceric acid, and lactic acid, act as important metabolites of which some are in direct contact with glucose metabolism.
   Nucleotides form more stable complexes with vanadyl cation than nucleosides.
5. A mixture of complexes are formed by the reaction of vanadyl cations with peptides. Strong complexes with serum albumin protein is a very good example [70].

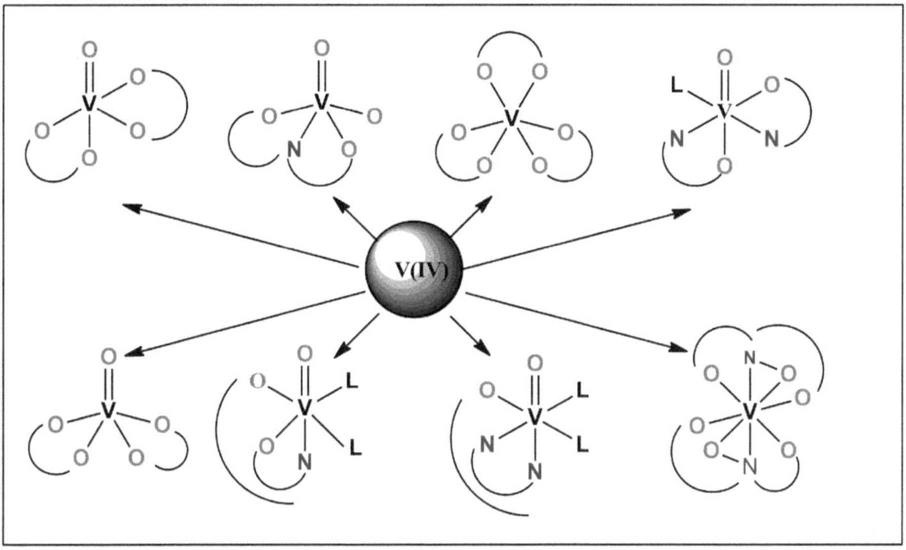

FIGURE 5.2 Coordination environments for V(IV) with N- and O- donor functionalities. (Adapted and modified from Ref. [28]).

6. A number of vanadium (IV) complexes formed by coordinating ligands such as bis (pyrrolidine-N-carbodithioate, cysteine, cystine, picolinic acid, *N,N*-ethylenediamine diacetic acid etc.) show promising insulin-mimetic activities. A vanadium hydroxypyridone complex also shows better insulin-mimetic properties.

5.4.2.2 Aqueous Interaction of V(V) in Aqueous Solutions with Ligands

With inorganic ligands

Crans et al. [28] has discussed the aqueous coordination behavior of V(V) compounds with inorganic ligands. However, some key points are depicted here:

1. V(V) easily forms complexes with ligands having N-, O- and S-donor active functionalities.
2. Weak interactions of vanadate with carbonates, phosphates, chromates, arsenate, and similar ligands are observed.
3. Vanadate–hydrogen peroxide (O_2^{2-}) complexes are of keen interest for biochemical and insulin-mimetic communities. Their complex behavior is very well studied.
4. Peroxo complexes of V(V) peroxovanadium' are very well-known complexes of vanadium. Oxidative mechanisms involved in their formation are very well discussed. They possess excellent oxidant properties and in the absence of

reducing agents, they are hydrolytically quite stable in neutral and slightly alkaline pH range.

5. Several oxoperoxovanadium(V) and oxodiperoxovanadium(V) complexes have been characterized at the crystallographic level (Table 5.2).
6. Peroxide is a triangular bidentate in all the cases known (Figure 5.3). The monomeric oxoperoxo complexes have one or two peroxides bound in the equatorial plane relative to the axial oxo ligand, resulting in pentagonal bipyramidal geometry in most of the structures. A few cases shows pentagonal pyramidal geometry; however, the oxo ligand of another molecule may well serve as the seventh ligand as a long axial bond.

TABLE 5.2
X-ray-characterized peroxovanadium (V) complexes

Oxoperoxo complexes	$[VO(O_2)(pic)(bipy)].H_2O$
	$K_3[VO(O_2)(ox)_2].0.5H_2O$
	$NH_4[VO(O_2)(H_2O)(dipic)]1.3H_2O$
	$NH_4([CO(O_2)IDA]$
	$[VO(O_2)(bipy)_2]ClO_4$
	$[VO(O_2)(phen)_2]ClO_4$
	$[VO(O_2)(bipy)_2]ClO_4$
	$VO(O_2)(pic)(H_2O)_2$
	$VO(O_2)(PAN)(pyr)$
Oxodiperoxo complexes	$NH_4[VO(O_2)_2(bipy)].4H_2O$
	$K_3[VO(O_2)_2CO_3]$
	$K_3[VO(O_2)_2(ox)].H_2O_2$
	$K_3[VO(O_2)_2(ox)].H_2O$
	$K_2[VO(O_2)_2(pic)].2H_2O$
	$K_2[VO(O_2)_2(OHpic)].3H_2O$
	$(NH_4)_3[VF_2O(O_2)_2]$
	$Cs_2[VFO(O_2)_2]$
	$K_2[VFO(O_2)_2]$
	$NH_4[VO(O_2)_2(NH_3)]$
	$(Hbipy)[VO(O_2)_2(bipy)\text{-}(3+x)\ H_2O_2.(2-x)H_2O$
Oxoperoxo dimers	$K_2[VO(O_2)(cit)]_2.H_2O$
	$K_3[F(O_2)\{VO(O_2)F\}_2]$
Oxodiperoxo dimers	$(NH_4)_3[HO\{VO(O_2)\}_2].\ H_2O$
	$(NH_4)_4[O\{VO(O_2)_2\}_2$
	$K_2[V_2O_2(O_2)_4(H_2O)].3H_2O$
	$[N(CH_3)_4]_2[V_2P_2(O_2)_4(H_2O)].2H_2O$
Triperoxo complex	$(NH_4)_2[V(O_2)3F]$
Alkylperoxo complex	$(dipic)VO(OOtBu).\ H_2O$

Note: bipy, 2,2-bipyridine; cit, citrato; dipic, pyridine 2,6-dicarboxylato; Hbipy, 2,2′ bipyridinium; IDA, iminodiacetato; OHpic, 3-hydroxypyridine -2-carboxylato; ox, oxalate; pic, picolinato, pyridine - 2-carboxylato; PAN, 1-(2-pyridylazo)-2-naphthol; phen, 1,10-phenanthroline.

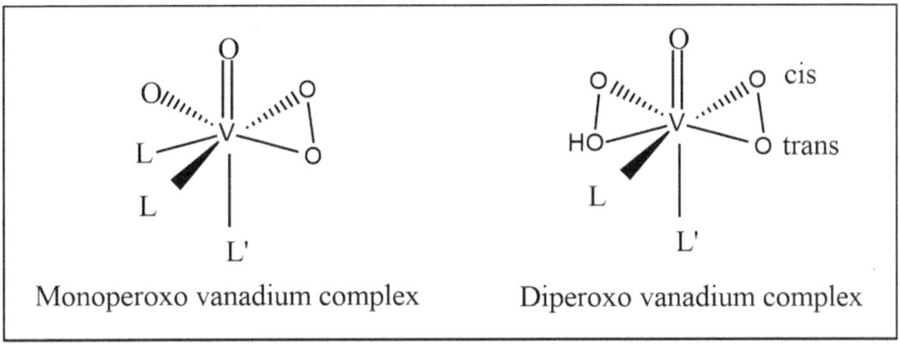

FIGURE 5.3 Coordination behavior in peroxovanadium complexes.

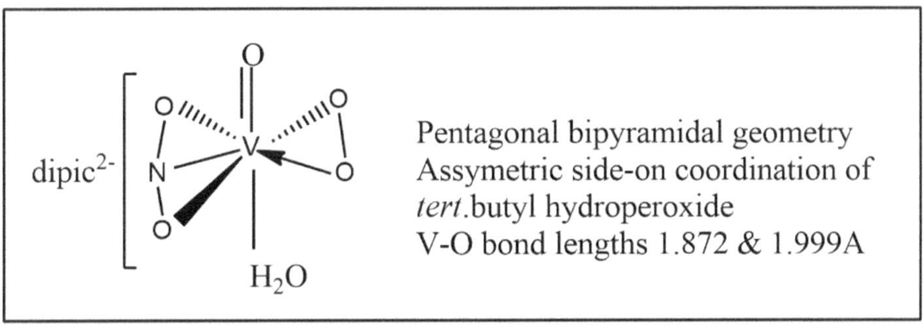

FIGURE 5.4 Structure of VO(dipic)(OOtBu).

7. In general, the O–O peroxo bond is longer in diperoxo complexes than in monoperoxo complexes. In the monoperoxo complexes, the peroxide is symmetrically coordinated, while in the diperoxo complexes, the $V–O_{trans}$ bond is slightly longer than the $V–O_{cis}$ bond. $V–O_{trans}$ and $V–O_{cis}$ bonds are defined in Figure 5.3.
8. VO(OOtBu)(dipic)(H_2O) is the first alkylperoxovanadium(V) complex to be crystallographically characterized [71]. The crystal structure reveals a pentagonal bipyramidal geometry with asymmetric, side-on coordination of *tert*-butyl hydroperoxide (Figure 5.4).
9. Peroxovanadates in aqueous medium coordinate with auxiliary ligands including substituted carboxylic acids, amino acids, peptides, and nucleosides, resulting in mono- and di-peroxovanadium complexes. These complexes show favorable catalytic and potent insulin-mimetic properties.
10. Hydroxylamine complexes of V(V) have also been studied [72]. These complexes are kinetically more inert than peroxovanadium complexes. Hydroxylamine NH_2OH is isoelectronic with H_2O_2 and readily coordinates

with vanadium. Complexes containing the hydroxylamide ligand [NH_2O^-] have been characterized. Interestingly, the ligand present in amavadin is the bis(propionate) derivative of hydroxylamide. It is a non-oxo vanadium (IV) complex isolated from the mushroom *Amanita muscaria*.

11. A series of mixed ligand oxovanadium(V) complexes of the type [$VO(NH_2O)_2(aa)$], containing the hydroxylamide ligand and simple amino acids coordinated through N (amino) and O(carboxylate) atoms (amino acids such as glycine, serine, threonine alanine valine, leucine, isoleucine, phenylalanine, tryptophan, cysteine and methionine) were prepared and characterized. Seven of these complexes, with the amino acid valine, leucine, isoleucine, phenylalanine, tryptophan, cysteine, and methionine, were described for the first time. Their infrared and Raman spectra, together with those of the previously investigated similar species with the amino acids glycine, serine, threonine, and alanine, and those of the related [$VO(NH_2O)_2(imidazole)_2$] Cl complex, were recorded and assigned. The spectroscopic behavior of those complexes whose structures has been determined by X-ray crystallography suggests an identical structure for the full series of complexes. The spectroscopic results also give a better characterization of the vibrational behavior of the interesting O–N– "side-on" bonded hydroxylamide ligand (Figure 5.5) [72].
12. The crystal and molecular structure of [$VO(DLmeth)(NH_2O)_2$].H_2O complex has been determined by X-ray diffraction methods (monoclinic, *P*21/*c* with $a = 5.8138(2)$, $b = 27.3496(8)$, $c = 7.5663(2)$ A°, $\beta = 103.059(3)$°, and $Z = 4$). The vanadium(V) cation shows pentagonal bipyramidal environment axially coordinated to the oxo group and equatorially to two hydroxylamido groups in a side-on fashion. The fifth equatorial binding site and the other apical position are occupied by an N and an O atom, respectively, from the methioninate anion acting as a bidentate ligand [73].

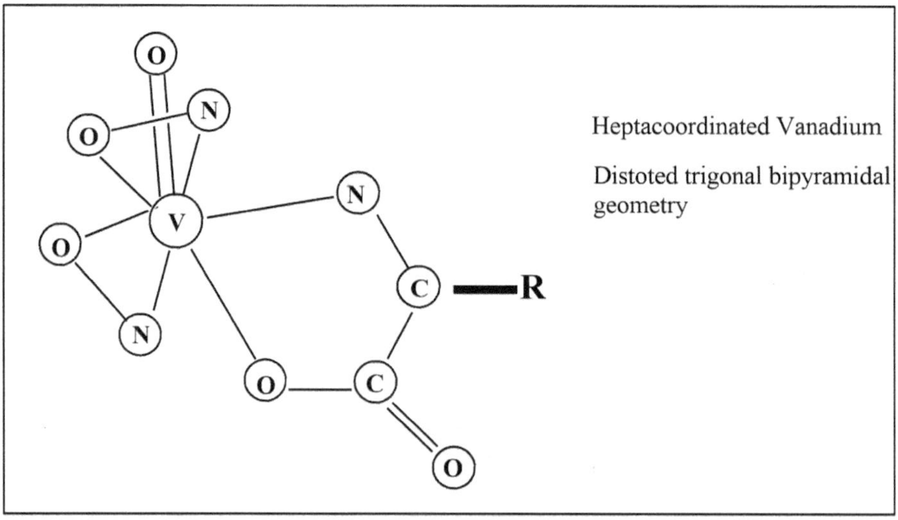

FIGURE 5.5 Structural characterization of [$VO(NH_2O)_2(aa)$] complexes.

With organic ligands

Complexes of vanadium(V) with organic ligands (monodentate, bidentate, and multidentate) have been studied [28]. Figure 5.6 depicts the selected examples of Possible coordination environments for vanadium (V) with ligands containing O and N donor *functionalities*. The presence of singly and doubly protonated vanadate ions, which are prevalent under physiological conditions, leads to their active reaction with hydroxylic compounds such as alcohols and acids. This interaction produces complexes often identified as esters or acid anhydrides, drawing a parallel to the vanadate–phosphate model. It is widely accepted that these complexes in aqueous solutions exhibit a four-coordinate structure. The stability constants for the formation of alkyl esters with aliphatic and aromatic alcohols are relatively low, usually around 0.2 M^{-1} for alkyl esters and approximately 1.0 M^{-1} for phenyl and tyrosyl vanadate esters. When carboxylic acids interact with vanadate, the coordination number increases, resulting in complexes with five or six coordination sites. These vanadium–carboxylate complexes are occasionally referred to as vanadate anhydrides due to their resemblance to organic acid anhydrides. Interestingly, there has been limited research on the formation of vanadium complexes with monodentate amide-containing ligands, likely due to low formation constants, which may also account for the lack of observed complexes with aliphatic amines. In contrast, aromatic amines such as pyridine and imidazole have been shown to form weak complexes with vanadate. However, in the presence of additional ligands, stable ternary complexes involving both aliphatic and aromatic amines can be formed, as demonstrated in studies involving Tris buffer and imidazole [28].

Extensive research on vanadium(V) systems, which often encompasses structural details, has significantly enhanced our comprehension of vanadium(V) complexes in aqueous solutions. The findings from these studies have been summarized below:

1. Vanadate interacts readily with 1,2-diols, producing multiple products, the predominant one being a 2:2 complex.
2. Interest in these complexes began in the 1970s when it was discovered that a vanadate–nucleoside complex effectively inhibited ribonuclease A [74]. The structure of the 2:2 vanadium–nucleoside complex has been extensively studied, making it one of the most well-characterized vanadium complexes in solutions. While simple vanadate–nucleoside esters can be formed, their stability is considerably lower than that of the 2:2 complex. Although 2:2 complexes form readily, ligands like tris(hydroxymethyl)aminomethane and imidazole have been shown to effectively compete, resulting in the formation of ternary complexes involving vanadate, nucleosides, and the competing ligands. The reasons behind the heightened stability of these ternary complexes remain unclear.
3. Several investigations regarding the interaction of multidentate thiols with vanadate have been conducted. It is noteworthy that vanadium(V) forms stable complexes with thiols containing ligands, some of which have been structurally elucidated.

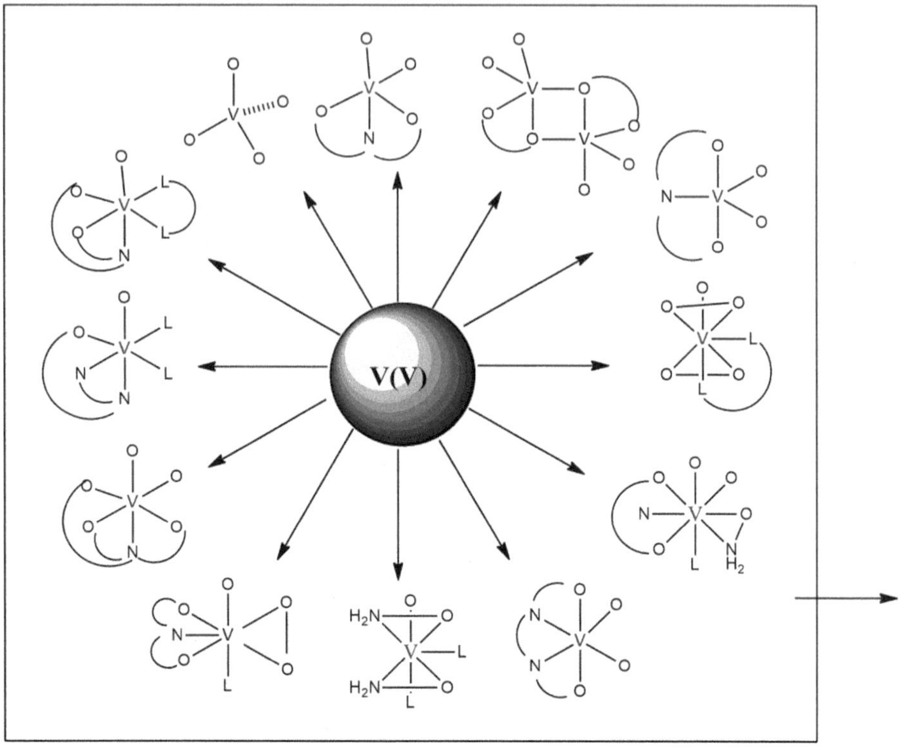

FIGURE 5.6 Possible coordination atmosphere for V(V) with ligands containing O- and N- donor functionalities.

4. A recent solution study of vanadate with Dithiothreitol (DTT) showed the study of these complexes is quite amenable even in aqueous solution [75].
5. The significance of thiol accessibility is illustrated by studies where vanadate was added to glycerol-3-phosphate dehydrogenase; despite the thiol presence on the active site, the inhibition caused by vanadate was reversible and not linked to redox processes.
6. Numerous reports envisage the interactions between amino acids, peptides, and peptide-like ligands with vanadate [76–80]. Amino acids generally form weak complexes with vanadate, with the only well-defined systems being stabilized by additional ligands such as Schiff's bases, peroxo groups, or hydroxylamido groups. Dipeptides can coordinate with vanadate through carboxylate, amine functionalities, and the deprotonated nitrogen of the peptide bond [28].
7. These studies indicate that oligopeptides are likely to form stable vanadate complexes through their amide backbones, with side chains such as serine, lysine, or glutamate coordinating to vanadate, as evidenced by studies with simple model peptides.

5.4.3 Vanadate–Phosphate Analogy

The vanadate–phosphate analogy highlights the structural and functional similarities between vanadate (V^{5+}) and phosphate (P^{5+}), particularly in biochemical reactions and enzymatic processes. Both compounds share similar tetrahedral geometries, allowing vanadate to mimic phosphate in various contexts. Vanadate interacts with enzymes that typically use phosphate substrates and can inhibit these enzymes similarly to how phosphate operates. It also influences energy transfer and signalling pathways, exhibiting insulin-mimetic effects that impact glucose metabolism.

Given the essential function of phosphates in biological systems, we can anticipate a variety of impacts from two categories of vanadium compounds: vanadate esters (1) and vanadate anhydrides (2). Phosphate esters are crucial cellular metabolites; their phosphate groups enhance solubility while their anionic nature boosts the recognition between cellular substrates and enzymes. In adenosine triphosphate (ATP), the phosphate group is responsible for storing approximately 7 kcal/mol of energy due to the formation of an anhydride bond, in contrast to the 3 kcal/mol stored by a phosphate ester bond. Phosphate is vital for both the structural integrity and functional roles in RNA and DNA. The analogy between vanadium and phosphate (V-P) has been used successfully with enzymes that facilitate the reactions of phosphate esters and anhydrides (Figure 5.7).

The effects of vanadate and various derivatives such as phosphate analogues have been reviewed in detail by Crans et al. [29] describing both the ground state analogy and the transition state analogy with phosphorus compounds. Notably, five-coordinate vanadium compounds resemble the transition state of phosphate ester hydrolysis, which has been recognized for many years. These compounds are often termed transition state analogues for phosphate ester hydrolysis reactions, explaining why many vanadium compounds serve as potent inhibitors of enzymes such as ribonucleases, phosphatases, ATPases, and other phosphorylases.

Various structural model studies of vanadate esters have been investigated to study the structurally similarity aspects of trialkoxo oxovanadium(V) complexes (3) and triphosphate esters. The simplest vanadium tris-alkoxides was reported in 1913. Since then, several vanadium tris-alkoxides have been reported. Studies focusing on structural characterization reveal the formation of vanadium compounds with five-coordinated V(V) center (4) with a wide range of alkoxide ligands due to dimerization even in the absence of other coordinating ligands. Dimerization is observed in solutions with 10 mM or more of the oxovanadium tris-alkoxides. As the concentration decreases, the fraction of presumed four-coordinated vanadium tris-alkoxides increases. Although the most common coordination geometry for trisalkoxide oxovanadium complexes is five-coordinate, there are notable exceptions when sterically bulky alkoxides are used. Various oxovanadium(V) complexes featuring t-butoxide, *tert*-butylsiloxide, norbornene oxide, adamantanoxide, 2,6-diisopropylphenoxide, or bidentate, sterically restricted di-alkoxides exhibit four-coordinate vanadium. This four-coordinate geometry is observed both in solution and in the solid state. Among these, a structurally characterized example is tris(2,6-diisopropylphenoxide) oxovanadium(V)

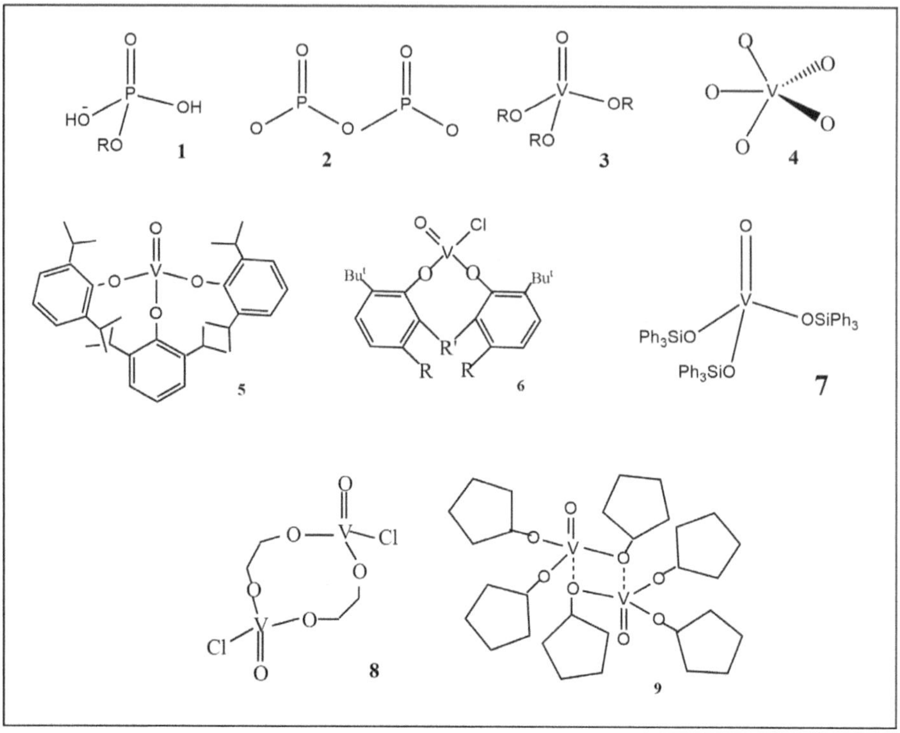

FIGURE 5.7 (V-P) analogy. Select examples of four-coordinated vanadium alkoxides. (Adapted from Ref. [29].

(5) confirming the existence of the four-coordinate V-P analogy. Additionally, several four-coordinate V(V) complexes supported by alkoxide, phenoxide, or siloxide have been reported, including those involving biphenolic diols(6), tri-*tert*-butyl oxide (7), and ethylene diol (8).

Theoretical evaluations investigating the influence of ligands on vanadium coordination geometry in simple systems indicate that electronic effects are not the primary cause of the observed coordination differences. Instead, subtle ligand effects, such as the gem-dimethyl effect, play a significant role. Among the various structurally characterized complexes, the one featuring the cyclopentanol ligand exhibited noticeable variation in V–O bond lengths—up to approximately 0.3 Å (9)—depending on whether the alkoxide group was directly coordinated to the vanadium atom or merely associated with its dinuclear counterpart. This structural behavior underscores the favorable association of the alkoxide, even in the presence of a sterically bulky ligand like cyclopentanol.

Vanadate and Phosphate Esters Analogues

A detailed study has been reported by Crans et al. [29]. However, to summarize the Crans content, vanadate esters, or alkoxides, are formed easily in aqueous solutions

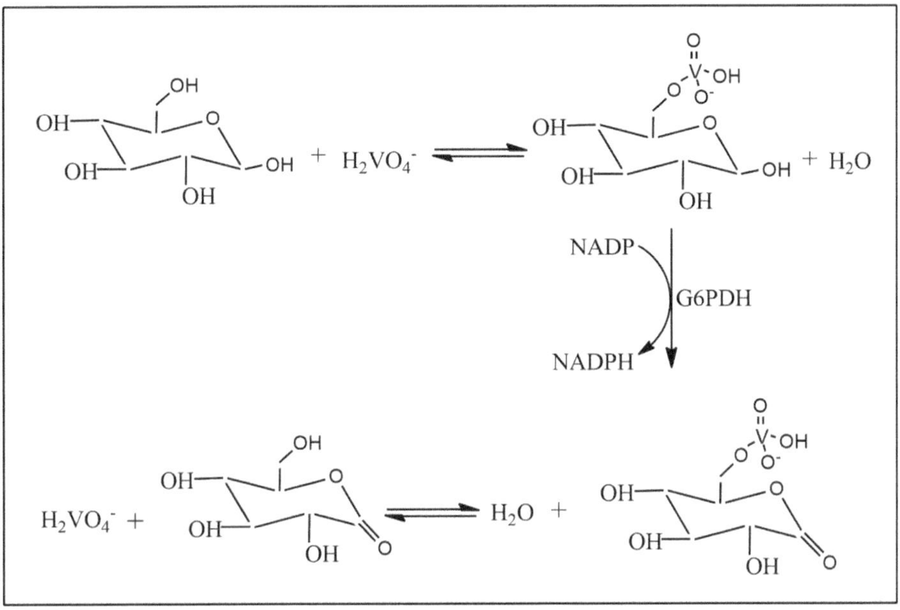

FIGURE 5.8 Formation of D-glucose-6-vanadate (a substrate for glucose-6-phosphate dehydrogenase). [Adapted from Ref. 29.]

and serve as substrates for various enzymes like dehydrogenases, isomerases, and aldolases. Their formation constants range from 0.1 to 0.2 M^{-1} for aliphatic esters and are three to five times higher for aromatic esters. Due to their higher lability in water compared to phosphate esters, reactions to isolate vanadate esters often occur in organic solvents. In aqueous solutions, these esters can rapidly decompose, with the potential for alcohol exchange if conditions are favorable. The reaction rates can be controlled, especially with sterically hindered alcohols that react slowly.

Despite their lower formation constants, the ability of vanadate esters to form bonds quickly at room temperature—unlike phosphate esters, which take years to equilibrate—suggests their significant potential in biological processes. A specific example is the reaction of D-glucose with vanadate, where glucose-6-vanadate is formed and recognized by the enzyme glucose-6-phosphate dehydrogenase, indicating its biological relevance (Figure 5.8).

Vanadate and phosphate Anhydride Analogues

Vanadium has oligomeric vanadate species analogous to pyrophosphate bonds, including dimers, trimers, tetramers, and pentamers. Solid-state studies have explored the properties of these V–O materials. Specifically, the tetramer and pentamer adopt cyclic forms in solution that do not structurally resemble the biologically relevant linear phosphates. Their existence may be attributed to greater stability of these cyclic derivatives over linear and cyclic trimers, or due to electronic repulsion preventing their formation. Additionally, a cyclic trimer has been prepared and characterized structurally, alongside known tetra- and pentanuclear analogues. A linear vanadate trimer was recently identified as a complex with the phosphatase PhoE from *Bacillus stearothermophilus*. Previously, research on these homovanadates focused primarily on the relationship between the vanadate dimer and pyrophosphate in inorganic systems. There is currently no data on systems featuring four-coordinate vanadium within an $O_3POV(O_3)$ unit, but structural information exists regarding these units in larger complexes or clusters where vanadium is five- or six-coordinate. A detailed study on V-P anhydride analogues has been reported [29].

5.4.4 Enzymatic Action of Vanadium

Vanadium is known to participate in a variety of enzymatic functions in biological systems. Its ability to exist in multiple oxidation states (+2, +3, +4, and +5) allows it to interact with enzymes and influence various biochemical processes. Reports reveal that the presence of established naturally occurring vanadium-dependent enzymes are either haloperoxidases or nitrogenases. Two primary types of vanadium-containing enzymes have been discovered: vanadium nitrogenases and vanadium-dependent haloperoxidases (V-HPOs) [81]. Vanadium nitrogenases are employed by nitrogen-fixing bacteria to convert dinitrogen gas into ammonia. While this metalloenzyme system is typically associated with a molybdenum–iron cofactor, certain bacteria have been found to produce distinct nitrogenases that utilize either vanadium–iron or iron–iron centers. These vanadium nitrogenases have been isolated from a variety of nitrogen-fixing microorganisms and are produced when molybdenum is scarce. In contrast, V-HPOs are more widely distributed across various organisms, including macroalgae, fungi, and bacteria. These enzymes, which feature a coordinated vanadate ion, facilitate the oxidation of halide ions into their respective hypohalous acids using hydrogen peroxide, and they are classified based on the most electronegative halide they can oxidize. For instance, vanadium chloroperoxidases (V-ClPOs) can oxidize chloride, bromide, and iodide, while vanadium bromoperoxidase (VBrPOs) are limited to oxidizing only bromide and iodide. Detailed studies on enzymatic behavior of vanadium has been done by various research groups [29, 81]. Therefore only information on vanadium enzymes is outline below.

Vanadium-Dependent Haloperoxidases

Vanadium-dependent haloperoxidases are a class of enzymes that use vanadate (VO_4^{3-}) or vanadyl (VO^{2+}) as a cofactor to catalyze the halogenation of organic compounds. These enzymes are found in marine algae, fungi, and bacteria and are of following types:

Vanadium chloroperoxidase (VClPO): Catalyzes the oxidation of chloride ions (Cl^-) to hypochlorous acid (HOCl) using hydrogen peroxide (H_2O_2).

Vanadium bromoperoxidase (VBPO): Catalyzes the oxidation of bromide ions (Br^-) to hypobromous acid (HOBr) using H_2O_2.

Vanadium iodoperoxidase (VIPO): Catalyzes the oxidation of iodide ions (I^-) to hypoiodous acid (HOI) using H_2O_2.

VClPOs possess a sufficiently high oxidation potential to oxidatively attack chloride along with iodide and bromide.

The mechanism for the formation of haloperoxidases proceeds via three steps:

Step 1: Activation: Vanadium in the active site binds to hydrogen peroxide (H_2O_2), forming a peroxovanadium intermediate.

Step 2: Halogenation: The peroxovanadium intermediate reacts with halide ions (Cl^-, Br^-, I^-), leading to the formation of hypohalous acids (HOCl, HOBr, HOI).

Step 3: Product release: The hypohalous acids are released, and the vanadium returns to its original oxidation state, ready for another catalytic cycle.

Phosphatases and Vanadate Inhibition

Vanadate (VO_4^{3-}) is structurally like phosphate (PO_4^{3-}) and can act as a competitive inhibitor for phosphatases, enzymes that hydrolyze phosphate esters.

- Protein tyrosine phosphatases (PTPs): Vanadate inhibits PTPs by mimicking the transition state of phosphate, thereby preventing dephosphorylation of tyrosine residues in proteins.
- Alkaline phosphatases (APs): Vanadate can inhibit alkaline phosphatases, which are involved in removing phosphate groups from various substrates, affecting bone mineralization and other metabolic processes.

Vanadium-Dependent Nitrogenases

In some nitrogen-fixing bacteria, vanadium is an essential component of a specialized nitrogenase enzyme that catalyzes the reduction of atmospheric nitrogen (N_2) to ammonia (NH_3). This enzyme contains a vanadium–iron (VFe) cofactor in its active site. It functions under conditions where the more common molybdenum nitrogenase is less efficient. At first, electrons are transferred to the vanadium–iron cofactor. Then nitrogen binds to the active site, and the enzyme reduces it to ammonia (NH_3) through a series of electron and proton transfers.

Vanadium-Dependent Oxidoreductases

Vanadium can also be a cofactor in certain oxidoreductases, enzymes that catalyze oxidation–reduction reactions. Vanadium-dependent aldehyde oxidase enzyme catalyzes the oxidation of aldehydes to carboxylic acids, with vanadium playing a role in electron transfer processes.

Vanadium-Dependent Catalase and Peroxidase

Vanadium compounds can mimic the activity of catalases and peroxidases, enzymes involved in the decomposition of hydrogen peroxide (H_2O_2). Vanadium binds to hydrogen peroxide, forming reactive intermediates. These intermediates facilitate the breakdown of H_2O_2 into water (H_2O) and oxygen (O_2).

In conclusion, vanadium possesses a unique set of biological properties that have created increasing interest in biomedical research. Its role as an essential trace element in certain organisms, coupled with its potential therapeutic applications, highlights its significance in biological systems. Vanadium has been shown to influence various metabolic pathways, particularly in glucose metabolism and insulin signaling, suggesting its possible utility in managing diabetes and metabolic disorders.

To add, vanadium compounds exhibit anti-inflammatory and antioxidant effects, which may contribute to their protective roles in cellular health. However, while the beneficial aspects of vanadium are promising, it is crucial to consider the potential toxicological effects associated with excessive exposure or inappropriate dosage. Much more work is being done by the scientific community to better understand the mechanisms of vanadium's biological actions with an easy approach, optimize its therapeutic potential, and minimize risks. Future studies should focus on elucidating the molecular pathways involved, as well as investigating the efficacy and safety of vanadium-based interventions in clinical settings. Overall, vanadium presents a fascinating intersection of chemistry and biology, with significant implications for health and disease management.

REFERENCES

1. Tripathi, D., Mani, V., & Pal, R. P. (2018). Vanadium in biosphere and its role in biological processes. *Biological Trace Element Research*, *186*(1), 52–67. doi:10.1007/s12011-018-1289.
2. Underwood, E. J., & Suttle, N. F. (2010) *The Mineral Nutrition of Livestock.* New York: CABI.
3. Moskalyk, R. R., & Alfantazi, A. M. (2003). Processing of vanadium: A review. *Minerals Engineering*, *16*(9), 793–805. doi:10.1016/s0892-6875(03)00213-9
4. Verma, S., Cam, M. C., & McNeill, J. H. (1998). Nutritional factors that can favorably influence the glucose/insulin system: Vanadium. *Journal of the American College of Nutrition*, *17*(1), 11–18. doi:10.1080/07315724.1998.10718730
5. Treviño, S., Díaz, A., Sánchez-Lara, E., Sanchez-Gaytan, B. L., Perez-Aguilar, J. M., & González-Vergara, E. (2019). Vanadium in biological action: Chemical, pharmacological aspects, and metabolic implications in diabetes mellitus. *Biological Trace Element Research*, *188*(1), 68–98. doi:10.1007/s12011-018-1540-6
6. Moskalyk, R. R., & Alfantazi, A. M. (2003). Processing of vanadium: A review. *Minerals Engineering*, *16*(9), 793–805. doi:10.1016/s0892-6875(03)00213-9
7. Perron, L. (2001) Vanadium, natural resources canada, minerals and resources sector. *Canada Minerals Yearbook*, *59*, 59.1–59.7.
8. Lu, W., Li, X., & Zhang, H. (2018). The next generation vanadium flow batteries with high power density—a perspective. *Physical Chemistry Chemical Physics*, *20*(1), 23–35. doi:10.1039/c7cp07456e
9. Rehder, D. (2008) *Bioinorganic Vanadium Chemistry*. A Wiley Series of Advanced Textbooks, 1st edn. Wiley, New York, pp. 1–4.
10. Rehder, D. (2015). The role of vanadium in biology. *Metallomics: Integrated Biometal Science*, *7*(5), 730–742. doi:10.1039/c4mt00304g
11. Almeida, M., Filipe, S., Humanes, M., Maia, M. F., Melo, R., Severino, N., & Wever, R. (2001). Vanadium haloperoxidases from brown algae of the Laminariaceae family. *Phytochemistry*, *57*(5), 633–642. doi:10.1016/s0031-9422(01)00094
12. Nriagu, J.P. (1998). *Vanadium in the Environment, Part 2: Health Effects*, Chichester: John Wiley and Sons.

13. Mukherjee, B., Patra, B., Mahapatra, S., Banerjee, P., Tiwari, A., & Chatterjee, M. (2004). Vanadium—An element of atypical biological significance. *Toxicology Letters*, *150*(2), 135–143. doi: 10.1016/j.toxlet.2004.01.009
14. Barceloux, D. G., & Barceloux, D. (1999). Vanadium. *Journal of Toxicology: Clinical Toxicology*, *37*(2), 265–278. doi:10.1081/clt-100102425
15. Badmaev, V., Prakash, S., & Majeed, M. (1999). Vanadium: A review of its potential role in the fight against diabetes. *Journal of Alternative and Complementary Medicine (New York, N.Y.)*, *5*(3), 273–291. doi:10.1089/acm.1999.5.273
16. Willsky, G. R., Chi, L.-H., Godzala, M., 3rd, Kostyniak, P. J., Smee, J. J., Trujillo, A. M., ... Crans, D. C. (2011). Anti-diabetic effects of a series of vanadium dipicolinate complexes in rats with streptozotocin-induced diabetes. *Coordination Chemistry Reviews*, *255*(19–20), 2258–2269. doi:10.1016/j.ccr.2011.06.015
17. Rehder, D. (2012). The potentiality of vanadium in medicinal applications. *Future Medicinal Chemistry*, *4*(14), 1823–1837. doi:10.4155/fmc.12.103
18. Costa Pessoa, J., & Tomaz, I. (2010). Transport of therapeutic vanadium and ruthenium complexes by blood plasma components. *Current Medicinal Chemistry*, *17*(31), 3701–3738. doi:10.2174/092986710793213742
19. Rehder, D. (1999). The coordination chemistry of vanadium as related to its biological functions. *Coordination Chemistry Reviews*, *182*(1), 297–322. doi:10.1016/s0010-8545(98)00194-5
20. Chasteen, N. D. (1983) *Structure and Bond 53*, 105.
21. Gresser, M. J., Tracey, A. S., & Stankiewicz, P. J. (1987). *Advances in Protein Phosphatases 4*, 35.
22. Gresser, M. J. & Tracey, A. S. (1990). In *Vanadium in Biological Systems*. Chasteen, N. D., Ed. Boston: Kluwer Academic.
23. Chasteen, N. D., Ed. (1990). *Vanadium in Biological Systems*, Dordrecht, The Netherlands: Kluwer Academic.
24. Rehder, D. (1991). The bioinorganic chemistry of vanadium. *Angewandte Chemie: International Edition in English*, *30*(2), 148–167. doi:10.1002/anie.199101481
25. Crans, D.C. (1994). Aqueous chemistry of labile oxovandates: Relevance to biological studies. *Comments on Inorganic Chemistry 16*(1–2), 1–33.
26. Rehder, D. (1995). Inorganic considerations on the function of vanadium in biological systems. *Metal Ions in Biological Systems*, *31*, 1–1.
27. Crans, D., Keramidas, A., & Drouza, C. (1996). Organic vanadium compounds—Transition state analogy with organic phosphorus compounds. *Phosphorus, Sulfur, and Silicon and the Related Elements*, *109*(1), 245–248. doi:10.1080/10 42650 960 8046244
28. Tracey, A. S., & Crans, D. C. (Eds.). (1999). *Vanadium Compounds*. ACS Publications. doi:10.1021/bk-1998-0711
29. Crans, D. C., Smee, J. J., Gaidamauskas, E., & Yang, L. (2004). The chemistry and biochemistry of vanadium and the biological activities exerted by vanadium compounds. *Chemical Reviews*, *104*(2), 849–902. doi:10.1021/cr020607t
30. Amaral, L. M. P. F., Moniz, T., Silva, A. M. N., & Rangel, M. (2023). Vanadium compounds with antidiabetic potential. *International Journal of Molecular Sciences*, *24*(21), 15675. doi:10.3390/ijms242115675
31. Winter, J. M., & Moore, B. S. (2009). Exploring the chemistry and biology of vanadium-dependent haloperoxidases. *The Journal of Biological Chemistry*, *284*(28), 18577–18581. doi:10.1074/jbc.R109.001602
32. Jandhyala, B. S., & Hom, G. J. (1983). Physiological and pharmacological properties of vanadium. *Life Sciences*, *33*(14), 1325–1340. doi:10.1016/0024-3205(83)90816-0

33. Skou, J. C. (1965). Enzymatic basis for active transport of Na+ and K+ across cell membrane. *Physiological Reviews*, *45*(3), 596–618. doi:10.1152
34. Gamyee, A., & Larmuth, J. (1876). *Journal of Anatomy and Physiology*, *II*, 235–250.
35. Jackson, D. E. (1912). *Journal of Pharmacology and Experimental Therapeutics*, *4*(1), 59–74.
36. Charney, A.N., Silva, P., & Epstein, F.H. (1975). *Journal of Applied Physiology*, *39*, 156–158.
37. Cantley, Jr. L.C., Josephson, L., & Warner, R. (1977). *Journal of Biological Chemistry*, *252* , 7421–7423.
38. Cantley, Jr., L.C., Resh, M. & Guidotti, G. (1978). *Nature*, *272*, 552–554.
39. Rehder, D. (2003). Biological and medicinal aspects of vanadium. *Inorganic Chemistry Communications*, *6*(5), 604–617. doi:10.1016/s1387-7003(03)00050-9
40. Butler, A. (1998). Acquisition and utilization of transition metal ions by marine organisms. *Science*, *281*(5374), 207–209.
41. Aihemaiti, A., Gao, Y., Meng, Y., Chen, X., Liu, J., Xiang, H., & Jiang, J. (2020). Review of plant-vanadium physiological interactions, bioaccumulation, and bioremediation of vanadium-contaminated sites. *The Science of the Total Environment*, *712*(135637), 135637. doi: 10.1016/j.scitotenv.2019.135637
42. Kordowiak, A.M., & Holko, P. (2009). Vanadium derivatives as compounds of high biological significance. Part I. Antidiabetic activity. *Postepy Biologii Komorki*, *36*, 361–376.
43. Aureliano, M., & Gândara, R. M. C. (2005). Decavanadate effects in biological systems. *Journal of Inorganic Biochemistry*, *99*(5), 979–985. doi: 10.1016/j.jinorgbio.2005.02.024
44. Lin, T. S., Chang, C. L., & Shen, F. M. (2004). Whole blood vanadium in Taiwanese college students. *Bulletin of Environmental Contamination and Toxicology*, *73*(5), 781–786. doi:10.1007/s00128-004-0495-9
45. Erman, L. Y., Mindrul, V. F., Noskov, V. G., Fadeev, V. N., & Kurochkin, V. K. (1995). Synthesis and molecular and electronic structure of dibenz(b,f)-1,4-oxazepine. *Journal of Structural Chemistry*, *36*(1), 175–178. doi:10.1007/bf02577767
46. Zhang, Z., Huang, C., Li, J., Leonard, S. S., Lanciotti, R., Butterworth, L., & Shi, X. (2001). Vanadate-induced cell growth regulation and the role of reactive oxygen species. *Archives of Biochemistry and Biophysics*, *392*(2), 311–320. doi:10.1006/abbi.2001.2464
47. Tripathi, D., Mani, V., & Pal, R. P. (2018). Vanadium in biosphere and its role in biological processes. *Biological Trace Element Research*, *186*(1), 52–67. doi:10.1007/s12011-018-1289-y
48. Allan, D. L., & Jarrell, W. M. (1989). Proton and copper adsorption to maize and soybean root cell walls. *Plant Physiology*, *89*(3), 823–832. doi:10.1104/pp.89.3.823
49. Saco, D., Martín, S., & San José, P. (2013). Vanadium distribution in roots and leaves of Phaseolus vulgaris: morphological and ultrastructural effects. *Biologia Plantarum*, *57*(1), 128–132. doi:10.1007/s10535-012-0133-z
50. Rauser, W. E. (1999). Structure and function of metal chelators produced by plants. *Cell Biochemistry and Biophysics*, *31*(1), 19–48. doi:10.1007/bf02738153
51. García-Jiménez, A., Trejo-Téllez, L. I., Guillén-Sánchez, D., & Gómez-Merino, F. C. (2018). Vanadium stimulates pepper plant growth and flowering, increases concentrations of amino acids, sugars and chlorophylls, and modifies nutrient concentrations. *PLoS One*, *13*(8), e0201908. doi: 10.1371/journal.pone.0201908
52. Nawaz, M. A., Jiao, Y., Chen, C., Shireen, F., Zheng, Z., Imtiaz, M., … Huang, Y. (2018). Melatonin pretreatment improves vanadium stress tolerance of watermelon

seedlings by reducing vanadium concentration in the leaves and regulating melatonin biosynthesis and antioxidant-related gene expression. *Journal of Plant Physiology, 220*, 115–127. doi: 10.1016/j.jplph.2017.11.003
53. Gokul, A., Cyster, L. F., & Keyster, M. (2018). Efficient superoxide scavenging and metal immobilization in roots determines the level of tolerance to Vanadium stress in two contrasting Brassica napus genotypes. *Suid-Afrikaanse Tydskrif Vir Plantkunde [South African Journal of Botany], 119*, 17–27. doi: 10.1016/j.sajb.2018.08.001
54. Aihemaiti, A., Jiang, J., Blaney, L., Zou, Q., Gao, Y., Meng, Y. & Xu, Y. (2019). The detoxification effect of liquid digestate on vanadium toxicity to seed germination and seedling growth of dog's tail grass. *Journal of Hazardous Materials, 369*, 456–464. doi: 10.1016/j.jhazmat.2019.01.091
55. Imtiaz, M., Mushtaq, M. A., Rizwan, M. S., Arif, M. S., Yousaf, B., Ashraf, M., & Tu, S. (2016). Comparison of antioxidant enzyme activities and DNA damage in chickpea (*Cicer arietinum* L.) genotypes exposed to vanadium. *Environmental Science and Pollution Research International, 23*(19), 19787–19796. doi:10.1007/s11356-016-7192-1
56. Kubes, J., Skalicky, M., Tumova, L., Martin, J., Hejnak, V., & Martinkova, J. (2019). Vanadium elicitation of Trifolium pratense L. cell culture and possible pathways of produced isoflavones transport across the plasma membrane. *Plant Cell Reports, 38*(5), 657–671. doi:10.1007/s00299-019-02397-y
57. Yang, J., Wang, M., Jia, Y., Gou, M., & Zeyer, J. (2017). Toxicity of vanadium in soil on soybean at different growth stages. *Environmental Pollution (Barking, Essex: 1987), 231*(Pt 1), 48–58. doi: 10.1016/j.envpol.2017.07.075
58. Yang, J.Y., & Tang, Y. (2015). Accumulation and biotransformation of vanadium in *Opuntia microdasys. Bulletin of Environmental Contamination and Toxicology, 94*, 448–452.
59. Jiang, B., Xing, Y., Zhang, B., Cai, R., Zhang, D., & Sun, G. (2018). Effective phytoremediation of low-level heavy metals by native macrophytes in a vanadium mining area, China. *Environmental Science and Pollution Research International, 25*(31), 31272–31282. doi:10.1007/s11356-018-3069-9
60. Jiang, Y., Zhang, B., He, C., Shi, J., Borthwick, A. G. L., & Huang, X. (2018). Synchronous microbial vanadium (V) reduction and denitrification in groundwater using hydrogen as the sole electron donor. *Water Research, 141*, 289–296. doi: 10.1016/j.watres.2018.05.033
61. Arnon, D. I., & Wessel, G. (1953). Vanadium as an essential element for green plants. *Nature, 172*(4388), 1039–1040. doi:10.1038/1721039a0
62. Sigel, A., & Sigel, H. Eds. (1995) *Vanadium and Its Role for Life. Metal Ions in Biological Systems, 3.* New York: Marcel Dekker.
63. Chasteen, N. D. (1981). Vanadyl (IV) EPR spin probes inorganic and biochemical aspects. In L. J. Berliner et al. (Eds.), *Biological Magnetic Resonance* (pp. 53–119). New York: Plenum Press. doi:10.1007/978-1-4613-3201-5_2
64. Boyd, D.W., & Kustin, K. (1984). Vanadium: A versatile biochemical effector with an elusive biological function. *Advances in Inorganic Biochemistry, 6*, 311–365.
65. Mustafi, D., & Makinen, M.W. (1988). ENDOR-determines solvation structure of vanadyl (2+) in frozen solutions. *Inorganic Chemistry, 27*(19), 3360–3368.
66. Crans, D. C., Mahroof-Tahir, M., & Keramidas, A. D. (1995). Vanadium chemistry and biochemistry of relevance for use of vanadium compounds as antidiabetic agents. *Molecular and Cellular Biochemistry, 153*(1–2), 17–24. doi:10.1007/bf01075914

67. Pettersson, L., Hedman, B., Nenner, A.-M., & Andersson, I. (1985). Multicomponent polyanions: 36. Hydrolysis and redox equilibria of the H^+-HVO_4^{2-} system in 0.6M NaCl. A complementary potentiometric and ^{51}VNMR study at low vanadium concentrations in acid solution. *Acta Chemica Scandinavica, A 39,* 499–506.
68. Selbin, J. (1965). The chemistry of oxovanadium (IV). *Chemical Reviews, 65*(2), 153–175. doi:10.1021/cr60234a00
69. Sanna, D., Micera, G., Buglyó, P., & Kiss, T. (1996). Oxovanadium (IV) complexes of ligands containing phosphonic moieties. *Journal of Chemical Society, Dalton Transactions*, (8), 1779–1779. doi:10.1039/dt9960001779
70. Chasteen, N. D., & Francavilla, J. (1976). An electron paramagnetic resonance study of vanadyl (IV)–Serum albumin complexes. *The Journal of Physical Chemistry, 80*(8), 867–871. doi:10.1021/j100549a02
71. Butler, A., Clague, M.J., & Meister, G.E. (1994). *Chemical Review, 94,* 625–638.
72. Arrambide, G., Gambino, D., & Baran, E. J. (2009). Synthesis and spectroscopic characterization of hydroxylamido/amino acid complexes of oxovanadium(V). *Journal of Coordination Chemistry, 62*(1), 63–74. doi:10.1080/00958970802474821
73. Piroa, O.E., Gustavo A. E., Arrambide, G., Gambino, D., & Baran, E. J. (2011). *Zeitschrift für Naturforschung A . 66b*, 752–754.
74. Lindquist, R. N., Lynn, J. L., & Lienhard, G. E. (1973). Possible transition-state analogs for ribonuclease. Complexes of uridine with oxovanadium (IV) ion and vanadium(V) ion. *Journal of the American Chemical Society, 95*(26), 8762–8768. doi:10.1021/ja00807a043
75. Paul, P.C., & Tracey, A.S. (1997). Aqueous interactions of vanadate and peroxovanadate with dithiothreitol. Implications for the use of this redox buffer in biochemical investigations. *Journal of Biological Inorganic Chemistry, 2*, 644–651.
76. Rehder, D. (1988). Interaction of vanadate ($H2VO_4^-$) with dipeptides investigated by vanadium-51 NMR spectroscopy. *Inorganic Chemistry, 27*(23), 4312–4316.
77. Crans, D. C., Bunch, R. L., & Theisen, L. A. (1989). Interaction of trace levels of vanadium (IV) and vanadium (V) in biological systems. *Journal of American Chemical Society,* 1989, *111*(19), 7597–7607.
78. Jaswal, J. S., & Tracey, A. S. (1991). Stereochemical requirements for the formation of vanadate complexes with peptides. *Canadian Journal of Chemistry, 69*(10), 1600–1607.
79. Tracey, A. S., Jaswal, J. S., Nxumalo, F., & Angus-Dunne, S.J. (1995). Condensation reactions between vanadate and small functionalised peptides in aqueous solution. *Canadian Journal of Chemistry, 73*(4), 489–498.
80. Crans, D.C., Holst, H., Keramida, A.D., & Rehder, D. (1995). A slow exchanging vanadium (V) peptide complex: vanadium (V)-glycine-tyrosine. *Inorganic Chemistry, 34*(10), 2524–2534.

6 Biological Activity of Chromium

Garima Pandey, Monika Kamboj, Sangeeta Bajpai and Bimal Krishan Banik

6.1 INTRODUCTION

Chromium(III) has held significant biological importance since its discovery six decades ago [1]. Initially identified alongside vitamin E and selenium for its role in preventing liver necrosis in experimental animals, chromium(III) was found to enhance insulin activity, leading to its classification as an essential trace element, albeit required in extremely low concentrations. Marketed primarily as chromium picolinate, this nutritional supplement carries an ironic bright red hue. However, numerous questions regarding chromium's biological intricacies remain unanswered [2,3]. While several countries acknowledge chromium(III) as an essential micronutrient, recent controversies have arisen due to the lack of clarity surrounding chromium's biology. The European Food Safety Authority (EFSA) panel in 2009 concluded insufficient evidence to support chromium(III)'s essentiality in animal nutrition [4]. Conversely, the Office of Dietary Supplements, National Institutes of Health, recognized chromium as a mineral required by humans in trace amounts in 2018 [5]. In contrast, chromium(VI) is known to be toxic, with most environmental chromate being anthropogenic or formed through the oxidation of chromium(III) by trace amounts of MnO_2 in water [6]. Chromate is genotoxic and classified as a Group I carcinogen, particularly linked to inhalation and lung cancer [7]. Laboratory experiments have demonstrated its uptake by cells and its ability to cause DNA strand breaks. The extent to which chromate enters cells in living organisms and reaches the nucleus for DNA interaction remains uncertain, given cells' reducing environment that favors chromium(III) [8,9].

The paradox of chromium's dual nature—beneficial as chromium (III) and toxic as chromium (VI)—poses a significant challenge for scientists worldwide. Bridging the gap between its nutritional, pharmacological, and toxicological effects necessitates adequate funding for experimental research elucidating the biochemical actions of biologically relevant chromium complexes within and on cells [10].The analogy of traffic lights—a continuum between red and green—underscores the ongoing uncertainty in chromium biology. While amber lights provide a transitional phase, the

DOI: 10.1201/9781032615110-8

duration of this phase in chromium biology remains uncertain. This uncertainty raises questions about missed therapeutic opportunities and potential risks to populations. Thus, resolving the dichotomy of chromium's effects is imperative for advancing both health interventions and environmental safety measures [11].

It's essential to differentiate between the two variants of chromium found in nature. Trivalent chromium, capable of forming bonds with up to three other atoms, is our primary focus—it's the biologically active form present in food. Conversely, hexavalent chromium, the other variant, is vastly dissimilar. Known for its high toxicity, hexavalent chromium is primarily discussed concerning pollution resulting from industrial activities. Trivalent chromium gained prominence in the 1950s due to its pivotal role in insulin metabolism and its participation in various enzyme reactions within organisms [12]. It serves as a crucial component of the glucose tolerance factor (GTF), positively influencing mechanisms regulating blood sugar levels [13]. Chromium acts synergistically with insulin, aiding in glucose uptake into cells and contributing to blood sugar level regulation. Insufficient chromium can hamper insulin effectiveness, leading to elevated glucose levels. Trivalent chromium is indispensable for normal human and animal development [14].

6.2 CHROMIUM AS BIOELEMENT

Chromium, along with trace minerals like zinc, copper, and selenium, demonstrates notable antioxidant properties. These minerals assist in scavenging free radicals from the body, thus shielding internal cells and tissues against oxidative harm [15,16]. Chromium plays a vital role in metabolic processes, insulin regulation, neurological function, energy metabolism, and cardiovascular health. Adequate intake of chromium through diet or supplements is essential to support these crucial physiological functions and promote overall well-being [17,18].

6.2.1 Sources

A variety of plant and animal foods serve as rich sources of chromium, helping to meet the body's dietary needs for this essential trace mineral. Chromium is chiefly acquired through dietary sources [19]. Chromium-rich foods encompass entrails, meat, mollusks, lobsters, vegetables, and unrefined sugar, with smaller amounts present in fish, vegetable oil, and fruits [20,21]. These sources of chromium are listed in Table 6.1.

Dietary supplements offer chromium in various forms, ranging from chromium polynicotinate, chromium picolinate, chromium polynicotinate, chromium nicotinate, chromium histidinate, and chromium chloride [22]. Chromium compounds contain varying percentages of elemental chromium, for example, chromium picolinate contains 12.4 percent of the weight of elemental chromium [23]. The quantity of elemental chromium rather than the weight of the whole chromium compound is specified on dietary supplement labels, and studies show that the absorption rates of these different forms of chromium are similar.

6.2.2 Dosage

Morris et al. (2016) conducted a study on plasma chromium levels in both healthy volunteers and diabetes patients, comparing the variations between the two groups. The study discovered a negative correlation between serum chromium levels in healthy individuals and the magnitude of insulin spikes following a glucose challenge. The significant and rapid decline in chromium levels, reaching up to 50 percent over a span of 45 min, was not caused by the excretion of chromium through urine. Chromium levels in diabetic participants did not exhibit any variation in response to insulin [24]. The 1997 study conducted by Anderson et al. revealed the most notable results on the decrease in cholesterol levels resulting from the addition of chromium. The study involved 180 participants with type 2 diabetes who were given either 200 mcg/day or 1000 mcg/day of Cr(III) in the form of chromium picolinate. The study found no notable impact of chromium supplementation on HDL cholesterol or triglyceride levels [25].

Upon ingestion, the majority of chromium binds to proteins within the body, while only about 5 percent circulates freely as ions in the bloodstream. This bound chromium is predominantly stored and concentrated in organs such as the liver, spleen, soft tissues, and bones. Any excess chromium that the body doesn't utilize is excreted through urine after absorption (Figure 6.1).

The biological roles of chromium (Cr) remain a subject of debate. While compounds containing Cr(VI) are classified as human carcinogens, Cr(III) is generally considered an essential micronutrient that aids in insulin activation, although this viewpoint is contested [26]. Despite its recognition as a micronutrient, specific biomolecules dependent on Cr(III) have not been definitively identified. The present perceptive of Cr(VI)-induced genotoxicity is based on the uptake-reduction model proposed by Wetterhahn, which outlines the vigorous transfer of Cr(VI) into the cells, intracellular reduction leading to potentially DNA-damaging intermediates, and the creation of kinetically non-reacting Cr(III) complexes [27]. Recent studies have suggested the formation of stable Cr(V) species through Cr(VI) reduction, adding to the complexity of the model. Research on Cr(III) involvement in intra-cellular

TABLE 6.1
Sources of Chromium

Plant Sources	Animal Sources	Others	Supplements
Broccoli, grapes (including grape juice), celery, tomatoes (including tomato juice), oranges, lettuce, apples, green beans, potatoes, Brewer's yeast, banana, Brazil nuts, whole wheat flour, tomato, white rice	Liver (from beef and lamb), Ham turkey breast, seafood, particularly mussels and other shellfish, egg	Barbecue sauce, whole wheat bread, ketchup, American cheese, peanut butter, spaghetti	Multivitamin/mineral supplements contain chromium (35 to 120 mcg), standalone chromium supplements (from 200 mcg to 500 mcg)

TABLE 6.2
Metabolic Activities of Chromium

Nature of Support	Activity
Metabolism support	Breaking down of macronutrients like proteins, carbohydrates, and fats Facilitating the absorption of micronutrients by cells and tissues
Insulin regulation	Insulin synthesis, binds to an oligopeptide to form chromodulin after ingestion, regulating glucose metabolism and maintaining blood sugar levels.
Cholesterol balance	Promotes optimal cholesterol synthesis Helps regulate fatty acid levels Contributing to cardiovascular health by supporting lipid profiles.
Neurological health	Crucial for interactions with neuronal cells, potentially impacting cognitive function and nervous system vitality.
Energy metabolism	Enhances energy levels, promoting stamina, sustained vigor, and improved physical performance by aiding metabolic processes
Cardiovascular Support	Regulates blood sugar and lipids profile to supports heart health, contributing to overall cardiovascular wellness.

redox reactions is limited but suggests its potential role in oxidative damage to DNA and proteins [28,29]. The mechanism behind Cr(III)'s antidiabetic effects, potentially through the formation of Cr(VI) and its insulin-mimetic properties, remains unclear. Overall, the interplay between Cr species and their biological effects is not fully understood. Chromium serves a pivotal role by supporting various metabolic activities in human physiology, as listed in Table 6.2.

6.3 BIOLOGICAL FUNCTIONS OF CHROMIUM

Chromium, an essential trace element, is crucial for numerous metabolic processes in the human body. Recent studies have investigated its roles in glucose metabolism, lipid regulation, protein synthesis, antioxidant properties, body composition, and neurological functions. Chromium is found in several chemical states in non-biological systems, but only Cr(III) and Cr(VI) are significant in biological contexts.

Trivalent chromium, an essential bioelement, holds significant sway in metabolic functions. Its presence is conspicuous in the tissues of human fetuses and infants, dwindling as individuals age, except for a slight uptick in lung chromium content from age 10, likely attributed to inhaled chromium deposits. The pinnacle accumulation of chromium is typically observed in hair, ranging from 0.2 to 2 mg/kg.

Chromium, an essential trace element, is crucial for numerous metabolic processes in the human body. Recent studies have investigated its roles in glucose metabolism, lipid regulation, protein synthesis, antioxidant properties, body composition, and neurological functions. This review synthesizes contemporary research findings,

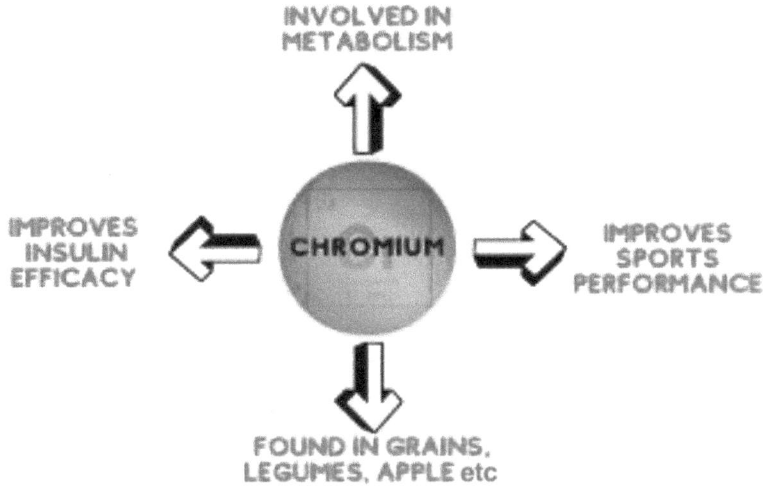

FIGURE 6.1 Features of chromium.

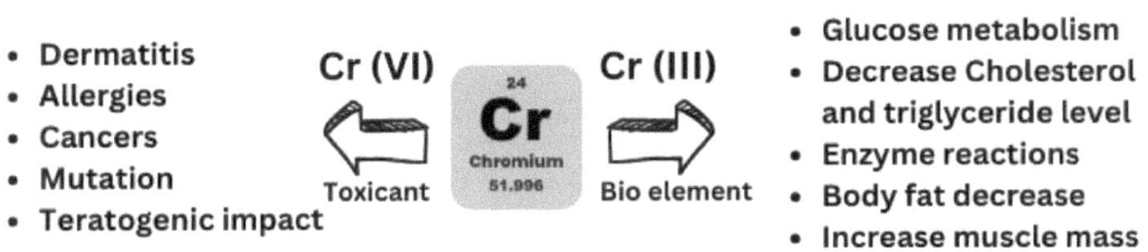

FIGURE 6.2 Breakdown of the effects of Cr(III) and Cr(VI) on humans.

focusing on the biological activities of chromium, its mechanisms, health benefits, and potential risks associated with its supplementation.

6.3.1 Glucose Metabolism

The function of chromium in glucose metabolism continues to be a primary focus of recent research on chromium. Chromium enhances insulin action, facilitating glucose uptake by cells and improving insulin sensitivity. In individuals with type 2 diabetes, the addition of chromium supplements resulted in a decrease of 0.6 percent (95% CI –0.9 to –0.2) in glycosylated hemoglobin levels and a decrease of 1.0 mmol/l (–1.4 to –0.5) in fasting glucose levels. However, there was no significant effect on lipids. No advantage was observed for persons not having diabetes. There were discernible dose–response relationships and variations seen among different chromium formulations [30]. Supplementation with chromium has been shown to effectively reduce fasting blood glucose levels in individuals diagnosed with type 2 diabetes. It notably decreased insulin levels in individuals diagnosed with type 2 diabetes.

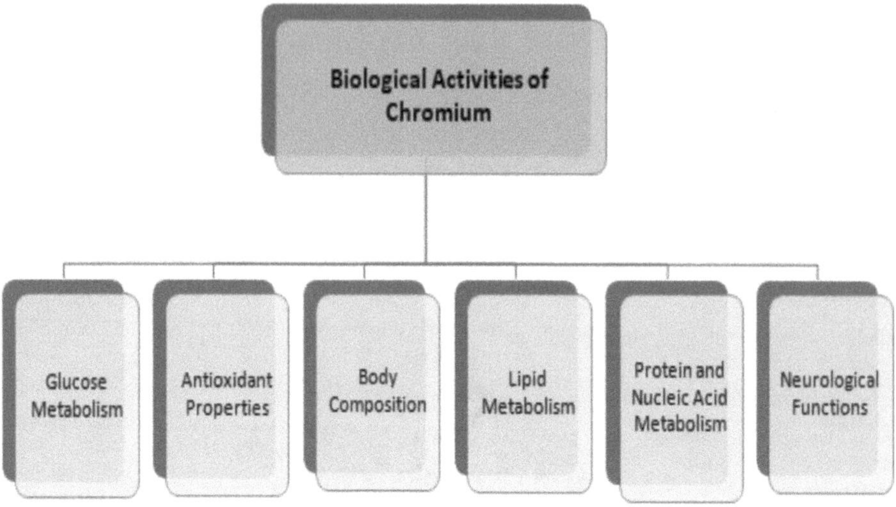

FIGURE 6.3 Biological activities of chromium.

Supplementing with chromium led to a considerable decrease in hemoglobin A1C levels in individuals diagnosed with type 2 diabetes[31].

6.3.2 Lipid Metabolism

The role of chromium in lipid metabolism and cardiovascular health has been widely studied. It was reported that chromium supplementation considerably reduced the levels of total cholesterol and low-density lipoprotein (LDL) while increasing the levels of high-density lipoprotein (HDL). Recent studies have investigated the characteristics of the oligopeptide low-molecular-weight chromium-binding substance (LMWCr). These studies indicate that LMWCr may play a role in the metabolism of carbohydrates and lipids by enhancing insulin signaling. This discovery has potential implications for the treatment of diabetes and related conditions [32]. Abdollahi et al. found both the inorganic and organic oral chromium supplements significantly augmented quantifiable body reservoirs of chromium in hair and red blood cells by approximately 25 percent. Nevertheless, none of the treatments had a significant impact on fasting plasma glucose and lipids, as well as the glycemic response to both the normal meal and tolbutamide [33].

6.3.3 Protein and Nucleic Acid Metabolism

The impact of chromium on protein and nucleic acid metabolism has also been a subject of recent research. Researchers highlighted involvement of chromium in amino acid metabolism and protein synthesis, noting its impact on protein turnover and muscle maintenance. Additionally, chromium is essential for nucleic acid synthesis, influencing DNA and RNA synthesis and repair mechanisms, which is critical for maintaining cellular integrity and function.

6.3.4 Antioxidant Properties

A study has demonstrated that chromium(III) has no impact on the levels of reduced glutathione (GSH), but it does lower the ferric-reducing ability of plasma (FRAP). Additionally, it impacts the enzyme superoxide dismutase (SOD) in a manner that is dependent on the dosage, but it does not have any impact on the activity of the enzyme glutathione peroxidase (GPx). Cr(VI), in contrast, induces a decline in SOD activity, enhances GPx activity, diminishes the amount of GSH, and, like Cr(III), exhibits an adverse impact on FRAP[34].The combination of Cr(III) and Fe(III) had a synergistic impact in generating reactive oxygen species (ROS) and reducing antioxidant activity. The study showed the protective effects of Cr(III) in antioxidant activity and its ability to reduce intracellular ROS generation when paired with Ni(II) or Mo(III) [35]. Alongside other trace minerals like zinc, copper, and selenium, chromium also possesses significant antioxidant properties. It plays a vital role in scavenging free radicals from the body, thereby safeguarding internal cells and tissues against oxidative damage. These findings underscore the potential of chromium in mitigating oxidative stress-related damage and its role in chronic disease prevention.

6.3.5 Impact on Body Composition

Researchers found that chromium picolinate (CP) supplementation did not result in any substantial changes in body composition or resting metabolic rate (RMR). There were no notable disparities observed in the levels of fasting plasma glucose, serum insulin, plasma glucagon, serum C-peptide, serum lipid concentrations, or iron and zinc indices between the two groups during the duration of the study. The serum total cholesterol concentration exhibited a substantial decrease ($p = 0.0016$) over time for all subjects collectively, likely due to the activity training [36]. In another study, supplementing with CrPic significantly raised blood Cr concentration and urine Cr excretion compared to picolinic acid and placebo ($p < 0.0001$). CrPic had no impact on body weight or fat despite the fact that all groups saw significant weight and fat loss ($p < 0.05$). Additionally, CrPic did not influence fat-free, mineral-free mass or assessments of iron status [37].

6.3.6 Neurological Functions

The neurotoxicity of Cr(VI) in rat primary hippocampal neurons was investigated using potassium dichromate (K_2Cr_2O7). The MTT assay was employed to assess the neuronal vitality. The findings indicated that the viability of neurons was progressively hindered by the dose and duration of Cr(VI) treatment. The analysis of the mechanism revealed that treatment with Cr(VI) leads to mitochondrial malfunction by influencing the expression of the Bcl-2 family. Additionally, treatment with Cr(VI) also stimulates the production of reactive oxygen species (ROS) within cells, causes

DNA damage, and activates MAPKs in neurons [38]. The pituitary and temporal lobe exhibited the highest observed levels of Cr in the human brain. Animal studies demonstrate that Cr(VI) causes extensive neurodegeneration and oxidative damage [39].

6.3.7 Chromium Supplementation

Chromium supplementation is also widespread for diverse purposes, including weight management. A plethora of chromium products are available in health food stores and pharmacies, encompassing body-building powders to slimming shakes and tablets. Chromium supplements are believed to aid weight loss by enhancing insulin sensitivity and appetite control. Interestingly, they have been linked to reducing weight while simultaneously increasing lean body mass. Metal ions have been categorized into distinct classes based on their biological function, specifically as essential, helpful, abiological, or deleterious. Interestingly, other authors have classified Cr(III) into each of these categories. Cr(III) is well recognized by nutritionists as a crucial micronutrient for human health. Animal tests conducted by Mertz and colleagues in the 1950s provided the initial data indicating the participation of Cr(III) in glucose metabolism [40,41]. Chromium (III) is regarded as an important trace element for humans. Prior work has revealed that Cr(III) serves as a second messenger in modulating the activity of enzymes, the effects of insulin, glucose tolerance factors (GTF), as well as peptide and nucleic acid stability. It also improves aberrant arterial smooth muscle cell activities linked to hyperglycemia-related vascular problems [42]. Typical blood chromium levels in babies are around 9.0 nM, dropping to 5.0–6.0 nM in grownups and 3.8 nM in type II diabetes individuals [43]. Cr(III) consumption should range from 25 and 200 g per day. However, the dietary composition in recent times is high in refined and processed meals frequently fall below of necessary chromium levels. Cardiovascular illness and diabetes can be caused by a lack of chromium, which manifests as impaired glucose tolerance, higher circulating insulin levels, fewer insulin receptors, and high cholesterol as well as triglyceride levels [44]. To remedy the shortage, supplements containing chromium such as chromium picolinate, nicotinate chromium, amino acid chromium, and chromium yeast can be found in the marketplace [45]. It can help improve the serotonergic characteristics of the brain in diabetics [46].

6.4 CONCENTRATIONS IN BIOLOGICAL FLUIDS AND TISSUES

Tipton and Cook (1963, 1965) reported presence of variable amount of chromium in different parts of human body. A comparatively high amount of Cr in lungs and a lower amount in other organs both in the United States and in other countries were established by them. It is also reported that the concentrations also vary with the geographical areas Table 6.3.

Earlier reports indicated normal chromium in blood levels to be between 3 and 30 μg/L and urinary excretion up to 10 μg/day, but these values were excessively high. Later studies found plasma Cr levels to be 0.14 μg/L. In 1989, Pääkkö et al. used plasma emission spectroscopy to analyze lung tissue Cr levels in 45 deceased

TABLE 6.3
Concentrations of Chromium in Biological Tissues

Organs and Tissues	Weight (Wet Weight)	Reference
hair	200 to 2,000 μg/kg	[47]
Lungs (New York and Chicago)	700 μg/kg	[48]
Liver	270 μg/kg	[48]
Kidney	90 μg/kg	[48]
Lung tissue of lung cancer (Smokers)	6.4 μg/g	[49]
Non-smokers	2.2 μg/g	[49]
Lung (in Denver area)	140 μg/kg	[48]
Liver (in Denver area)	30 μg/kg	[48]
Kidney (in Denver area)	40 μg/kg	[48]
Lung (in Japan)	300 μg/kg	[50]
Liver (in Japan)	40 μg/kg	[50]
Kidney (in Japan)	15 μg/kg	[50]
Median lung tissue concentration smelter workers in northern Sweden that contracted lung cancer	390 μgCr/kg	[51]
Median lung tissue concentration in deceased rural referents contracted lung cancer	110 μgCr/kg	[51]
Blood	3–30 μg/L	[52]
Urinary excretion	10 μg/day	[53][54]
Plasma	0.14 μg//L	[55][56]
Lung tissue in deceased subjects (in northern Finland)	1.3 and 4.8 μg/g dry and wet tissue, respectively	[57]

individuals from northern Finland, finding average concentrations of 1.3 μg/g in dry tissue and 4.8 μg/g in wet tissue. Cr levels in all tissues decrease from birth until around age 10, after which there is a slight increase in lung Cr levels but a continued decrease in other organs. This suggests that lung Cr is primarily from inhaled air, while dietary intake is the main source for Cr in other organs.

After this period, lung Cr concentrations show a slight increase, while levels in other organs continue to decrease [36]. This suggests that Cr accumulation in the lungs is primarily due to inhalation, whereas dietary intake is the main source of Cr in other organs.

[Ox]: Biological oxidants, for example H_2O_2 or peroxides of proteins and lipids
[Red]: Biological reductants, such as glutathione, ascorbate, or NAD(P)H [46].

Cr(VI) appears mainly as dichromate ($Cr_2O_7^{2-}$) in acidic aqueous solutions or chromate (CrO_4^{2-}) in alkali aqueous solutions at its maximum oxidation state [59]. Because of its capacity to easily enter the membranes of cells, Cr(VI) causes a notably greater hazard threat to individuals than Cr(III). It is classed as a mutagenic and carcinogenic ion due its intensive reduction activity in cells that generate free radicals,

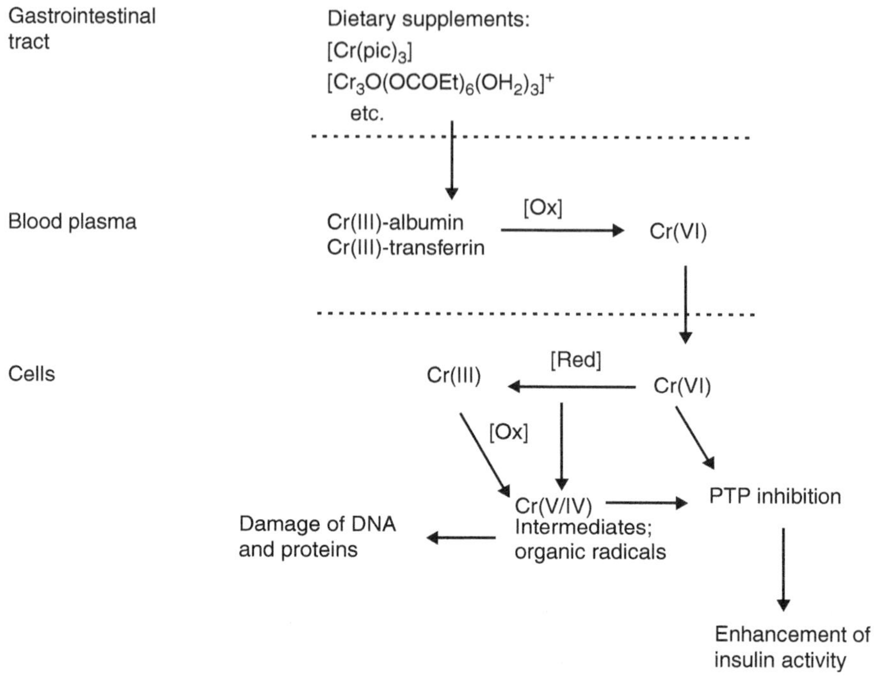

FIGURE 6.4 Proposed oxidative mechanism of biological activities of Cr(III).

causing considerable DNA damage. As a consequence, even slight disturbances in Cr homeostasis can cause acute and chronic illnesses such as headaches, ulcers, vomiting, airway irritations, ocular damage, gastrointestinal troubles, hematological, hepatic, and kidney issues, and possibly cancer [60].

6.5 MECHANISM OF BIOLOGICAL REACTIONS OF CHROMIUM

Despite numerous investigations, the precise mechanisms through which chromium ions impact carbohydrate and lipid metabolism remain unclear. Significant advancements in understanding the mechanisms of chromium (Cr) reactions over the past decade have included the adoption of global kinetic analysis (GKA), utilization of techniques such as electro-spray mass spectrometry (ESMS), X-ray absorption spectroscopy, and electron paramagnetic resonance spectroscopy for characterizing active intermediates and refinement of methods for isolating or quantitatively generating such intermediates through kinetic analyses. GKA, initially developed for enzymological studies, has found applications in elucidating inorganic reaction mechanisms, particularly notable in research concerning ruthenium (Ru) complexes as redox catalysts. Recent advancements in GKA involve the incorporation of additional dimensions into datasets, such as the temperature control, balanced concentration of reagents, or pH values, enhancing the depth of analysis. In summary, the recent advancements in mechanistic studies have significantly enhanced our understanding of chromium's behavior across different chemical and biological systems, shedding light on its reactivity, kinetics, and potential implications for the environment and human health.

6.5.1 Mechanism of Carbohydrate Metabolism

One of the prevailing theories suggests the participation of an oligopeptide known as chromodulin (LMWCr), a low-molecular-weight chromium-binding substance. LMWCr, comprised of specific amino acids, has the capacity to bind four chromium atoms, exhibiting its full biological activity only in this form. Studies by Vincent et al. proposed a mechanism wherein insulin binding triggers the activation of chromodulin, facilitating the subsequent activation of insulin receptor kinase and insulin signal intensification. This activation pathway ultimately leads to the translocation of GLUT4 protein to the cell membrane, facilitating glucose transport [61]. However, the exact structure of LMWCr remains elusive, though recent research has shed light on its sequence and chromium-binding properties. Alternatively, experiments with chromium supplementation may also result from nonspecific binding to amino acid sequences in proteins found in liver tissues. Furthermore, while chromodulin activity in stimulating insulin receptor tyrosine kinase is observed, it is noted that other metal ions and xenobiotics can also induce similar effects. Another theory proposes direct activation of the insulin receptor by chromium ions, with studies indicating the induction of reactive oxygen species (ROS) synthesis and modulation of membrane phosphatases activity. Additionally, the impact of chromium on membrane fluidity, potentially affecting glucose uptake, has been suggested, along with its ability to interact with lipid bilayers. Furthermore, the estrogenic properties of chromium may influence insulin-stimulated processes through estrogen receptor activation pathways [62], though the exact molecular mechanisms remain incompletely understood. Overall, these various hypotheses contribute to our evolving understanding role of chromium role in metabolic processes.

The mechanism begins with the conversion of the inactive form of the insulin receptor (IR) to its active form upon binding with insulin (I) (Figure 6.5). This binding prompts the movement of chromium, likely in the form of Cr-transferrin (Cr-Tf), from the blood into insulin-dependent cells. Inside the cells, chromium binds to apochromodulin (represented as a triangle), forming holochromodulin (represented as a square). Holochromodulin then binds to the insulin receptor, further enhancing its kinase activity. Apochromodulin alone cannot bind to the insulin receptor or activate its kinase activity. When insulin levels decrease, holochromodulin is released from the cell, thereby stopping its activation effects [64].

After absorption, chromium enters the body and encounters chromodulin, an oligopeptide synthesized in the liver that exhibits an exceptional affinity for chromium. Studies indicate that chromodulin activates the insulin receptor, thereby enhancing insulin signaling mechanisms of the body. When activated, this receptor facilitates the

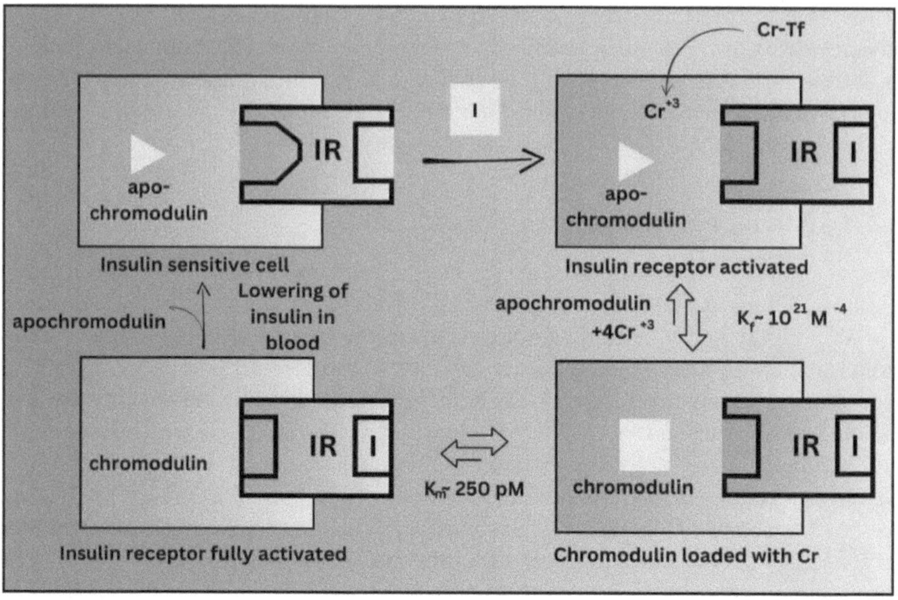

FIGURE 6.5 Proposed mechanism for the activation of insulin receptor by chromodulin [I]: Insulin, [IR]: Insulin receptor [63].

efficient uptake of glucose into cells and helps regulate blood sugar levels. However, this process relies on the presence of insulin. To some extent, chromodulin mimics insulin by initially binding with chromium and subsequently engaging with subunits of the insulin receptor to activate it. This interaction not only imitates the presence of insulin but also augments it—research suggests that chromodulin collaborates with insulin to boost insulin signaling by up to eight times the baseline activity. In essence, chromium forms a bond with chromodulin to enhance insulin sensitivity, which can significantly impact glucose levels and various other physiological processes.

6.5.2 Mechanism of Lipid Metabolism

Engaging with pathways commonly associated with insulin, chromodulin bonded with chromium also enhances AMPK activity in muscle cells. AMPK, also known as adenosine monophosphate kinase, is crucial for maintaining energy balance by metabolizing fatty acids and stored glucose for energy [65]. This process helps regulate body weight and promotes the development of lean muscle by fostering an anabolic state. The significance of these mechanisms cannot be overstated. Insulin signaling directly influences critical aspects of health such as insulin sensitivity and blood sugar levels, while AMPK plays pivotal roles in energy balance and insulin response. Both pathways are central to maintaining a healthy body weight, underscoring why proper chromium supplementation focuses not only on insulin response but also on weight management.

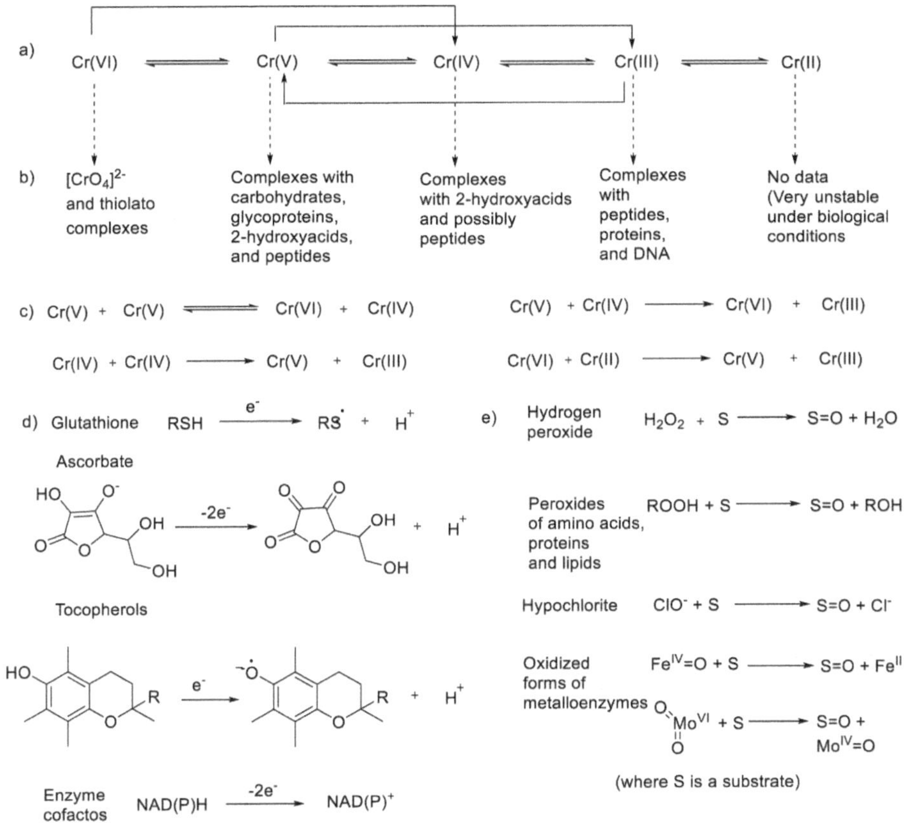

Scheme 6.1 Outline of redox chemistry of Cr in biological systems.
(a) Proposed redox transformations of Cr; (b) moderately stable complexes; (c) probable reactions between various Cr oxidation states; (d) key reductants; (e) key oxidants [66,67].

6.6 ABSORPTION AND METABOLISM

Cr(III) is frequently ingested in the form of carboxylato complexes, such as oxalato or citrato, or amino acid complexes, through plant- or animal-based food. However, there is limited information available regarding the specific forms of Cr(III) in various dietary sources [68]. The gastrointestinal absorption of Cr(III) in humans and experimental animals is minimal, ranging from 0.4 to 2 percent. It appears that the absorption levels of Cr(III) are influenced by organic acid ligands, such as nicotinate or ascorbate. However, certain investigations have not shown a significant correlation between the type of Cr(III) complexation and the absorption levels [69]. Although numerous models of Cr(III) pharmacokinetics in mammals have been established, there is limited understanding regarding the speciation of Cr(III) in organisms. Transferrin, a prominent protein responsible for transporting various metal ions, is very probable to serve as the primary carrier of Cr(III) in the bloodstream [70,71].

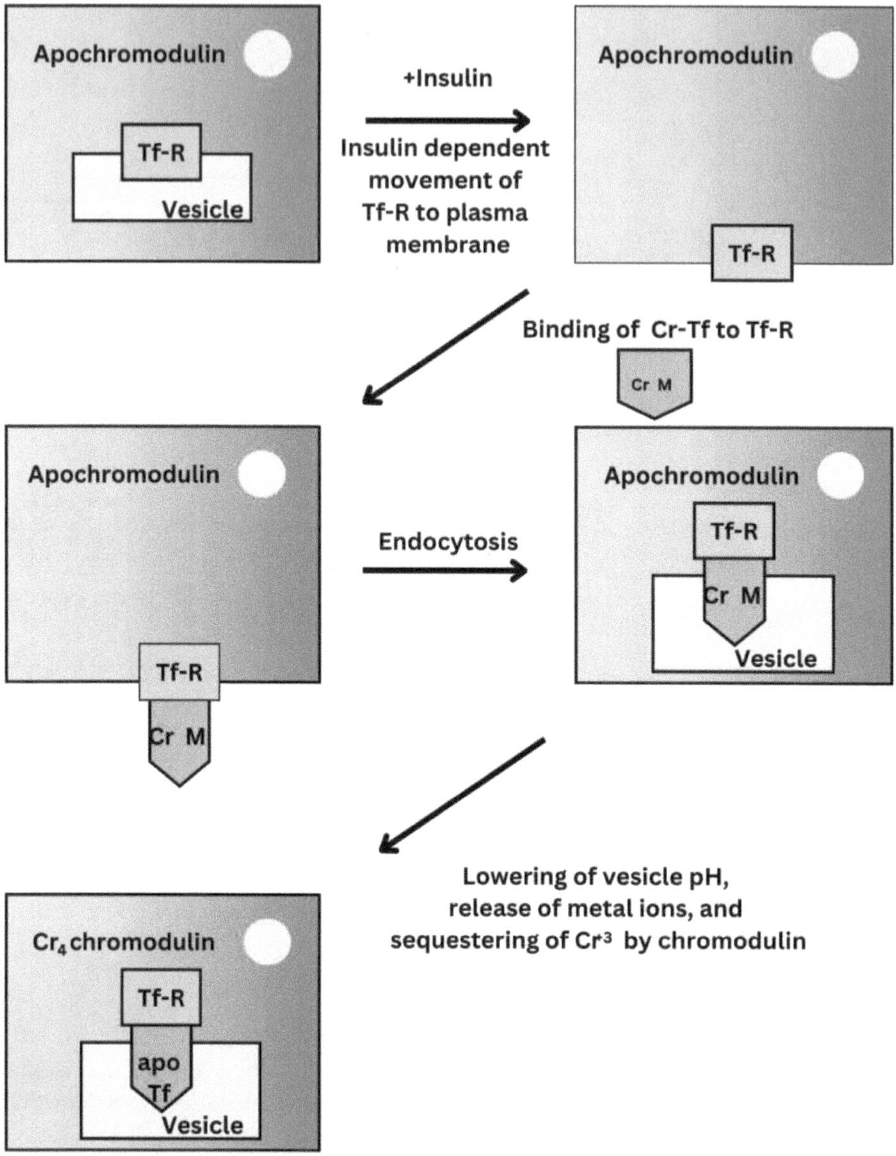

FIGURE 6.6 A proposed mechanism for the activation of insulin receptor by chromodulin [51]. [I]: Insulin, [IR]: Insulin receptor [63].

In rejoinder to the increase in concentrations of plasma insulin, the transferrin receptors (Tf-R) in insulin-sensitive cells drift from vesicles to the plasma membrane. Transferrin (represented as a pentagon), which carries two metal ions (one chromic ion and one other metal cation, M), binds to these receptors and is internalized by endocytosis. Within the resulting vesicle, ATP-driven proton pumps lower the pH, causing the metal ions to be released from transferrin. The released chromium is then sequestered by apochromodulin (depicted as an open circle) to form chromium-loaded chromodulin (represented as a dark circle) [63,72].

6.7 TOXICOKINETICS

The toxicokinetics of chromium compounds are significantly influenced by the valence state of chromium and the nature of its ligands. The distinct behaviors of chromium(III) and chromium(VI) in the body highlight the importance of considering the valence state in toxicokinetic studies and health risk assessments [73]. Chromium(III) primarily binds to plasma proteins such as transferrins, which facilitates its transport in the bloodstream. Chromium(VI) is taken up by erythrocytes (red blood cells) in the bloodstream, where it is reduced to chromium(III) [74]. In this reduced form, it mainly binds to hemoglobin. Once chromium(VI) enters the blood, it is rapidly reduced to chromium(III) by antioxidants like glutathione and ascorbate. This reduction process helps to mitigate the potential toxic effects of chromium(VI) (Scheme 6.1). Chromium entering the bloodstream, particularly through the lungs, is primarily excreted by the kidneys and liver [75]. The kidneys absorb chromium from the blood through the renal cortex and subsequently release it into the urine. This makes urine sampling a useful method for monitoring exposure to certain water-soluble chromium compounds, such as those found in welding fumes. This can help in early detection and intervention to prevent adverse health effects. These findings emphasize the complexity of chromium toxicokinetics and the necessity for targeted research and monitoring to protect against its potential health risks. Owing to the limited data availability on the subject, the toxicokinetics of chromium in biological systems is not well-documented. However, certain assumptions can be made about the factors influencing its uptake, particularly in the respiratory system. The rate of chromium uptake in the airways is primarily influenced by two factors: the size distribution of the particles being inhaled and the water solubility of the concerned compound. These aspects play a crucial role in determining how chromium is absorbed and processed inside the body. Chromium can exist in multiple oxidation states, with the most common being trivalent (Cr(III)) and hexavalent (Cr(VI)). These are the primary states in which chromium is found in various environments, including industrial settings and biological systems. Chromium can occasionally be present in the quadrivalent (Cr(IV)) or pentavalent (Cr(V)) states, although these forms are less common in biological contexts. Understanding the toxicokinetics of chromium, particularly its absorption, distribution, metabolism, and excretion, requires more extensive research. The available data highlight the importance of particle size and solubility in chromium uptake, which could inform future studies and potential health risk assessments related to chromium exposure [76].

Research indicates that chromium concentrations in human lungs tend to increase with age. The absorption of chromium through inhalation occurs particularly quickly for water-soluble chromium(VI) compounds. However, accurately quantifying the extent of this uptake is challenging. Preliminary estimates suggest that within a few hours after the instillation of chromium(III) chloride about 5 percent is absorbed in the lungs. Investigations into the fate of Cr (VI) and Cr (III) following intratracheal administration provide some insights into their pulmonary retention and absorption. Chromium(III) is retained in the lungs to a greater extent than chromium(VI). Additionally, chromates with low water solubility are typically cleared to the gastrointestinal tract, whereas more soluble chromates are absorbed into the bloodstream. These findings underline the complexity of biokinetics of chromium and highlight areas needing further research to better understand its health impacts, particularly regarding long-term exposure and cancer risk. Oral exposure to chromium results in low absorption rates in the gastrointestinal tract, estimated to be 5 percent or less. The uptake of chromium(VI) compounds in the gastrointestinal tract is significantly influenced by their water solubility.

6.8 BIOLOGICAL FATE OF CHROMIUM

The gastrointestinal absorption of chromium following oral exposure is low, with an estimated absorption rate of 5 percent or less the rate of which is largely governed by the water solubility of these compounds. More soluble compounds are more readily absorbed. In vitro studies indicate that gastrointestinal juices can reduce chromium(VI) to chromium(III). There is insufficient in vivo data to confirm whether this reduction process effectively eliminates differences in absorption between ingested Cr(VI) and Cr (III). Pulmonary cells have some capacity to reduce hexavalent chromium, but this capacity is significantly lower compared to liver cells. This suggests that while the lungs can partially reduce Cr(VI), the liver plays a more crucial role in this reduction process. The lower reduction capacity of pulmonary cells compared to liver cells highlights the importance of systemic detoxification mechanisms in managing chromium exposure risks. More in vivo studies are needed to fully understand the extent to which gastrointestinal and pulmonary reduction processes influence overall chromium absorption and toxicity. Understanding these mechanisms is essential for assessing the health risks associated with chromium exposure, particularly in occupational and environmental settings where inhalation and ingestion of chromium compounds may occur.

6.8.1 Transport and Retention

Studies on feminine investigational animals have demonstrated that both the Cr(III) and Cr(VI) could be transported to the fetus to a limited extent. However, the available data are insufficient for making quantitative estimates of fetal exposure. Chromium(III) and chromium(VI) can cross the placental barrier to a limited extent, but precise quantitative estimates of fetal exposure are not available. Chromium concentrations inside the tissues of newborns are higher than in later stages of life. A considerable decline in chromium concentration in the tissues of children is observed till the 10 years. After this period, lung tissue concentration slightly increases, while a minor decrease is observed in other tissues. Chromium absorbed into the bloodstream is dispersed to various organs. The most significant retention of chromium is found in the spleen, liver, and bone marrow [48].

6.8.2 Implications for Health and Development

Even though the data on fetal exposure are not quantitative, the ability of chromium to cross the placenta raises concerns about potential developmental impacts. The higher concentrations in newborns and the decline in childhood suggest possible changes in metabolism, excretion, or tissue binding capacities with age. The significant retention in critical organs such as the spleen, liver, and bone marrow highlights the need for understanding long-term health effects, as these organs are crucial for immune function, detoxification, and blood cell production.

6.8.3 Chromium Elimination

The elimination of absorbed chromium from the body occurs in two phases in both animals and humans: a rapid phase—representing the clearance of chromium from the blood and a slower phase—representing the clearance of chromium from tissues. The primary routes of chromium excretion are: (i) urinary excretion accounts for a little over 50 percent of eliminated chromium and (ii) fecal excretion accounts for approximately 5 percent of eliminated chromium. The remaining chromium not excreted is deposited into deep body parts, such as bone and soft tissue. Elimination from these tissues is very slow [77].

Half-time of whole-body chromium elimination: The estimated half-time for whole-body elimination of chromium(VI) is 22 days following intravenous administration. The estimated half-time for chromium(III) is 92 days following intravenous administration. The significantly longer half-time for chromium(III) compared to chromium(VI) suggests different metabolic and retention behaviors that could influence toxicity profiles. The prolonged half-times, especially for chromium(III), indicate that chromium can remain in the body for extended periods, potentially leading to cumulative toxic effects. Regular monitoring of chromium levels in individuals exposed to chromium, especially in occupational settings, is important to manage and mitigate potential health risks [77].

6.9 ANTIMICROBIAL STUDIES

Complexes of metals with the capacity to be employed as a therapeutic agent are in high demand these days and can be employed as therapeutic substances. These complexes show antibacterial, antifungal, anti-inflammatory, anti-cancer, nematicidal, pesticidal, antioxidant, and other applications in pharma chemistry [78]. Cr(III) was shown to bind with cysteine as well as histidine in microbe protein molecules, leading to the development of stable compounds. The quantity of Cr(III) associated with DNA estimated by pharmacokinetic modelling of Cr(III) buildup as an outcome of dietary supplementation is adequate to minimize variations in DNA replication efficiency and increasing mutation rate [79].

Researchers have synthesized Schiff base complexes of Cr, which are further evaluated for biological activities against bacteria like *Staphylococcus aureus* and *Escherichia coli* and fungi such as *Aspergillus flavus* and *Aspergillus niger*. In comparison with regular streptomycin, the various molecules exhibit poor to moderate

TABLE 6.4
Antimicrobial Studies of Chromium

S. No.	Complex	Strain	Activity	References
1.	[Cr(Glucose) (hydroquinoline)]	*Staphylococcus aureus* (SA), *Escherichia coli* (EC), *Aspergillus flavus* (AF), *Candida albicans* (CA), and *Pseudomonas aeruginosa* (PA)	SA= 70%, EC= 70%, AF, CA, PA= 25%	[85]
2.	$[Cr(L)(2H_2O)Cl]H_2O$ (L= ligand)	*Bacillus subtilis* (BS), *Staphylococcus aureus* (SA), *Escherichia coli* (EC), and *Pseudomonas aeruginosa* (PA)	Zone of inhibition (mm) BS= 24 SA= 22 EC= 24 PA= 22	[86]
3.	$Cr(CO)_2(C_{20}H_{14}N_2O_2Cl_2)$	*S. aureus* (SA), *E. coli* (EC), *A. flavus* (AF) and *A. niger* (AN)	Zone of inhibition (mm) SA= 10.0 ± 1.0 EC= 9.0 ± 0.70 AF= 19.0 ± 1.4 AN= 20.0 ± 1.8	[80]
4.	$[Cr_2(O)_4(dap)_2]$	*Escherichia coli* (EC), *Staphylococcus aureus* (SA), *Aspergillus niger* (AN), *Candida albicans* (CA)	Zone of inhibition (mm) EC= 18 SA= 21 AN= 18 CA= 21	[87]
5.	[Cr(*N,N′*-bis(5-methylsalicylaldimine)-*N,N′*-bis-(2-aminoetyl) ethane-1,2-diamine)]	*Staphylococcus aureus* (SA), *Escherichia coli* (EC), and *Pseudomonas aeruginosa* (PA)	Zone of inhibition (mm) SA= 45 ± 0.4 EC= 41 ± 0.4 PA= 36 ± 0.4	[88]

activity toward bacterial species and moderate to good effectiveness against fungal species [80]. In another study, researchers created chromium oxide nanoparticles utilizing a natural extract from *Eucalyptus globulus*. Biological activity tests have revealed that these nanoparticles exhibit a high inhibitory tendency toward a few bacteria [81]. Murcia et al. synthesized Cr complexes with triazole-based ligands and found complexes were more effective against fungus than bacteria. The biological effects of the Cr complexes against *Candida tropicalis* are very interesting because they displayed minimum inhibitory concentration values ranging from 7.8 to 15.6 [its correct] μgmL^{-1} [82]. Schiff bases consisting of NO as donor atoms were evaluated for antimicrobial activity. The biological function of the ligands showed a significant improvement in coordination with the presence of metal ions versus each of the examined bacterial/fungal strains, indicating that metal chelates were more potent than ligands [83]. When evaluated in comparison with amphotericin, the Cr(III) complex and ligand had virtually comparable action toward *Pencillium chrysogenum* strains [84].

6.10 CONCLUSION

The genotoxicity and carcinogenicity of chromium(VI) are thoroughly established, highlighting the potential hazards associated with this form of chromium. Conversely, chromium(III) compounds are generally considered to be nontoxic due to their limited bioavailability. However, it is crucial to acknowledge that within cells, chromium(III) complexes formed as a result of the uptake and reduction of chromium(VI) exhibit high genotoxicity. While the essentiality of chromium(III) for glucose and fat metabolism in humans remains uncertain, it is worth noting that certain chromium(III) complexes have been reported to positively impact insulin function through interactions with cellular insulin receptors. The understanding of the ways in which chromium affects cells at the cellular and sub-cellular levels has advanced alongside developments in cell and molecular biology. Although we have identified several cellular responses to Cr compounds, we still have much to learn about how Cr actually works within cells. Currently, there is a growing interest in understanding the genotoxic and carcinogenic effects of Cr(VI) beyond its potential health impacts in workplaces or environmental exposure, although these remain important issues.

The carcinogenic properties of Cr(VI) make it a key focus in studying how environmental factors contribute to cancer. By examining how Cr(VI) interacts with cells and biomolecules, we can gain valuable insights into the broader mechanisms of carcinogenicity. Soon, cutting-edge techniques such as gene expression microarray data analysis and quantitative proteome analysis are expected to be extensively employed in studying the genotoxic effects of Cr(VI).

REFERENCES

[1] A.M. Zayed, N. Terry, Chromium in the environment: factors affecting biological remediation, *Plant Soil.* 249 (2003) 139–156. [its correct] https://doi.org/10.1023/A:1022504826342.

[2] X. Yao, R. Liu, X. Li, Y. Li, Z. Zhang, S. Huang, Y. Ge, X. Chen, X. Yang, Zinc, selenium and chromium co-supplementation improves insulin resistance by preventing hepatic endoplasmic reticulum stress in diet-induced gestational diabetes rats, *J. Nutr. Biochem.* 96 (2021) 108810. https://doi.org/https://doi.org/10.1016/j.jnutbio.2021.108810.

[3] T. Pholen, B. Londe, Reproduced with permission of the copyright owner . Further reproduction prohibited without, *J. Allergy Clin. Immunol.* 130 (1998) 556. http://dx.doi.org/10.1016/j.jaci.2012.05.050.

[4] EFSA NDA Panel EFSA Panel on Dietetic Products. Scientific opinion on dietary reference values for chromium. *EFSA J.* 12 (2014) 3845. https://doi.org/https://doi.org/10.2903/j.efsa.2014.3845.

[5] R. Iskra, H. Antonyak, Chromium in health and longevity, in: M. Malavolta, E. Mocchegiani (Eds.), *Trace Elements and Minerals in Health and Longevity*. Springer International Publishing, Cham, 2018: pp. 133–162. https://doi.org/10.1007/978-3-030-03742-0_5.

[6] J. Gorny, G. Billon, C. Noiriel, D. Dumoulin, L. Lesven, B. Madé, Chromium behavior in aquatic environments: a review, *Environ. Rev.* 24 (2016) 503–516. https://doi.org/10.1139/er-2016-0012.

[7] D.M. Proctor, M. Suh, S.L. Campleman, C.M. Thompson, Assessment of the mode of action for hexavalent chromium-induced lung cancer following inhalation exposures, *Toxicology* 325 (2014) 160–179. https://doi.org/https://doi.org/10.1016/j.tox.2014.08.009.

[8] C.L. Mitchelmore, J.K. Chipman, DNA strand breakage in aquatic organisms and the potential value of the comet assay in environmental monitoring, *Mutat. Res. Mol. Mech. Mutagen.* 399 (1998) 135–147. https://doi.org/https://doi.org/10.1016/S0027-5107(97)00252-2.

[9] B. Poljsak, I. Pócsi, M. Pesti, Interference of chromium with cellular functions, in: G. Banfalvi (Ed.), *Cellular Effects of Heavy Metals*. Springer Netherlands, Dordrecht, 2011: pp. 59–86. https://doi.org/10.1007/978-94-007-0428-2_3.

[10] A. Levina, P.A. Lay, Chemical properties and toxicity of chromium(III) nutritional supplements, *Chem. Res. Toxicol.* 21 (2008) 563–571. https://doi.org/10.1021/tx700385t.

[11] T.J. O'Brien, S. Ceryak, S.R. Patierno, Complexities of chromium carcinogenesis: role of cellular response, repair and recovery mechanisms, *Mutat. Res. Mol. Mech. Mutagen.* 533 (2003) 3–36. https://doi.org/https://doi.org/10.1016/j.mrfmmm.2003.09.006.

[12] N. Tor, The carcinogenicity of chromium, *Environ. Health Perspect.* 40 (1981) 121–130. https://doi.org/10.1289/ehp.8140121.

[13] Y. Schneider, Endothelial cells in high glucose conditions: assessment of antioxidant enzymes, NO system, and insulin-signaling pathway with involvement of potential protective effect of GTF (glucose tolerance factor) on these parameters (University of Haifa (Israel)ProQuest Dissertations & Theses, 2018. 28762011). https://www.proquest.com/openview/191e1a61a43b22d7072aa91db915537c/1?cbl=2026366&diss=y&pq-origsite=gscholar

[14] G. Genchi, G. Lauria, A. Catalano, A. Carocci, M.S. Sinicropi, The double face of metals: The intriguing case of chromium, *Appl. Sci.* 11 (2021). https://doi.org/10.3390/app11020638.

[15] M.R. Islam, S. Akash, M.H. Jony, M.N. Alam, F.T. Nowrin, M.M. Rahman, A. Rauf, M. Thiruvengadam, Exploring the potential function of trace elements in human health: a therapeutic perspective, *Mol. Cell. Biochem.* 478 (2023) 2141–2171. https://doi.org/10.1007/s11010-022-04638-3.

[16] G. Martemucci, C. Costagliola, M. Mariano, L. D'andrea, P. Napolitano, A.G. D'Alessandro, Free radical properties, source and targets, antioxidant consumption and health, *Oxygen* 2 (2022) 48–78. https://doi.org/10.3390/oxygen2020006.

[17] M. Thirunavukkarasu, S. Penumathsa, B. Juhasz, L. Zhan, M. Bagchi, T. Yasmin, M.A. Shara, H.S. Thatte, D. Bagchi, N. Maulik, Enhanced cardiovascular function and energy level by a novel chromium (III)-supplement, *BioFactors* 27 (2006) 53–67.

[18] F.H. Nielsen, Importance of making dietary recommendations for elements designated as nutritionally beneficial, pharmacologically beneficial, or conditioinally essential, *J. Trace Elem. Exp. Med.* 13 (2000) 113–129. https://doi.org/https://doi.org/10.1002/(SICI)1520-670X(2000)13:1<113::AID-JTRA13>3.0.CO;2-D.

[19] T.E. Tuormaa, Chromium, selenium, copper and other trace minerals in health and reproduction, *J. Orthomol. Med.* 15 (2000) 145–156.

[20] F. Deeba, N. Abbas, T. Shafique, S. Rehman, N. Hussain, Chromium in various food items as regulators of different metabolic activities, *Pak. J. Biochem. Mol. Biol.* 43 (2010) 15–18.

[21] EFSA Panel on Food Additives and Nutrient Sources added to Food (ANS), Scientific Opinion on ChromoPrecise® cellular bound chromium yeast added for nutritional purposes as a source of chromium in food supplements and the bioavailability of chromium from this source, *EFSA J.* 10 (2012) 2951. https://doi.org/https://doi.org/10.2903/j.efsa.2012.2951.

[22] R.A. DiSilvestro, E. Dy, Comparison of acute absorption of commercially available chromium supplements, *J. Trace Elem. Med. Biol.* 21 (2007) 120–124. https://doi.org/https://doi.org/10.1016/j.jtemb.2007.01.004.

[23] T.O. Berner, M.M. Murphy, R. Slesinski, Determining the safety of chromium tripicolinate for addition to foods as a nutrient supplement, *Food Chem. Toxicol.* 42 (2004) 1029–1042. https://doi.org/https://doi.org/10.1016/j.fct.2004.02.015.

[24] B.W. Morris, Chromium action and glucose homeostasis, *J. Trace Elem. Exp. Med.* 12 (1999) 61–70. https://doi.org/https://doi.org/10.1002/(SICI)1520-670X(1999)12:2<61::AID-JTRA3>3.0.CO;2-E.

[25] R.A. Anderson, N. Cheng, N.A. Bryden, M.M. Polansky, N. Cheng, J. Chi, J. Feng, Elevated intakes of supplemental chromium improve glucose and insulin variables in individuals with type 2 diabetes, *Diabetes* 46 (1997) 1786–1791. https://doi.org/10.2337/diab.46.11.1786.

[26] A. Levina, R. Codd, C.T. Dillon, P.A. Lay, Chromium in biology: toxicology and nutritional aspects, in: *Progress in Inorganic Chemistry* K.D. Karlin (Ed.), John Wiley & Sons, Inc 2002: pp. 145–250. https://doi.org/https://doi.org/10.1002/0471267287.ch2.

[27] A.M. Standeven, K.E. Wetterhahn, Chromium(VI) toxicity: uptake, reduction, and DNA damage, *J. Am. Coll. Toxicol.* 8 (1989) 1275–1283. https://doi.org/10.3109/10915818909009118.

[28] E. Rudolf, M. Červinka, The role of intracellular zinc in chromium(VI)-induced oxidative stress, DNA damage and apoptosis, *Chem. Biol. Interact.* 162 (2006) 212–227. https://doi.org/https://doi.org/10.1016/j.cbi.2006.06.005.

[29] J.M. Myers, W.E. Antholine, C.R. Myers, The intracellular redox stress caused by hexavalent chromium is selective for proteins that have key roles in cell survival and thiol redox control, *Toxicology* 281 (2011) 37–47. https://doi.org/https://doi.org/10.1016/j.tox.2011.01.001.

[30] E.M. Balk, A. Tatsioni, A.H. Lichtenstein, J. Lau, A.G. Pittas, Effect of chromium supplementation on glucose metabolism and lipids: A systematic review of randomized controlled trials, *Diabetes Care* 30 (2007) 2154–2163. https://doi.org/10.2337/dc06-0996.

[31] O. Asbaghi, N. Fatemeh, R.K. Mahnaz, G. Ehsan, E. Elham, N. Behzad, A.-L. Damoon, A.N. Amirmansour, Effects of chromium supplementation on glycemic

control in patients with type 2 diabetes: a systematic review and meta-analysis of randomized controlled trials, *Pharmacol. Res.* 161 (2020) 105098. https://doi.org/https://doi.org/10.1016/j.phrs.2020.105098.
[32] C.M. Davis, J.B. Vincent, Chromium in carbohydrate and lipid metabolism, *JBIC J. Biol. Inorg. Chem.* 2 (1997) 675–679. https://doi.org/10.1007/s007750050183.
[33] M.B. Rabinowitz, H.C. Gonick, S.R. Levin, M.B. Davidson, Effects of chromium and yeast supplements on carbohydrate and lipid metabolism in diabetic men, *Diabetes Care* 6 (1983) 319–327. https://doi.org/10.2337/diacare.6.4.319.
[34] A. Dlugosz, K.P. Rembacz, A. Pruss, M. Durlak, J. Lembas-Bogaczyk, Influence of chromium on the natural antioxidant barrier, *Polish J. Environ. Stud.* 21 (2012) 331–335.
[35] S. Terpilowska, A.K. Siwicki, Pro- and antioxidant activity of chromium(III), iron(III), molybdenum(III) or nickel(II) and their mixtures, *Chem. Biol. Interact.* 298 (2019) 43–51. https://doi.org/https://doi.org/10.1016/j.cbi.2018.10.028.
[36] S.L. Volpe, H.-W. Huang, K. Larpadisorn, I.I. Lesser, Effect of chromium supplementation and exercise on body composition, resting metabolic rate and selected biochemical parameters in moderately obese women following an exercise program, *J. Am. Coll. Nutr.* 20 (2001) 293–306. https://doi.org/10.1080/07315724.2001.10719050.
[37] H.C. Lukaski, W.A. Siders, J.G. Penland, Chromium picolinate supplementation in women: effects on body weight, composition, and iron status, *Nutrition* 23 (2007) 187–195. https://doi.org/https://doi.org/10.1016/j.nut.2006.12.001.
[38] T. Zhang, L. Feng, J. Cui, W. Tong, H. Zhao, T. Wu, P. Zhang, X. Wang, Y. Gao, J. Su, X. Fu, Hexavalent chromium induces neurotoxicity by triggering mitochondrial dysfunction and ROS-mediated signals, *Neurochem. Res.* 49 (2024) 660–669. https://doi.org/10.1007/s11064-023-04063-y.
[39] J.P. Wise, J.L. Young, J. Cai, L. Cai, Current understanding of hexavalent chromium [Cr(VI)] neurotoxicity and new perspectives, *Environ. Int.* 158 (2022) 106877. https://doi.org/https://doi.org/10.1016/j.envint.2021.106877.
[40] K. Schwarz, W. Mertz, A glucose tolerance factor and its differentiation from factor 3, *Arch. Biochem.* 72 (1957) 515–518.
[41] P.A. Lay, Chromium in biology: Toxicology and nutritional aspects, *Journal of Inorganic Biochemistry*, 96 (2003), Issue 1, pp 8, https://doi.org/10.1016/S0162-0134(03)80417-8.
[42] S. Liu, L. Zhang, H. Kim, J. Sun, J. Yoon, Recent advances and challenges in monitoring chromium ions using fluorescent probes, *Coord. Chem. Rev.* 501 (2024) 215575. https://doi.org/10.1016/j.ccr.2023.215575.
[43] E. Wysocka, M. Cymerys, G. Mielcarz, W. Bryl, S. Dzięgielewska, L. Torliński, The way of serum chromium utilization may contribute to cardiovascular risk factors in centrally obese persons, *Arch. Med. Sci.* 7 (2011) 257–263. https://doi.org/10.5114/aoms.2011.22076.
[44] M. Valko, H. Morris, T.D.M. Cronin, Metals, Toxicity and oxidative stress, *Curr. Med. Chem.* 12 (2005) 1161–1208. https://doi.org/http://dx.doi.org/10.2174/0929867053764635.
[45] W.T. Cefalu, Z.Q. Wang, X.H. Zhang, L.C. Baldor, J.C. Russell, Oral chromium picolinate improves carbohydrate and lipid metabolism and enhances skeletal muscle glut-4 translocation in obese, hyperinsulinemic (JCR-LA corpulent) rats, *J. Nutr.* 132 (2002) 1107–1114. https://doi.org/https://doi.org/10.1093/jn/132.6.1107.
[46] J.R. Komorowski, M. Tuzcu, N. Sahin, V. Juturu, C. Orhan, M. Ulas, K. Sahin, Chromium picolinate modulates serotonergic properties and carbohydrate

metabolism in a rat model of diabetes, *Biol. Trace Elem. Res.* 149 (2012) 50–56. https://doi.org/10.1007/s12011-012-9393-x.

[47] W. Mertz, Chromium occurrence and function in biological systems, *Physiol. Rev.* 49 (1969) 163–239. https://doi.org/10.1152/physrev.1969.49.2.163.

[48] H.A. Schroeder, J.J. Balassa, I.H. Tipton, Abnormal trace metals in man—Chromium, *J. Chronic Dis.* 15 (1962) 941–964. https://doi.org/https://doi.org/10.1016/0021-9681(62)90114-5.

[49] S. Anttila, P. Kokkonen, P. Pääkkö, P. Rainio, P. -L Kalliomäki, J. Pallon, K. Malmqvist, P. Pakarinen, V. Näntö, S. Sutinen, High concentrations of chromium in lung tissue from lung cancer patients, *Cancer* 63 (1989) 467–473. https://doi.org/10.1002/1097-0142(19890201)63:3<467::AID-CNCR2820630313>3.0.CO;2-Z.

[50] K. Hyodo, S. Suzuki, N. Furuya, K. Meshizuka, An analysis of chromium, copper, and zinc in organs of a chromate worker, *Int. Arch. Occup. Environ. Health.* 46 (1980) 141–150. https://doi.org/10.1007/BF00378192.

[51] L. Gerhardsson, G.F. Nordberg, Lung cancer in smelter workers--interactions of metals as indicated by tissue levels, *Scand. J. Work. Environ. Health.* 19 Suppl 1 (1993) 90–94.

[52] F.J. Feldman, G.D. Christian, W.C. Purdy, Coulometric determination of chromium in biologic materials, *Am. J. Clin. Pathol.* 49 (1968) 826–833. https://doi.org/10.1093/ajcp/49.6.826.

[53] R. Cornelis, A. Speecke, J. Hoste, Neutron activation analysis for bulk and trace elements in urine, *Anal. Chim. Acta.* 78 (1975) 317–327. https://doi.org/https://doi.org/10.1016/S0003-2670(00)00152-5.

[54] W. Mertz, Effects and metabolism of glucose tolerance factor, *Nutr. Rev.* 33 (1975) 129–135. https://doi.org/https://doi.org/10.1111/j.1753-4887.1975.tb07105.x.

[55] C. Veillon, W.R. Wolf, B.E. Guthrie, Determination of chromium in biological materials by stable isotope dilution, *Anal. Chem.* 51 (1979) 1022–1024. https://doi.org/10.1021/ac50043a056.

[56] H. Nomiyama, M. Yotoriyama, K. Nomiyama, Normal chromium levels in urine and blood of japanese subjects determined by direct flameless atomic absorption specectrophotometry, and valency of chromium in urine after exposure to hexavalent chromium, *Am. Ind. Hyg. Assoc. J.* 41 (1980) 98–102. https://doi.org/10.1080/15298668091424438.

[57] P. Pääkkö, P. Kokkonen, S. Anttila, P.-L. Kalliomäki, Cadmium and chromium as markers of smoking in human lung tissue, *Environ. Res.* 49 (1989) 197–207. https://doi.org/https://doi.org/10.1016/S0013-9351(89)80065-9.

[58] A. Levina, P.A. Lay, *Redox Chemistry and Biological Activities of Chromium(III) Complexes,* 2nd ed., Elsevier B.V., 2019. https://doi.org/10.1016/b978-0-444-64121-2.00009-x.

[59] M. Szabó, J. Kalmár, T. Ditrói, G. Bellér, G. Lente, N. Simic, I. Fábián, Equilibria and kinetics of chromium(VI) speciation in aqueous solution – A comprehensive study from pH 2 to 11, *Inorganica Chim. Acta.* 472 (2018) 295–301. https://doi.org/https://doi.org/10.1016/j.ica.2017.05.038.

[60] J. Briffa, E. Sinagra, R. Blundell, Heavy metal pollution in the environment and their toxicological effects on humans, *Heliyon* 6 (2020) e04691. https://doi.org/10.1016/j.heliyon.2020.e04691.

[61] J.B. Vincent, Elucidating a biological role for chromium at a molecular level, *Acc. Chem. Res.* 33 (2000) 503–510. https://doi.org/10.1021/ar990073r.

[62] S. Lewicki, R. Zdanowski, M. Krzyzowska, A. Lewicka, B. Debski, M. Niemcewicz, M. Goniewicz, The role of chromium III in the organism and its possible use in

diabetes and obesity treatment, *Ann. Agric. Environ. Med.* 21 (2014) 331–335. https://doi.org/10.5604/1232-1966.1108599.

[63] J.B. Vincent, The biochemistry of chromium, *J. Nutr.* 130 (2000) 715–718. https://doi.org/https://doi.org/10.1093/jn/130.4.715.

[64] A.R. Saltiel, The paradoxical regulation of protein phosphorylation in insulin action, *FASEB J.* 8 (1994) 1034–1040. https://doi.org/https://doi.org/10.1096/fasebj.8.13.7926368.

[65] A. Gruzman, G. Babai, S. Sasson, Adenosine monophosphate-activated protein kinase (AMPK) as a new target for antidiabetic drugs: A review on metabolic, pharmacological and chemical considerations, *Rev. Diabet. Stud.* 6 (2009) 13–36. https://doi.org/10.1900/RDS.2009.6.13.

[66] A. Levina, L. Zhang, P.A. Lay, Structure and reactivity of a chromium(V) glutathione complex, *Inorg. Chem.* 42 (2003) 767–784. https://doi.org/10.1021/ic020621o.

[67] A. Levina, P.A. Lay, Mechanistic studies of relevance to the biological activities of chromium, *Coord. Chem. Rev.* 249 (2005) 281–298. https://doi.org/10.1016/j.ccr.2004.02.017.

[68] D.E. Kimbrough, Y. Cohen, A.M. Winer, L. Creelman, C. Mabuni, A Critical assessment of chromium in the environment, *Crit. Rev. Environ. Sci. Technol.* 29 (1999) 1–46. https://doi.org/10.1080/10643389991259164.

[69] B. Gammelgaard, K. Jensen, B. Steffansen, In vitro metabolism and permeation studies in rat jejunum: organic chromium compared to inorganic chromium, *J. Trace Elem. Med. Biol.* 13 (1999) 82–88. https://doi.org/https://doi.org/10.1016/S0946-672X(99)80028-5.

[70] T.H. Lim, T. Sargent, N. Kusubov, Kinetics of trace element chromium(III) in the human body, *Am. J. Physiol. Integr. Comp. Physiol.* 244 (1983) R445–R454. https://doi.org/10.1152/ajpregu.1983.244.4.R445.

[71] H. Sun, H. Li, P.J. Sadler, Transferrin as a metal ion mediator, *Chem. Rev.* 99 (1999) 2817–2842. https://doi.org/10.1021/cr980430w.

[72] B.W. Morris, S. MacNeil, C.A. Hardisty, S. Heller, C. Burgin, T.A. Gray, Chromium homeostasis in patients with type II (NIDDM) diabetes. *J. Trace Elem. Med. Biol.* 13 (1999) 57–61. https://doi.org/https://doi.org/10.1016/S0946-672X(99)80024-8.

[73] E.J. O'Flaherty, Chromium toxicokinetics BT – toxicology of metals: biochemical aspects, in: R.A. Goyer, M.G. Cherian (Eds.), *Handbook of Experimental Pharmacology*, vol 115, Springer, Berlin, 1995: pp. 215–228. https://doi.org/10.1007/978-3-642-79162-8_10.

[74] P.C. Bragt, E.A. Van Dura, Toxicokinetics of hexavalent chromium in the rat after intratracheal administration of chromates of different solubilities, *Ann. Occup. Hyg.* 27 (1983) 315–322. https://doi.org/10.1093/annhyg/27.3.315.

[75] S.A. Katz, H. Salem, The toxicology of chromium with respect to its chemical speciation: A review, *J. Appl. Toxicol.* 13 (1993) 217–224. https://doi.org/https://doi.org/10.1002/jat.2550130314.

[76] J.P. Goullé, E. Saussereau, J. Grosjean, C. Doche, L. Mahieu, J.M. Thouret, M. Guerbet, C. Lacroix, Accidental potassium dichromate poisoning. Toxicokinetics of chromium by ICP-MS-CRC in biological fluids and in hair, *Forensic Sci. Int.* 217 (2012) e8–e12. https://doi.org/https://doi.org/10.1016/j.forsciint.2011.10.020.

[77] M.G. Bemben, D.A. Bemben, M.J. Hartman, Chromium , in Wolinsky, I., & Driskell, J.A. (Eds.). *Sports Nutrition: Vitamins and Trace Elements, Second Edition* (2nd ed.). CRC Press. 2005, pp. 267–274. https://doi.org/10.1145/566654.566639.

[78] O.P. Gurjar, S. Kumari, P. Saini, S. Bugalia, A short review on biological applications of macrocyclic complexes of chromium, *J. Chil. Chem. Soc.* 67 (2022) 5716–5721. https://doi.org/10.4067/s0717-97072022000405716.

[79] M.C. Vlasiou, K.S. Pafiti, Chromium coordination compounds with antimicrobial activity: synthetic routes, structural characteristics, and antibacterial activity, *Open Med. Chem. J.* 14 (2020) 1–25. https://doi.org/10.2174/1874104502014010001.

[80] A.A. Abdel Aziz, Synthesis, spectroscopic characterization, thermal studies, catalytic epoxidation and biological activity of chromium and molybdenum hexacarbonyl bound to a novel N_2O_2 Schiff base, *J. Mol. Struct.* 979 (2010) 77–85. https://doi.org/10.1016/j.molstruc.2010.06.004.

[81] T.M. Thakur, R.S. Lokhande, M.M. Thigle, V.R. Patil, Nanoparticles of chromium oxide by green synthesis using Eucalyptus globulous leaves extract; characterization and biological activity studies, *Mater. Today Proc.* 79 (2023) 100–106. https://doi.org/10.1016/j.matpr.2022.09.368.

[82] R.A. Murcia, S.M. Leal, M. V. Roa, E. Nagles, A. Muñoz-Castro, J.J. Hurtado, Development of antibacterial and antifungal triazole chromium(III) and cobalt(II) complexes: Synthesis and biological activity evaluations, *Molecules* 23 (2018) 1–16. https://doi.org/10.3390/molecules23082013.

[83] N. Fahmi, S. Shrivastava, R. Meena, S.C. Joshi, R. V. Singh, Microwave assisted synthesis, spectroscopic characterization and biological aspects of some new chromium(iii) complexes derived from NO donor Schiff bases, *New J. Chem.* 37 (2013) 1445–1453. https://doi.org/10.1039/c3nj40907d.

[84] G. Sridhar, I. Mohammed Bilal, D. Easwaramoorthy, S. Kutti Rani, B. Siva Kumar, C.S. Manohar, Synthesis, characterization and antimicrobial activities of copper, nickel, cobalt, chromium complexes derived from (Z)-4-fluoro-N-(2,7-dimethylhept-6-enylidene) benzenamine, *J. Braz. Chem. Soc.* 28 (2017) 756–767. https://doi.org/10.21577/0103-5053.20160224.

[85] S.E. t. Ashoor, R.A. Abokhater, L. Belkhiri, S.A. Gadir, Spectroscopy and DFT studies on Cr(III) complexes with saccharides, 8-hydroxyquinoline and their biological activity, *J. Mol. Struct.* 1281 (2023) 135134. https://doi.org/10.1016/j.molstruc.2023.135134.

[86] S. Alghool, H.F.A. El-Halim, A. Dahshan, Synthesis, spectroscopic thermal and biological activity studies on azo-containing Schiff base dye and its cobalt(II), chromium(III) and strontium(II) complexes, *J. Mol. Struct.* 983 (2010) 32–38. https://doi.org/10.1016/j.molstruc.2010.08.025.

[87] A.A. Soliman, S.A. Ali, A.H. Marei, D.H. Nassar, Synthesis, characterization and biological activities of some new chromium molybdenum and tungsten complexes with 2,6-diaminopyridine, *Spectrochim. Acta – Part A, Mol. Biomol. Spectrosc.* 89 (2012) 329–332. https://doi.org/10.1016/j.saa.2011.12.061.

[88] S. Praveen Kumar, R. Suresh, K. Giribabu, R. Manigandan, S. Munusamy, S. Muthamizh, V. Narayanan, Synthesis and characterization of chromium(III) Schiff base complexes: Antimicrobial activity and its electrocatalytic sensing ability of catechol, *Spectrochim. Acta – Part A Mol. Biomol. Spectrosc.* 139 (2015) 431–441. https://doi.org/10.1016/j.saa.2014.12.012.

Section III

Role of Trace Metals (Se, V and Cr) in Human Health

7 Effect of Selenium and Its Compounds on Human Health

Monika Kamboj, Sangeeta Bajpai, Garima Pandey and Bimal Krishan Banik

7.1 INTRODUCTION

Selenium is a trace element and essential micronutrient of biological importance in humans [1,2]. It got the status of an essential element in 1954 [3]. Till then it was considered a toxic element as it was associated with the disease in grazing cattle. It is required by the body to maintain good health. Selenium is required to produce thyroid hormones, boost the immune system, and act as an antioxidant [2,4]. It plays a significant role in the synthesis of selenoprotein. In our body, through metabolism, selenium is converted to selenoprotein, which plays a remarkable role in various biological functions and maintaining good health. The toxicity of selenium in animals was first observed and described by Marco Polo in the thirteenth century [5]. However, in 1897, eight decades after the discovery of selenium, Modica reported the toxic effect of Se in research laboratory animals like rabbits [6]. Selenium is a topic of discussion due to its health effects on humans. It is considered an essential poison due to the narrow line of demarcation between its toxic and deficiency levels [7]. Adequate amounts of selenium in the diet can be lethal and inadequate levels can lead to chronic disorders. This creates confusion among the health authorities about the position of Se as a poisonous or a healthy element. In the early 1930s, the toxic effect of selenium was recognized. To date, plenty of research papers and reviews demonstrate the benefits and toxic effects of selenium. Many factors play a significant role in understanding selenium toxicity such as the age of a person, gender, nutritional status, physical activity, variation of selenoprotein in genes, mutation, and distribution of selenium in diet. Selenium poisoning is more likely seen in people having an undernourished protein diet and ailing from any liver disease. Rousseau et al. reported how exercise also alters the selenium level in young and elderly people [8]. Intoxication due to selenium depends upon its chemical composition, mode of administration, and species. Table 7.1 shows the toxicity of different selenium forms. It has been observed that oral consumption of selenium compounds causes less toxicity. Intake of high selenium may lead to selenosis. The toxic effect of selenium can be acute, subacute, and chronic. Table 7.2 represents the different types of selenosis.

DOI: 10.1201/9781032615110-10

TABLE 7.1
Toxicity of different selenium forms

Chemical Composition of Selenium	Toxicity
Selenium salts (Na_2SeO_3 and Na_2SeO_4)	Highest
Selenoamino acids	Moderate
Elemental Se	Least
Diphenyl selenide, hydrogen selenide	Least
Na_2Se	Least
SeS_2	Least

TABLE 7.2
Types and symptoms of selenosis [9]

Type of Selenosis	Dose	Duration of Exposure	Symptoms	Recommendation
Acute selenosis	Consuming plants with high Se contents in large amounts or due to incorrect formulation of supplements	Short period	• Breath with a garlic odor • Tiredness • Shortness of breath • Salivation, nausea, and vomiting along with stomach pain • Muscular and respiratory pain • Liver disease, endocarditis, and myocarditis in extreme cases.	• Reduce intake of selenium in the diet • No supplement
Subacute selenosis	Exposure to large doses of selenium. Observed in grazing animals that have consumed Se-accumulated plants.	Long period	• Blindness • Ataxia • Disorientation • Respiratory pain	• Reduce intake of selenium in the diet • No supplement

7.2 HEALTH BENEFITS OF SELENIUM

Selenium is a widely distributed, naturally occurring micronutrient. It is consumed by humans and other organisms orally through the ingestion of food or water. It also enters the body through the process of inhalation through air. Cereals, wheat grain,

TABLE 7.3
Relationship between serum selenium status and disease risk [10]

Disease Type	Serum Se Concentration (μg/L)	Effect
HIV	<85 –	Increased HIV morbidity and mortality
Cancer	<130 >150	All-cause mortality and increased cancer mortality
Prostate Cancer	<130 >170	Increased incidence and mortality of prostate cancer
Cardiovascular disease (CVD)	<55 >145	Increased morbidity and mortality from CVD
Irritable bowel syndrome (IBD)	– –	Increased incidence, severity, duration, and colon cancer risk of IBD

"–" indicates that no data is available.

Taken from ref [10] with permission from authors.

and other Se-rich foods (Brazil nuts, eggs, fish, chicken, crab, nuts, onion, garlic, potato, broccoli) are the main sources of Se through which it is absorbed. Selenium is good for health if the appropriate level of selenium in the diet is maintained. The changes in dietary intake may otherwise result in selenium deficiency diseases. The World Health Organization (WHO) had recommended the Se level for different age groups as described in Table 4.2 (Chapter 4). Maintaining the selenium level in the diet is crucial for good health benefits.

The increase of Se level in food, water, and air occurs due to the following reasons:

- The level of selenium in farm food increases if grown in soil using Se-enriched fertilizers. Se-contaminated water used for irrigation is another factor for increasing selenium in food.
- Exposure of people to soil, water, and air contaminated with hazardous waste containing selenium.
- Selenium recovery and paint industry workers are exposed to selenium through breathing. Exposure to selenium through inhalation causes dizziness, tiredness, and irritation of mucous membranes. In extreme cases, bronchitis occurs.

Low selenium levels in the diet may lead to Se deficiency, resulting in heart and muscular disorders. Intake of high selenium may lead to selenosis as shown in Table 7.2.

Figure 7.1 depicts the different health benefits of selenium. Selenium, due to its antioxidative properties, is responsible for reducing oxidative stress, thereby

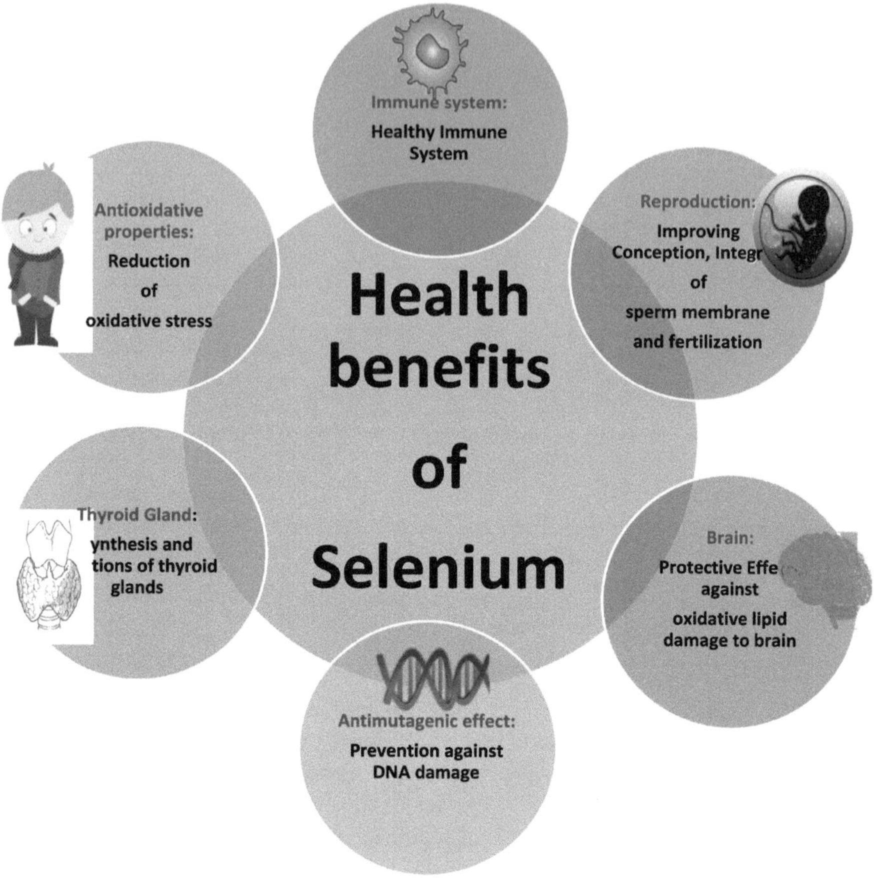

FIGURE 7.1 Health benefits of selenium.

protecting the cell damage. Selenium is also a super nutrient that helps in fighting cancer and boosts the immune system.

Selenoaminoacids are mainly considered a powerful antioxidant and act as scavengers for free radicals which are responsible for cell damage. Selenium is obtained from food, absorbed in the digestive system, and finally taken by the liver to transform it into selenoprotein, which performs various physiological functions. So, selenoprotein plays an important role in metabolism and antioxidative properties. The health effects of different selenoproteins are represented in Table 7.4.

7.3 DEFICIENCY OF SELENIUM

Selenium is an essential micronutrient for growth and performing normal physiology in humans. So, it is recommended in daily diet. It plays a crucial role in maintaining good health by controlling body functions. It acts as an immunity booster, regulating metabolism and reducing stress. Our body requires selenium for protein synthesis,

TABLE 7.4
Health benefits of selenoprotein [11]

Selenoprotein	Health Benefits	Reference
GPX1	Cancer, chondrogenic differentiation, Type 2 diabetic mellitus, depression, Keshan disease, cataracts, macular degeneration	12-18
GPX2	Cancer, intestinal inflammation	19,20
GPX3	Cancers, myocardial fibrosis, ventricular remodeling	21-23
GPX4	Osteoarthritis, cancer, cardiomyopathy, ischemia-reperfusion injury, brain function	24-28
GPX6	Huntington's disease	29
TXNRD1	Keshan disease, Idiopathic pulmonary arterial hypertension, hepatocellular carcinoma, osteoarthritis, genetic generalized epilepsy	16, 30-33
TXNRD2	Primary open-angle glaucoma, cardiovascular diseases, cancers, glaucoma	34-38
TXNRD3	Male reproduction, colitis, carcinogenesis	39,40
DIO1	Thyroid hormone metabolism, inhibit hepatosteatosis, renal cancer	41- 43
DIO2	Osteoarthritis, obesity, mental retardation	44-46
DIO3	Osteoarthritis, brain development, sepsis, and septic shock	47-49
MSRB1	Antioxidant activity, anti-inflammatory effect, regulate the immune system	50,51
SEPHS2	Cancer	52
SELENOF	Cancer, glycose metabolism disorder	53,54
SELENOH	Colorectal cancer	55
SELENOI	Hereditary spastic paraplegia 81	56
SELENOK	AD, cervical cancer	57,58
SELENOM	Glioblastoma, non-alcoholic fatty liver disease, synaptic deficits and cognitive dysfunction	59-61
SELENON	Myopathies	62
SELENOO	Thyroid cancer	63
SELENOP	Cancer, seizures and ataxia, Cardiovascular diseases	64-68
SELENOS	Hashimoto's thyroiditis, cardiovascular diseases	69-70
SELENOT	Glioblastoma, AD, cardiovascular diseases	71-73
SELENOV	Not known	-
SELENOW	Osteoporosis, anemia	74,75

regulating the thyroid gland, fighting infection, and reproduction. The deficiency of selenium in humans is represented schematically in Figure 7.2.

The deficiency of selenium is common worldwide. Deficiency is predicted when the person takes less than adequate selenium in their diet. The body requires selenium for the synthesis of selenoprotein, which is required for the normal functioning of the thyroid gland, immune system, and other vital organs. Selenium is obtained from food, absorbed in the digestive system, and finally taken by the liver to transform it

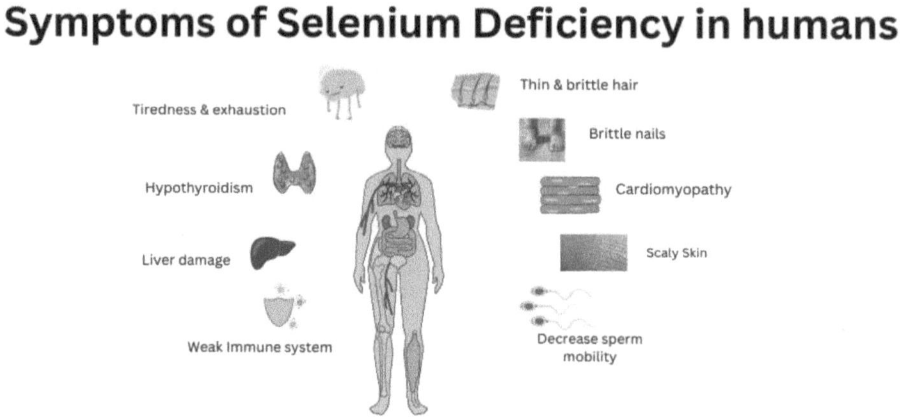

FIGURE 7.2 Symptoms due to selenium deficiency.

into selenoprotein, which performs various physiological functions. So, selenoprotein plays an important role in metabolism and antioxidative properties. A deficiency of selenium may also make the body prone to cancer. The deficiency of selenium may interfere with body functions and deteriorate health, resulting in several diseases of the brain, liver, and heart [76]. Hypothyroidism, tiredness, exhaustion liver damage, weak immune system, cardiomyopathy, scaly skin, decreased sperm mobility, thin and brittle hair, and nails are the common symptoms of selenium deficiency [77]. It is important to note that the effect of selenium on human health is not solely determined by its intake, but also by its form and chemical properties. Research has shown that Se deficiency is highly associated with over 40 different diseases. Some of the common issues and diseases related to selenium deficiency are discussed below.

7.3.1 Tiredness and Exhaustion

A deficiency of selenium makes a person fatigued as selenium plays a significant role in metabolism. It affects the functioning of the brain and results in loss of memory. Also, a person has poor concentration and mood swings.

7.3.2 Thin and Brittle Hair

Selenium deficiency affects hair growth, resulting in thin, brittle, and hair loss.

7.3.3 Scaly Skin

Selenium, an essential dietary micronutrient, is required for skin function, its growth, and protection against UV radiation. The function of selenium is attributed to selenoprotein, which protects the skin by acting as an antioxidant and nourishes the growth of keratinocytes present in the epidermis [78]. The person with selenium deficiency has dry skin with lesions [78]. Intake of dietary supplements is the treatment for scaly skin. Selenium has the property of anti-carcinogenesis. It protects the skin damage by (1) preventing DNA damage due to oxidation, (2) inhibiting immunosuppressive cytokines, and (3) boosting immunity [79].

7.3.4 Keshan Disease (KD)

A form of cardiomyopathy, Keshan disease is an endemic cardiovascular disease (CVD). The deficiency of selenium increases the risk of CVD. The disease was first reported from Keshan County of China, hence its name. It is most prevalent among children and women (childbearing age). Keshan disease is a myocardial disease, characterized by an enlarged heart, dilation of heart chambers, weak heart muscles, thin heart wall, irregular heartbeat, stroke, and heart failure. It has a high mortality rate if not given proper treatment. The different stages of KD are given in Table 7.5. Exposure to Coxsackievirus or chemicals may further upsurge the severity of Keshan disease [2,80]. The morbidity rate of KD was 50 percent, which led to often death. To combat this disease, the use of supplements of selenium, and selenium enrich salts are recommended that will reverse this disease.

7.3.5 Hypothyroidism (Endocrine System Disorder)

Selenium is one of the nutrients required for the proper functioning of the thyroid gland. In the human body, the thyroid gland contains the highest concentration of selenium (0.2–2 μg/g). DIO1, DIO2, and DIO3 are selenoproteins in the thyroid that play a crucial role in the metabolism of thyroid hormones. DIO1 activates T4 (thyroxine) to T3 (triiodothyronine) by removing an atom of iodine from it. DIO2 activates thyroid hormones and produces T3 intracellularly. DIO3 deactivates them by producing rT3 from T4 and T2 from T3 [81,82]. Studies show the deficiency of selenium has an impact on less production of thyroid hormone. A deficiency of selenium reduces the selenoprotein function, which thereby decreases the production of thyroid hormones. The hypothalamic–pituitary complex is thereby stimulated to produce a high amount of thyroid-stimulating hormone (TSH). DIOs are stimulated by TSH to perform thyroid functions. In this process, H_2O_2 produced is not removed adequately and

TABLE 7.5
Stages of Keshan disease (KD)

Types	Symptoms
Acute KD	Nausea, feeling dizzy, chills, shortness of breath, and loss of appetite. Patients may also experience irregular heartbeat, Cardiogenic shock, and heart failure
Subacute KD	Congestive heart failure, cardiogenic shock, heart expansion
Chronic KD	Dilated heart chambers, congestive or chronic heart failure. shortness of breath, fainting spells, or hemoptysis
Latent KD	dizziness, heart palpitations, and fatigue, abnormal electrocardiogram (EKG)

accumulates in thyroid gland tissue, resulting in fibrosis [83]. Myxedematous cretinism is a condition brought on by insufficient selenium and iodine intake. Thyroid atrophy, hypothyroidism, aberrant skeletal growth, and mental impairment are the hallmarks of this condition. The syndrome gets its name from the extensive buildup of mucin. The absence of fusion of the cranial plates, uneven tooth implantation, dry, scaly skin, infantile facial features, lengthy bone growth distortion resulting in small stature, delayed sexual maturation, and hypothyroidism linked to thyroid atrophy are some other signs.

7.3.6 Kashin–Beck Disease (KBD)

KBD is a type of chronic osteoarthrosis that affects the joints. It is an endemic disease, predominant in areas with low selenium content in water, soil, and growing crops. The origin of this disease is multifaceted, among which selenium deficiency is an implicating cause.

Maintaining a sufficient level of selenium in the environment can be the major factor in preventing and controlling the disease [84]. The KDS and selenium deficiency relationship was first reported in 1970 in China. This disease is highly prevalent in the population of China and other parts of Asia, affecting children and youth [85,86]. The preliminary signs of KBD are swollen joints, retarded growth, pain in joints, morning stiffness, fingers and toes shortened, deformity, weak muscle strength, and finally disability of the joint with limited mobility [87,88].

The prevalence of this disease has been reduced in China by the government's efforts to prevent and control it for decades [89, 90].

7.3.7 Nervous System Disease (Neurological)

The antioxidant properties of selenium in the nervous system are mediated through selenoprotein. Selenium is a cofactor for the enzyme (GPX) glutathione peroxidase, crucial for the functioning of the nervous system [91]. Selenium deficiency leads to neurological disorders due to a low supply of selenium to perform normal functions of the brain. Alzheimer's disease, Parkinson's disease, and epilepsy are some of the neurodegenerative disorders due to selenium deficiency [92,93]. Selenium supplements can overcome these deficiency effects. Cardoso and his research associates conducted a trial study of 40 patients with Alzheimer's disease. They showed that 30 mg/day supplementation of sodium selenate increases selenium levels in the central nervous system and stops the neurodegenerative effect in Alzheimer's patients [94]. Song et al. demonstrated the effects and mechanism of selenium compounds (Se-methylselenocysteine, selenomethionine, and sodium selenate) in improving the pathology of Alzheimer's patients [95].

7.3.8 Immunological Disorders

The invasion of pathogenic microorganisms in the human body is restricted by the barrier of the immune system. The immune system maintains physiological balance and preserves homeostasis. Selenium plays a significant role in regulating the functions of the immune system. Deficiency of selenium results in immunological disorders due to improper functioning of the immune system. The incompetency of the immune system due to selenium deficit makes the body vulnerable to infections. Low selenium levels in plasma enhance the susceptibility to cancer and other chronic inflammation [96].

Selenium deficiency also impairs innate and adaptive immunity. Selenium levels in plasma increase with selenium supplements. The cellular immune system response is amplified with the selenium supplement by enhancing the secretion of interferon-gamma and the expression of cytokines. It also boosts T and helper cells' production [97]. The immunological health issues and disorders due to selenium deficiency can be mitigated with selenium supplementation. Leukocytes in human blood are activated by consuming selenium-fortified food [97]. Organic forms of Se have higher tissue retention and bioavailability than mineral forms.

7.3.9 Muscle Pain

Numerous studies imply that muscular pain is linked to the deficiency of selenium [98]. Selenium accumulates in the liver and muscle after absorption in the form of selenomethionine. Selenium acts as an antioxidant, required for the normal functioning of muscles through its biological form, selenoprotein. Intake of inadequate amounts of selenium is related to poor muscular function in humans due to lack of energy. Selenoprotein is the biological form of selenium that is crucial for performing the muscle's functions. Selenoprotein (SELENON) plays a significant role in performing the function of muscle activity by calcium signaling and oxido-reductive homeostasis. Mutation in the SELENON gene affects muscular activity and causes SELENON-related myopathy. Muscle pain, weakness, stiffness, and fatigue are the symptoms that occur due to a mutation in the selenocysteine tRNA that fails to mature and translate to selenoprotein [99]. The muscle pain can be reduced by increasing the selenium supplements (100 g/day of selenomethionine). Inadequate selenium and vitamin E intake causes muscular dystrophy condition that affects the muscles of the heart and skeleton. This condition can be reversed by supplements of selenium and vitamin E.

7.3.10 Reproductive System

Selenium is vital for growth and reproduction in human beings as it plays important and active roles in various cellular metabolic pathways. The health of the male and female reproductive system depends upon selenium. Selenium is required in the normal progression of testicular growth, spermatogenesis, and proper function and mobility of spermatozoa. Selenium safeguards the sperm cells in the growth phase to prevent it from oxidative damage or mutation. The sperm cell membrane contains high levels of polyunsaturated fatty acids, which are vulnerable to oxidation by free radicals. This process produces reactive oxygen species in the cytoplasm of sperm cells and from leukocytes within semen. These reactive species compromise the integrity and morphology of the spermatozoa membrane, ultimately leading to its death. SELENOP is secreted from the liver and is transported around the body. This is crucial for male reproductive health as selenium deficiency affects the testes [100]. Male fertility requires a series of consecutive processes and any defect can lead to infertility and difficulty in the fusion process leading to failed fertilization [101]. It has been found that a low level of selenium was present in both the serum and semen of infertile males. Fertility can be improved by taking selenium supplements that provide the necessary nutrients for conception. The role of selenium in female fertility is not clearly understood due to the limited information available. The oocytes that came out of the follicles failed to fertilize because the follicles had low GPx activity due to selenium deficiency. Women with polycystic ovary syndrome (PCOS) have lower levels of serum selenium and increased oxidative stress levels [102].

7.3.11 Cancer

Cancer is a global issue with a high mortality rate. There is no effective cure for this deadly ailment. Selenium has antioxidant properties that protect cells from damage, enhance the immune system, and suppress the growth of tumors. A research study conducted by M.R. Lener and co-researchers involved 169 patients from Estonian and Polish populations who suffered from colorectal cancer. The study found a strong correlation between a low level of selenium in serum and an increased risk of colorectal cancer. As a result, the patients were advised to take selenium supplements with a dosage of 100 μg/l [103]. The relationship between selenium and cancer is complex, as it possesses both cancer-promoting and anti-cancer properties. It is also a therapeutic agent for cancer [103]. Studies conducted in the 1990s suggested that selenium may have an anti-cancer effect on different types of cancer cells [104]. There is limited evidence regarding the use of selenium supplements as preventive measures against cancer. Research shows that single-nucleotide polymorphism (SNP) in the gene for selenoprotein GPX1 and SELENOF are related to increased risk of mortality due to cancer [104]. The potential of selenium for curing prostate cancer in males has been investigated [104]. Selenium has an anticancer effect in humans. A daily dose of 100–200 μg of selenium inhibits genetic damage and cancer development. However, a higher dose of selenium, more than 400μg/day, can induce cancer [104]. Selenium nanoparticles have potential cytotoxic effects and are potent against prostate cancer cells [105]. Advanced research on selenium nanoparticles has recently revealed their potential as an anticancer agent and autophagy regulator [105]. Recently, nanofabrication has been a fascinating option for treating several types of cancer [106].

7.3.12 Type 2 Diabetes Mellitus

Type 2 diabetes mellitus is a serious health issue due to metabolic disorders and is linked to obesity, sedentary lifestyle, and aging. It is characterized by insulin deficiency due to dysfunction of pancreatic beta cells. Selenium plays a crucial role in the normal functioning of the endocrine system. This simple essential mineral has a significant impact on our health. Without this essential element, the endocrine system could not perform its vital functions properly. Therefore, it is important to ensure that our body receives an adequate amount of selenium to keep the endocrine system in an excellent position and improve our overall well-being. Recent studies show that selenoprotein plays a vital role in insulin biosynthesis and secretion, and excessive selenium intake can cause hyperglycemia and hyperinsulinemia, which causes serious health consequences [106]. Selenium primarily exerts its biological effects through selenoprotein. Glutathione peroxidase 1 (GPx1), selenoprotein P (SELENOP), selenoprotein S (SELENOS), selenoprotein V (SELENOV), and iodothyronine deiodinases (Dios) are selenoprotein associated with glucose and lipid homeostasis [106]. The health effects of various selenoprotein on Type 2 diabetes mellitus are as follows:

- *GPX1*: It regulates the concentration of hydrogen peroxide and reduces the amount of reactive oxygen species within the cell.
- *SELENOP*: Promotes insulin resistance in Type 2 diabetes mellitus [107].
    - *SELENOS*: It protects the pancreas with its antioxidant properties, helps regulate glucose utilization, and increases glucose output in the liver. In skeletal muscle, it positively correlates with serum amyloid A (SAA), while in adipose tissue, it positively correlates with HOMA-IR. However, it negatively correlates with fasting plasma glucose (FPG) in serum. Additionally, it inhibits oxidative stress and ER stress, thereby protecting the muscles from apoptosis [107].
- *SELENOV*: Prevents fat accumulation and boost energy expenditure. Protects tissues from oxidative damage and ER stress [108].
- *Iodothyronine deiodinases*: It regulates energy homeostasis. Energy homeostasis is the process of maintaining a balance in the energy supply and demand of the body. Any disruption in this balance can lead to the development of diabetes. Figure 7.3 represents the relationship between selenium status and the risk of diabetics [108].

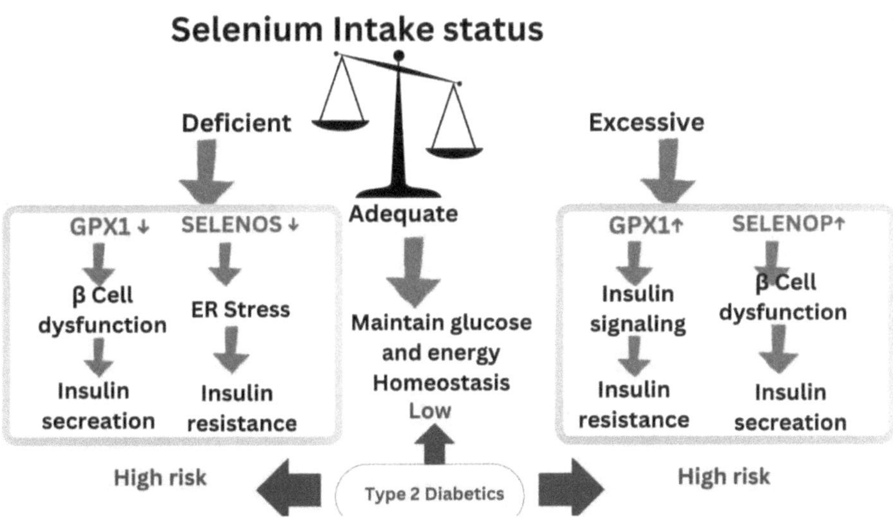

FIGURE 7.3 Selenium status and risk of type 2 Diabetes mellitus. (Taken from Ref [108] with permission from authors.)

7.4 TOXICITY OF SELENIUM AND SELENIUM COMPOUNDS

The toxicity of selenium is dependent on its chemical species. Selenium is a chemical element that can exhibit various oxidation states such as –2, 0, +4, and +6 in its salts. Selenite (SeO_3^{-2}) and selenate (SeO_4^{-2}) are the most commonly found selenium salts in water bodies. The sources of selenium include selenite (SeO_3^{-2}), selenate (SeO_4^{-2}), and organic compounds such as selenomethionine (SeMet), selenocysteine, and methyl selenocysteine. They all have nutritional value but excessive intake of these forms of selenium may lead to toxicity. Although selenium is an essential trace element for humans, its acute toxicity level is above 1.4 mg/l [109]. It is worth noting that the toxicity of Se(IV) salts is higher than Se(VI) salts, and both can contribute to increasing pollution in aquatic environments. It is important to include a small amount of selenium in your daily diet to maintain good health. However, too little or too much selenium can lead to health problems. It has been reported that two selenium compounds, selenium sulfide (SeS) and selenium hydride (SeH_2), are toxic. Two selenium compounds, selenium sulfide (SeS) and selenium hydride (SeH_2), have been identified as toxic. SeH_2, in particular, is a highly toxic selenium compound that can cause similar harmful effects as arsenic. Short-term exposure to SeH_2 through inhalation can irritate the respiratory system's mucous membrane. This exposure can also lead to the accumulation of fluid in the lungs, as well as bronchitis and pneumonia. Selenomethionine is less toxic than inorganic selenium and is the most effective supplement for increasing selenium levels. This is due to its nonspecific incorporation into proteins and storage in organs and tissues [109]. Methylselenocysteine is a naturally occurring less toxic organoselenium compound. It is biologically more active as it is readily incorporated into proteins and after ingestion is easily accumulated. In aquatic ecosystems, selenite is more toxic than selenate but in a soil environment, selenate becomes more toxic [109]. Selenium compounds that react with thiols are toxic and those that do not react with thiols are nontoxic. When selenite and SeO_2 react with glutathione and other thiols, they produce toxic superoxide and hydrogen peroxide. Similarly, selenocysteine and selenocystamine, which are both diselenides, are reduced to selenols in the presence of glutathione and other thiols, resulting in the production of toxic superoxide and hydrogen peroxide. However, selenate and selenoether are nontoxic, as they do not react with thiols to produce toxic superoxide and hydrogen peroxide. Nevertheless, in tissues, they can become toxic if they are reduced to selenite or selenols. The toxicity of selenium presents itself either acutely or chronically when the damage due to oxidation exceeds the body's antioxidant defense mechanisms. Selenium has an anticancer effect in humans. A daily dose of 100–200 µg of selenium inhibits genetic damage and cancer development. However, a higher dose of selenium, more than 400µg/day, can induce cancer. The effect of high doses of selenium is shown in Figure 7.4 [109].

7.5 EFFECT OF SELENIUM STATUS ON GUT MICROBIOTA

Selenium is a mineral that can be found in Brazil nuts, cereals, meat, fish, seafood, milk, and nuts. It can exist in two forms in our diet, organic seleno-amino acids and inorganic selenium, salts such as selenite and selenate. When we consume these forms of selenium, they are absorbed in the gastrointestinal tract and transported to the liver,

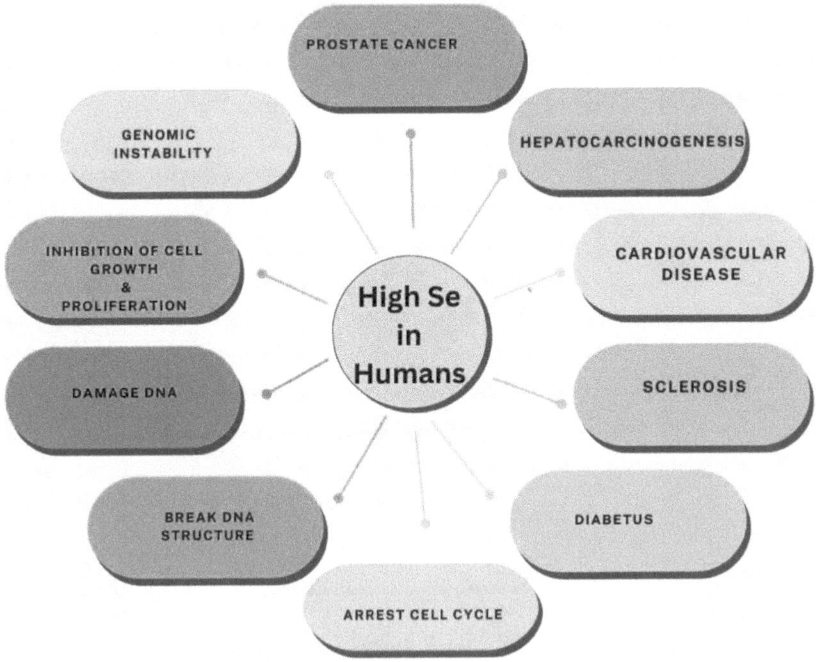

FIGURE 7.4 Effects of selenium toxicity in humans.

where they are metabolized and used to produce selenoprotein. Finally, the selenium is distributed to various tissues in the body. The gut microbiota is a complex population of microbes that live in our gastrointestinal tract. A review conducted by Cai et al. [110] found that inadequate intake of selenium can result in changes to the gut microbiota, which can make our body more susceptible to diseases. There is a symbiotic relationship between the host and the gut microbiota, where both parties benefit. However, this relationship can be compromised when there is not enough or too much selenium in our diet. Having an adequate supply of selenium in the body helps maintain homeostasis and promotes the biotransformation of selenium compounds. Selenium salts are metabolized into selenocysteine and selenomethionine. However, inadequate intake of selenium can lead to competition for selenium among bacteria such as *Escherichia coli*, Enterobacteria, and Clostridia. These bacteria may consume more selenium, leading to increased biotransformation of selenium compounds. This can further affect the expression of selenoprotein in the host, decreasing their levels. As a result, the immune system weakens, and there is an increased risk of inflammatory bowel disease and cancer. Having an adequate supply of selenium helps maintain homeostasis. However, when the host takes in enough selenium, the gut bacteria (such as lactic acid bacteria and Turicibacter) consume it faster. This leads to the transformation of these compounds and the excretion of volatile selenium compounds, which increases the excretion rate of selenium [110]. Figure 7.5 depicts how the gut microbiota is affected by the levels of selenium (Se) present in the body, and the biotransformation of selenium compounds [111].

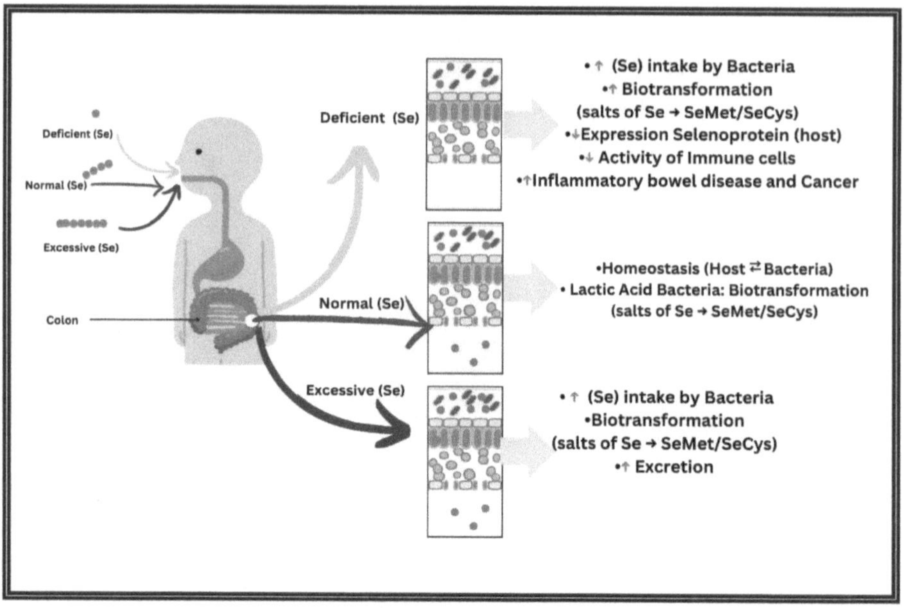

FIGURE 7.5 Regulation of the gut microbiota, Se status, and Biotransformation of Se compounds. (Taken from ref [111] with permission from authors.)

7.6 FACTORS RESPONSIBLE FOR LOW SELENIUM LEVELS

Several factors may cause low levels of selenium, including:

- *Low selenium level in soil*: Selenium is an essential nutrient for humans and animals, and is not evenly distributed geographically. The presence of selenium in soil depends on various factors such as location, precipitation, evaporation, and soil pH levels. The concentration of selenium in plants is directly proportional to selenium content in the soil. Therefore, crops grown in soil with varying levels of selenium, be it deficient or rich, will contain different levels of selenium. This can have a direct impact on the overall selenium intake of people who consume these crops.
- *Consumption of selenium-deficient diet*: People who do not consume selenium-rich food may suffer from selenium deficiency symptoms and diseases. The daily requirement of selenium is not met in the diet of such individuals, which has low levels of the mineral.

7.7 MEDICAL CONDITIONS

Patients with certain medical conditions such as kidney dialysis, HIV, and colitis have low selenium levels. The liver metabolizes selenium to selenide, which is further converted to selenoprotein. A patient with liver cirrhosis also has low selenium levels even if he gets adequate selenium from diet.

7.8 PREVENTIVE MEASURES FOR SELENIUM DEFICIENCY

Selenium deficiency can be prevented or treated by taking certain measures.

- *Consume selenium-rich foods*: To prevent selenium deficiency, it is advisable to consume food that is rich in selenium as the first line of action. This is the safe route to prevent low levels of selenium in the body. Brazil nuts, eggs, mushrooms, chia, and sunflower seeds are some of the selenium-rich food.
- *Taking dietary selenium supplements*: If the body lacks selenium, taking a selenium supplement can help prevent selenium deficiency. Selenium can be found in many multivitamins. However, supplementation should only be considered if your daily diet does not provide sufficient selenium. It is important to note that taking high doses of selenium supplements can lead to side effects and cause selenium toxicity due to its narrow window. Selenomethionine or selenite are the forms of selenium supplements. Selenomethionine has better absorption in the body.

7.9 ROLE OF SELENIUM NANOPARTICLES ON HUMAN HEALTH

In previous sections, we have discussed the beneficial effects of selenium and its compounds. Selenium is an essential micronutrient required for human health and can be obtained from various sources. It is incorporated into selenocysteine, which is used for the synthesis of selenoproteins. The therapeutic window of selenium is narrow, with excessive intake potentially leading to toxicity, and inadequate intake resulting in various disease symptoms. However, selenium nanoparticles have less toxicity than selenium salts and are therefore a better option for therapeutic use.

Advantages of selenium nanoparticles:

- Less toxic
- High biological activity (antioxidation, anticancer activity)
- High rate of absorption
- Directly absorbed in the body and converted to organic selenium
- Enhance immunity
- High catalytic efficiency

Some of the application of selenium nanoparticles is depicted in Figure 7.6 [112].

Pouri et al. have discovered a new method to synthesize spherical selenium nanoparticles. This method is safe, and pollution-free, and these nano-selenium found applications in medicine [113]. Research suggests that selenium nanoparticles are biologically safe for human health and are better absorbed than organic and inorganic forms. The efficiency of selenium at the nanoscale is the same as that of organic selenium but with low toxicity. Therefore, they can be used as antioxidants [113]. Studies have revealed that selenium nanoparticles also increase the activity of selenoenzyme glutathione peroxidase and thioredoxin reductase, with the same efficacy as organic

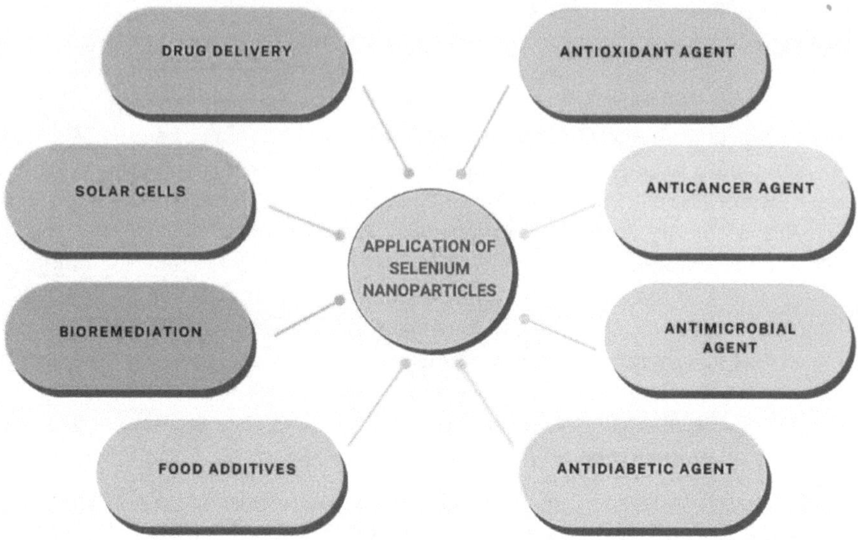

FIGURE 7.6 Application of selenium nanoparticles. (Taken from ref [112] with permission from authors.)

selenium. As the particle size decreases and the surface-to-volume ratio increases at the nanoscale, it boosts their biological activity and can be used as an antioxidant. Additionally, nano-selenium is effective against cancer cells. Nano-selenium encourages the cell to produce reactive oxygen species (ROS) that damage DNA and stop the cell cycle by dysfunctional mitochondria [113]. The effectiveness of these nano-selenium particles can be improved by modifying their surface.

Yang Fang and colleagues have reported a simple method for decorating the surface of selenium nanoparticles with Spirulina polysaccharide. They evaluated the anticancer properties of the decorated nanoparticles and found that they were more effective than undecorated nanoparticles. According to their research, the uptake of selenium nanoparticles by the cells was enhanced, increasing their anti-cancer activity [113]. Combined with radiotherapy, they have also proven effective against lung cancer cells [113]. Nanoparticles can be utilized for several therapeutic applications like cancer therapy, diabetes, Alzheimer's disease, microbial infection, and inflammation. Depending on the specific application, these nanoparticles can be administered orally or intravenously [113]. Selenium nanoparticles are synthesized and researched to minimize the toxicity of selenium and increase its therapeutic potential [113]. The potential role of selenium nanoparticles in the human body is depicted in Figure 7.7.

In a nutshell, the role of selenium in maintaining the function of our body is noteworthy. A selenium-rich healthy diet protects us from various diseases. Selenium and its compounds have supranutritional value and have found important applications as chemotherapeutic agents. Selenium has an impact on various health conditions. It is summarized in Figure 7.8 [114].

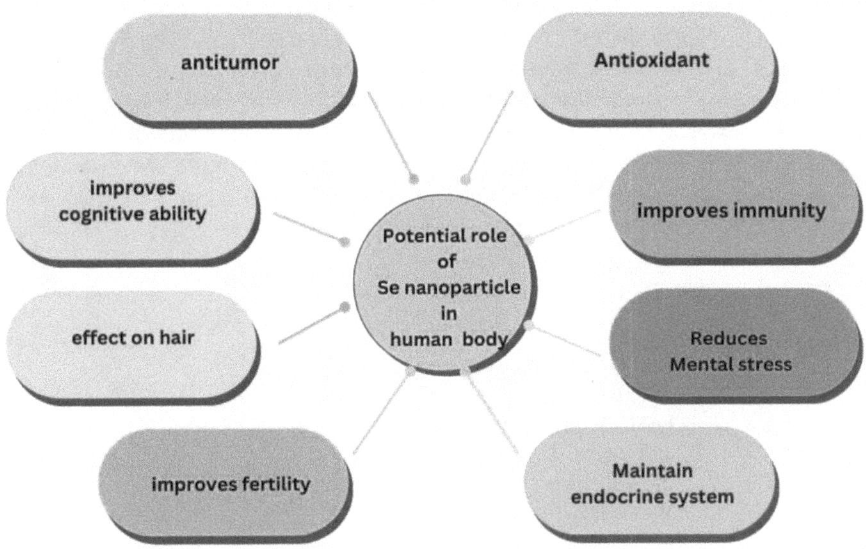

FIGURE 7.7 The potential role of selenium nanoparticles in the human body.

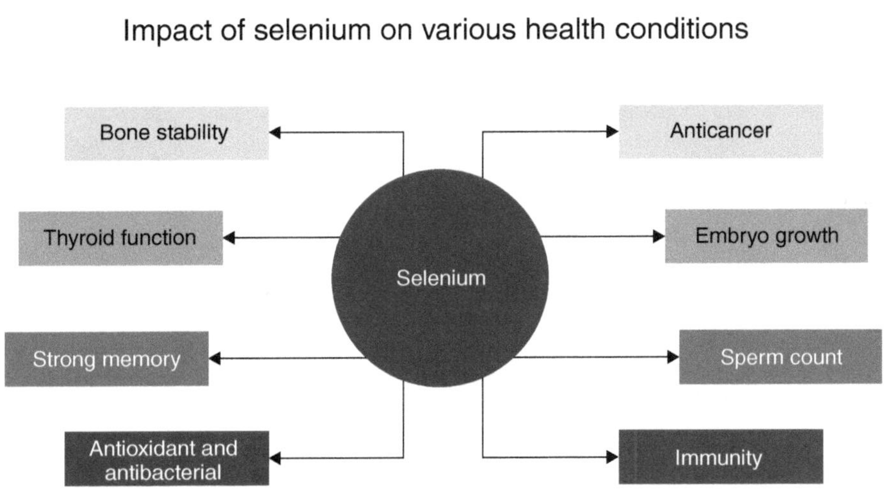

FIGURE 7.8 Impact of Selenium on various health conditions.

7.10 CONCLUSION

Selenium is an essential trace element found in many foods. It is a supra nutrient for human health and is available as a selenium supplement. Selenium and its compounds have an impact on the health conditions of humans. It has anti-inflammatory, antioxidant, antimicrobial, and anticancer properties. It plays a crucial role in maintaining the proper functions of the immune and reproductive systems. The therapeutic window of selenium is very narrow. Inadequate intake of selenium results in selenium deficiencies and disorders while excessive intake results in selenium toxicity. The biological function of selenium is expressed in terms of selenoproteins. Recently, research suggests that selenium nanoparticles are biologically safe for human health and are better absorbed than organic and inorganic forms. The efficiency of selenium at the nanoscale is the same as that of organic selenium but with low toxicity. The selenium nanoparticles have important therapeutic applications but with low toxicity. Further extensive research on the biomedical application of selenium nanoparticles could be an interesting area of future studies.

ABBREVIATIONS

Se	selenium
HIV	human immunodeficiency virus
CVD	cardiovascular diseases
IBD	inflammatory bowel disease
WHO	World Health Organization
UV	ultraviolet
KD	Keshan disease
ER	endoplasmic reticulum
EKG	electrocardiogram
FPG	fasting plasma glucose
HOMA-IR	homeostatic model assessment for insulin resistance
SeMet	selenomethionine
μg	microgram
SeO_2	selenium dioxide
SeCys	selenocysteine
ROS	reactive oxygen species
Na_2SeO_3	sodium selenite
Na_2SeO_4	sodium selenate
Na_2Se	sodium selenide
SeS_2	selenium disulfide
GPX1	glutathione peroxidase 1
GPX2	glutathione peroxidase 2
GPX3	glutathione peroxidase 3
GPX4	glutathione peroxidase 4
GPX6	glutathione peroxidase 6
TXNRD1	thioredoxin reductase 1
TXNRD2	thioredoxin reductase2
TXNRD3	thioredoxin reductase 3
DIO1	iodothyronine deiodinase 1
DIO2	iodothyronine deiodinase 2
DIO3	iodothyronine deiodinase3
MSRB1	methionine sulfoxide reductase Bl
SEPHS2	selenophosphate synthetase 2

SELENOF	selenoprotein F
SELENOH	selenoprotein H
SELENOI	selenoprotein I
SELENOK	selenoprotein K
SELENOM	selenoprotein M
SELENON	selenoprotein N
SELENOO	selenoprotein O
SELENOP	selenoprotein P
SELENOS	selenoprotein S
SELENOT	selenoprotein T
SELENOV	selenoprotein V
SELENOW	selenoprotein W

REFERENCES

1. Hatfield, D. L.; Gladyshev, V. N. How Selenium Has Altered Our Understanding of the Genetic Code. *Mol. Cell. Biol.* **2002**, *22* (11), 3565–3576. DOI: 10.1128/MCB.22.11.3565-3576.2002
2. Rayman, M. P. Selenium and Human Health. *Lancet* **2012**, *379* (9822), 1256–1268. DOI: 10.1016/S0140-6736(11)61452-9.
3. Gilinskaya, L.G.; Grigorieva, T. N.; Razvorotneva, L. I.; Trofimova, L. B. Composition and Physicochemical Properties of Natural Blue Clays **2008**, *16* (2), 143–153
4. Arthur, J. R.; McKenzie, R. C.; Beckett, G. J. Selenium in the Immune System. *J. Nutr.* **2003**, *133* (5) (Suppl. 1), 1457S–1459S. DOI: 10.1093/jn/133.5.1457S
5. Marco Polo, I. I. *Miliono*; Arnoldo Mondadori Editore, **1982**, 72.
6. Modica, Q. Azione Cronica del Selenio. *Arch. Farmacol. Ter.* **1897**, *5*, 61–84.
7. El-Ramady, H.; Domokos-Szabolcsy, É.; Shalaby, T.; Prokisch, J.; Fári, M. Selenium in Agriculture: Water; Air; Soil; Plants; Food. Animals and Nanoselenium. In *CO_2 Sequestration, Biofuels and Depollution. Environmental Chemistry for a Sustainable World*; Vol. 5. Lichtfouse, E., Schwarzbauer, J., Robert, D., Eds.; Springer. **2015**, DOI: https://doi.org/10.1007/978-3-319-11906-9_5
8. Rousseau, A. S.; Margaritis, I.; Arnaud, J.; Faure, H.; Roussel, A. M. Physical Activity Alters Antioxidant Status in Exercising Elderly Subjects. *J. Nutr. Biochem.* **2006**, *17* (7), 463–470. DOI: 10.1016/j.jnutbio.2005.10.001
9. Rosenfeld, I.; Beath, O. A. *Selenium: Geobotany, Biochemistry, Toxicity, and Nutrition*; Academic Press, 1964.
10. Sun, Y.; Wang, Z.; Gong, P.; Yao, W.; Ba, Q.; Wang, H. Review on the Health-Promoting Effect of Adequate Selenium Status. *Front. Nutr.* **2023**, 10, 1136458. DOI: 10.3389/fnut.2023.1136458
11. Zhang, F.; Li, X.; Wei, Y. Selenium and Selenoproteins in Health. *Biomolecules,* **2023**, *13* (5), 799. DOI: 10.3390/biom13050799.
12. Zhao, Y.; Wang, H.; Zhou, J.; Shao, Q. Glutathione Peroxidase GPX1 and Its Dichotomous Roles in Cancer. *Cancers (Basel)* **2022**, *14* (10), 2560. DOI: 10.3390/cancers14102560
13. Yan, J.; Guo, Y.; Fei, Y.; Zhang, R.; Han, Y.; Lu, S. GPx1 Knockdown Suppresses Chondrogenic Differentiation of ATDC5 Cells Through Induction of Reductive

Stress. *Acta Biochim. Biophys. Sin. (Shanghai)* **2017**, *49* (2), 110–118. DOI: 10.1093/abbs/gmw125
14. Kafeel, S.; Hashim, Z.; Fawwad, A.; Nawab, S. N. Predisposition of SOD1, GPX1, CAT Genetic Variants and Their Haplotypes in Cataractogenesis of Type 2 Diabetes Mellitus in Pakistan. *Acta Diabetol.* **2022**, *59* (5), 623–632. DOI: 10.1007/s00592-021-01832-5
15. Johnson, L. A.; Phillips, J. A.; Mauer, C.; Edwards, M.; Balldin, V. H.; Hall, J. R.; Barber, R.; Conger, T. L.; Ho, E. J.; O'Bryant, S. E. The Impact of GPX1 on the Association of Groundwater Selenium and Depression: A Project Frontier Study. *BMC Psychiatry* **2013**, *13*, 7. DOI: 10.1186/1471-244X-13-7.
16. Pei, J.; Fu, W.; Yang, L.; Zhang, Z.; Liu, Y. Oxidative Stress Is Involved in the Pathogenesis of Keshan Disease (an Endemic Dilated Cardiomyopathy) in China. *Oxid. Med. Cell. Longev.* **2013**, *2013*, 474203. DOI: 10.1155/2013/474203
17. Orhan, H.; Marol, S.; Hepşen, I. F.; Sahin, G. Effects of Some Probable Antioxidants on Selenite-Induced Cataract Formation and Oxidative Stress-Related Parameters in Rats. *Toxicology* **1999**, *139* (3), 219–232. DOI: 10.1016/s0300-483x(99)00128-6.
18. Sun, Y.; Zheng, Y.; Wang, C.; Liu, Y. Glutathione Depletion Induces Ferroptosis, Autophagy, and Premature Cell Senescence in Retinal Pigment Epithelial Cells. *Cell Death Dis.* **2018**, *9* (7), 753. DOI: 10.1038/s41419-018-0794-4
19. Brigelius-Flohé, R.; Kipp, A. P. Physiological Functions of GPx2 and Its Role in Inflammation-Triggered Carcinogenesis. *Ann. N. Y. Acad. Sci.* **2012**, *1259*, 19–25. DOI: 10.1111/j.1749-6632.2012.06574.x
20. Esworthy, R. S.; Yang, L.; Frankel, P. H.; Chu, F. F. Epithelium-Specific Glutathione Peroxidase, Gpx2, Is Involved in the Prevention of Intestinal Inflammation in Selenium-Deficient Mice. *J. Nutr.* **2005**, *135* (4), 740–745. DOI: 10.1093/jn/135.4.740.
21. Chang, C.; Worley, B. L.; Phaëton, R.; Hempel, N. Extracellular Glutathione Peroxidase GPx3 and Its Role in Cancer. *Cancers (Basel)* **2020**, *12* (8), 2197. DOI: 10.3390/cancers12082197.
22. Li, G.; Qin, Y.; Cheng, Z.; Cheng, X.; Wang, R.; Luo, X.; Zhao, Y.; Zhang, D.; Li, G. Gpx3 and Egr1 Are Involved in Regulating the Differentiation Fate of Cardiac Fibroblasts Under Pressure Overload. *Oxid. Med. Cell. Longev.* **2022**, *2022*, 3235250. DOI: 10.1155/2022/3235250.
23. Covington, T. A.; Pilz, P. M.; Mulhern, R. M.; Ngoy, S.; Loscalzo, A.; Liu, J.; Fisch, S.; Grune, J. GPx3 Deficiency Exacerbates Maladaptive RV Remodeling in Experimental Pulmonary Artery Banding. *Am. J. Physiol. Lung Cell. Mol. Physiol.* **2023**, *324* (4), L550–L556. DOI: 10.1152/ajplung.00379.2022.
24. Covington, T. A.; Pilz, P. M.; Mulhern, R. M.; Ngoy, S.; Loscalzo, A.; Liu, J.; Fisch, S.; Grune, J. GPx3 Deficiency Exacerbates Maladaptive RV Remodeling in Experimental Pulmonary Artery Banding. *Am. J. Physiol. Lung Cell. Mol. Physiol.* **2023**, *324* (4), L550–L556. DOI: 10.1152/ajplung.00379.2022.
25. Miao, Y.; Chen, Y.; Xue, F.; Liu, K.; Zhu, B.; Gao, J.; Yin, J.; Zhang, C.; Li, G. Contribution of Ferroptosis and GPX4's Dual Functions to Osteoarthritis Progression. *EBiomedicine* **2022**, *76*, 103847. DOI: 10.1016/j.ebiom.2022.103847.
26. Hangauer, M. J.; Viswanathan, V. S.; Ryan, M. J.; Bole, D.; Eaton, J. K.; Matov, A.; Galeas, J.; Dhruv, H. D.; Berens, M. E.; Schreiber, S. L.; McCormick, F.; McManus, M. T. Drug-Tolerant Persister Cancer Cells Are Vulnerable to GPX4 Inhibition. *Nature* **2017**, *551* (7679), 247–250. DOI: 10.1038/nature24297
27. Wang, Y.; Yan, S.; Liu, X.; Deng, F.; Wang, P.; Yang, L.; Hu, L.; Huang, K.; He, J. PRMT4 Promotes Ferroptosis to Aggravate Doxorubicin-Induced Cardiomyopathy

via Inhibition of the Nrf2/GPX4 Pathway. *Cell Death Differ.* **2022**, *29* (10), 1982–1995. DOI: 10.1038/s41418-022-00990-5

28. Zhang, J.; Bi, J.; Ren, Y.; Du, Z.; Li, T.; Wang, T.; Zhang, L.; Wang, M.; Wei, S.; Lv, Y.; Wu, R. Involvement of GPX4 in Irisin's Protection Against Ischemia Reperfusion-Induced Acute Kidney Injury. *J. Cell. Physiol.* **2021**, *236* (2), 931–945. DOI: 10.1002/jcp.29903
29. Reichert, C. O.; de Freitas, F. A.; Sampaio-Silva, J.; Rokita-Rosa, L.; Barros, P. L.; Levy, D.; Bydlowski, S. P. Ferroptosis Mechanisms Involved in Neurodegenerative Diseases. *Int. J. Mol. Sci.* **2020**, *21* (22), 8765. DOI: 10.3390/ijms21228765
30. Shema, R.; Kulicke, R.; Cowley, G. S.; Stein, R.; Root, D. E.; Heiman, M. Synthetic Lethal Screening in the Mammalian Central Nervous System Identifies Gpx6 as a Modulator of Huntington's Disease. *Proc. Natl. Acad. Sci. U. S. A.* **2015**, *112* (1), 268–272. DOI: 10.1073/pnas.1417231112
31. Lin, W.; Tang, Y.; Zhang, M.; Liang, B.; Wang, M.; Zha, L.; Yu, Z. Integrated Bioinformatic Analysis Reveals TXNRD1 as a Novel Biomarker and Potential Therapeutic Target in Idiopathic Pulmonary Arterial Hypertension. *Front. Med. (Lausanne)* **2022**, *9*, 894584. DOI: 10.3389/fmed.2022.894584.
32. Fu, B.; Meng, W.; Zeng, X.; Zhao, H.; Liu, W.; Zhang, T. TXNRD1 Is an Unfavorable Prognostic Factor for Patients with Hepatocellular Carcinoma. *Biomed. Res. Int.* **2017**, *2017*, 4698167. DOI: 10.1155/2017/4698167
33. Liang, J.; Wang, S.; Hu, J.; Hong, X.; Zhu, M.; Liu, X.; Alswadeh, M.; Mo, F.; Dai, M. Targeted Inhibition of TXNRD1 Prevents Cartilage Extracellular Matrix Degeneration by Activating Nrf2 Pathway in Osteoarthritis. *Biochem. Biophys. Res. Commun.* **2022**, *20* (635), 267–276. doi: 10.1016/j.bbrc.2022.10.059
34. Kudin, A.P.; Baron, G.; Zsurka, G.; Hampel, K.G.; Elger, C.E.; Grote, A.; Weber, Y.; Lerche, H.; Thiele, H.; Nürnberg, P.; Schulz, H.; Ruppert, A.K.; Sander, T.; Cheng, Q.; Arnér, E.S.; Schomburg, L.; Seeher, S.; Fradejas-Villar, N.; Schweizer, U.; Kunz, W.S. Homozygous Mutation in TXNRD1 is Associated with Genetic Generalized Epilepsy. *Free Radic. Biol. Med.* **2017**, *106*, 270–277. doi: 10.1016/j.freeradbiomed.2017.02.040.
35. Bailey, J.N.; Loomis, S.J.; Kang, J.H.; Allingham, R.R.; Gharahkhani, P.; Khor, C.C.; Burdon, K.P.; Aschard, H.; Chasman, D.I.; Igo, R.P. Jr.; Hysi, P.G.; Glastonbury, C.A.; Ashley-Koch, A.; Brilliant, M.; Brown, A.A.; Budenz, D.L.; Buil, A.; Cheng, C.Y.; Choi, H.; Christen, W.G.; Curhan, G.; De Vivo, I.; Fingert, J.H.; Foster, P.J.; Fuchs, C.; Gaasterland, D.; Gaasterland, T.; Hewitt, A.W.; Hu, F.; Hunter, D.J.; Khawaja, A.P.; Lee, R.K.; Li, Z.; Lichter, P.R.; Mackey, D.A.; McGuffin, P.; Mitchell, P.; Moroi, S.E.; Perera, S.A.; Pepper, K.W.; Qi, Q.; Realini, T.; Richards, J.E.; Ridker, P.M.; Rimm, E.; Ritch, R.; Ritchie, M.; Schuman, J.S.; Scott, W.K.; Singh, K.; Sit, A.J.; Song, Y.E.; Tamimi, R.M.; Topouzis, F.; Viswanathan, A.C.; Verma, S.S.; Vollrath, D.; Wang, J.J.; Weisschuh, N.; Wissinger, B.; Wollstein, G.; Wong, T.Y.; Yaspan, B.L.; Zack, D.J.; Zhang, K.; Study, E.N.; ANZRAG Consortium.; Weinreb, R.N.; Pericak-Vance, M.A.; Small, K.; Hammond, C.J.; Aung, T.; Liu, Y.; Vithana, E.N.; MacGregor, S.; Craig, J.E.; Kraft, P.; Howell, G.; Hauser, M.A.; Pasquale, L.R.; Haines, J.L.; Wiggs, J.L.; Genome-wide Association Analysis Identifies TXNRD2, ATXN2 and FOXC1 as Susceptibility Loci for Primary Open-angle Glaucoma. *Nat Genet.* **2016**,*48*(2):189–94. doi: 10.1038/ng.3482.
36. Sibbing, D.; Pfeufer, A.; Perisic, T.; Mannes, A.M.; Fritz-Wolf, K.; Unwin, S.; Sinner, M.F.; Gieger, C.; Gloeckner, C.J.; Wichmann, H.E.; Kremmer, E.; Schäfer, Z.; Walch, A.; Hinterseer, M.; Näbauer, M.; Kääb, S.; Kastrati, A.; Schömig, A.; Meitinger, T.; Bornkamm, G.W.; Conrad, M.; Beckerath, N.V. Mutations in the Mitochondrial

Thioredoxin Reductase Gene TXNRD2 Cause Dilated Cardiomyopathy. *Eur Heart J.* **2011,** *32* (9),1121–33. doi: 10.1093/eurheartj/ehq507

37. Kariž, S.; Mankoč, S.; Petrovič, D. Association of Thioredoxin Reductase 2 (TXNRD2) Gene Polymorphisms with Myocardial Infarction in Slovene Patients with Type 2 Diabetes Mellitus. *Diabetes Res Clin Pract.* **2015**,*108*(2), 323–8. doi: 10.1016/j.diabres.2015.01.038.
38. Gencheva, R.; Arnér, E.S.J. Thioredoxin Reductase Inhibition for Cancer Therapy, *Annu Rev Pharmacol Toxicol.* **2022,** *62,* 177–196. doi: 10.1146/annurev-pharmtox-052220-102509.
39. Aboobakar, I.F.; Kinzy, T.G.; Zhao, Y.; Fan, B.; Pasquale, L.R.; Qassim, A.; Kolovos, A.; Schmidt, J.M.; Craig, J.E.; Cooke Bailey, J.N.; Wiggs, J.L.; NEIGHBORHOOD Consortium. Mitochondrial TXNRD2 and ME3 Genetic Risk Scores Are Associated with Specific Primary Open-Angle Glaucoma Phenotypes. *Ophthalmology.* **2023**, *130*(7), 756–763. doi: 10.1016/j.ophtha.2023.02.018. Epub 2023 Feb 20. PMID: 36813040; PMCID: PMC10330404.
40. Dou, Q.; Turanov, A.A.; Mariotti, M.; Hwang, J.Y.; Wang, H.; Lee, S.G.; Paulo, J.A.; Yim, S.H.; Gygi, S.P.; Chung, J.J.; Gladyshev, V.N. Selenoprotein TXNRD3 Supports Male Fertility Via the Redox Regulation of Spermatogenesis. *J. Biol. Chem.* **2022**, *298*(8):102183. doi: 10.1016/j.jbc.2022.102183.
41. Liu, Q.; Du, P.; Zhu, Y.; Zhang, X.; Cai, J.; Zhang, Z. Thioredoxin Reductase 3 Suppression Promotes Colitis and Carcinogenesis Via Activating Pyroptosis and Necrosis. *Cell. Mol. Life Sci.* **2022**, *79*, 106. https://doi.org/10.1007/s00018-022-04155-y
42. França, M.M.; German, A.; Fernandes, G.W.; Liao, X.-H.; Bianco, A.C.; Refetoff, S.; Dumitrescu, A.M. Dumitrescu. Human Type 1 Iodothyronine Deiodinase (DIO1) Mutations Cause Abnormal Thyroid Hormone Metabolism. *Thyroid*, **2021**, *31*, 202–207.
43. Bruinstroop, E.; Zhou, J.; Tripathi, M.; Yau, W.W.; Boelen, A.; Singh, B.K.; Yen, P.M. Early Induction of Hepatic Deiodinase Type 1 Inhibits Hepatosteatosis During NAFLD Progression. *Mol. Metab.* **2021**, *53*, 101266.
44. Poplawski, P.; Rybicka, B.; Boguslawska, J.; Rodzik, K.; Visser, T.J.; Nauman, A.; Piekielko-Witkowska, A. Induction of Type 1 Iodothyronine Deiodinase Expression Inhibits Proliferation and Migration of Renal Cancer Cells. *Mol. Cell Endocrinol.* **2017**, *442*, 58–67.
45. Bomer, N.; Hollander, W.D.; Ramos, Y.F.M.; Bos, S.D.; van der Breggen, R.; Lakenberg, N.; Pepers, B.A.; van Eeden, A.E.; Darvishan, A.; Tobi, E.W.; Duijnisveld, B.J.; van den Akker, E.B.; Heijmans, B.T.; van Roon-Mom, W.M.; Verbeek, F.J.; van Osch, G.J.; Nelissen, R.G.; Slagboom, P.E.; Meulenbelt, I. Underlying Molecular Mechanisms of DIO2 Susceptibility in Symptomatic Osteoarthritis. *Ann. Rheum. Dis.* **2015**, *74*, 1571–1579. doi: 10.1136/annrheumdis-2013-204739.
46. Bradley, D.; Liu, J.; Blaszczak, A.; Wright, V.; Jalilvand, A.; Needleman, B.; Noria, S.; Renton, D.; Hsueh, W. Adipocyte DIO2 Expression Increases in Human Obesity but Is Not Related to Systemic Insulin Sensitivity. *J. Diabetes Res.* **2018**, *2018*, 2464652.
47. Guo, T.-W.; Zhang, F.-C.; Yang, M.-S.; Gao, X.-C.; Bian, L.; Duan, S.-W.; Zheng, Z.-J.; Gao, J.-J.; Wang, H.; Li, R.-L.; Feng, G.Y.; St Clair, D.; He, L. Positive Association of the DIO2 (Deiodinase Type 2) Gene with Mental Retardation in the Iodine-deficient Areas of China. *J. Med. Genet.* **2004**, *41*, 585–590. doi: 10.1136/jmg.2004.019190. Erratum in: J Med Genet. **2005**, *42(3)*, 288.

48. Meulenbelt, I.; Bos, S.D.; Chapman, K.; van der Breggen, R.; Houwing-Duistermaat, J.J.; Kremer, D.; Kloppenburg, M.; Carr, A.; Tsezou, A.; González, A.; Loughlin, J.; Slagboom, P.E. Meta-analyses of Genes Modulating Intracellular T3 Bio-availability Reveal a Possible Role for the DIO3 Gene in Osteoarthritis Susceptibility. *Ann. Rheum. Dis.* **2011**, *70* (1), 164–167. doi: 10.1136/ard.2010.133660.
49. Liu, Y.Y.; Brent, G.A. Thyroid Hormone and the Brain: Mechanisms of Action in Development and Role in Protection and Promotion of Recovery After Brain Injury. *Pharmacol. Ther.* **2018**, *186*, 176–185.
50. Vidart, J.; Axelrud, L.; Braun, A.C.; Marschner, R.A.; Wajner, S.M. Relationship among Low T3 Levels, Type 3 Deiodinase, Oxidative Stress, and Mortality in Sepsis and Septic Shock: Defining Patient Outcomes. *Int. J. Mol. Sci.* **2023**, *24*, 3935.
51. He, Q.; Li, H.; Meng, F.; Sun, X.; Feng, X.; Chen, J.; Li, L.; Liu, J. Methionine Sulfoxide Reductase B1 Regulates Hepatocellular Carcinoma Cell Proliferation and Invasion via the Mitogen-Activated Protein Kinase Pathway and Epithelial-Mesenchymal Transition. *Oxid. Med. Cell Longev.* **2018**, *2018*, 5287971.
52. Lee, H.-J.; Park, J.S.; Yoo, H.J.; Lee, H.M.; Lee, B.C.; Kim, J.H. The Selenoprotein MsrB1 Instructs Dendritic Cells to Induce T-Helper 1 Immune Responses. *Antioxidants* **2020**, *9*, 1021.
53. Carlisle, A.E.; Lee, N.; Matthew-Onabanjo, A.N.; Spears, M.E.; Park, S.J.; Youkana, D.; Doshi, M.B.; Peppers, A.; Li, R.; Joseph, A.B.; Smity, M.; Simin, K.; Zhu, L.J.; Greer, P.L.; Shaw, L.M.; Kim, D. Selenium detoxification is required for cancer-cell survival. *Nat. Metab.* 2020, *2*, 603–611. https://doi.org/10.1038/s42255-020-0224-7
55. Flowers, B.; Bochnacka, O.; Poles, A.; Diamond, A.M.; Kastrati, I. Distinct Roles of SELENOF in Different Human Cancers. *Biomolecules* **2023**, *13*, 486.
56. Li, M.; Zhang, Y.; Zhou, J.; Liu, H. Selenoprotein F Knockout Caused Glucose Metabolism Disorder in Young Mice by Disrupting Redox Homeostasis. *Antioxidants* **2022**, *11*, 2105.
57. Bertz, M.; Kühn, K.; Koeberle, S.C.; Müller, M.F.; Hoelzer, D.; Thies, K.; Deubel, S.; Thierbach, R.; Kipp, A.P. Selenoprotein H Controls Cell Cycle Progression and Proliferation of Human Colorectal Cancer Cells. *Free Radic. Biol. Med.* **2018**, *127*, 98–107.
58. Sarma, A.S.; Siddardha, B.; Pragna Lakshmi, T.; Ranganath, P.; Dalal, A. A Novel Homozygous Synonymous Splicing Variant in SELENOI Gene Causes Spastic Paraplegia 81. *J. Gene Med.* **2023**, *2023*, e3501.
59. Zhang, Z.H.; Song, G.L. Roles of Selenoproteins in Brain Function and the Potential Mechanism of Selenium in Alzheimer's Disease. *Front. Neurosci.* 2021, *15*, 646518.
60. Abdurahman, A.; Li, Y.; Jia, S.-Z.; Xu, X.-W.; Lin, S.-J.; Ouyang, P.; He, Z.J.; Zhang, Z.-H.; Liu, Q.; Xu, Y.; Song, G.L. Knockdown of the SELENOK Gene Induces Ferroptosis in Cervical Cancer Cells. *Metallomics* **2023**, *15*, mfad019. doi: 10.1093/mtomcs/mfad019.
61. Rogachev, V.V.; Goltyaev, M.V.; Varlamova, E.G.; Turovsky, E.A. Molecular Mechanisms of the Cytotoxic Effect of Recombinant Selenoprotein SELENOM on Human Glioblastoma Cells. *Int. J. Mol. Sci.* **2023**, *24*, 6469.
62. Cai, J.; Huang, J.; Yang, J.; Chen, X.; Zhang, H.; Zhu, Y.; Liu, Q.; Zhang, Z. The Protective Effect of Selenoprotein M on Non-alcoholic Fatty Liver Disease: The Role of the AMPKα1-MFN2 Pathway and Parkin Mitophagy. *Cell Mol. Life Sci.* **2022**, *79*, 354.
63. Lin, S.; Chen, C.; Ouyang, P.; Cai, Z.; Liu, X.; Abdurahman, A.; Peng, J.; Li, Y.; Zhang, Z.; Song, G.-L. SELENOM Knockout Induces Synaptic Deficits and

Cognitive Dysfunction by Influencing Brain Glucose Metabolism. *J. Agric. Food Chem.* **2023**, *71*, 1607–1619.

64. Pozzer, D.; Varone, E.; Chernorudskiy, A.; Schiarea, S.; Missiroli, S.; Giorgi, C.; Pinton, P.; Canato, M.; Germinario, E.; Nogara, L.; Blaauw, B.; Zito, E. A Maladaptive ER Stress Response Triggers Dysfunction in Highly Active Muscles of Mice with SELENON Loss. *Redox Biol.* **2019**, *20*, 354–366. doi: 10.1016/j.redox.2018.10.017.
65. Zhao, Y.; Chen, P.; Lv, H.-J.; Wu, Y.; Liu, S.; Deng, X.; Shi, B.; Fu, J. Comprehensive Analysis of Expression and Prognostic Value of Selenoprotein Genes in Thyroid Cancer. *Genet. Test. Mol. Biomarkers* 2022, *26*, 159–173.
66. Wang, C.; Yu, Q.; Song, T.; Wang, Z.; Song, L.; Yang, Y.; Shao, J.; Li, J.; Ni, Y.; Chao, N.; Zhang, Li.; Li, W. The Heterogeneous Immune Landscape Between Lung Adenocarcinoma and Squamous Carcinoma Revealed by Single-cell RNA Sequencing. *Signal. Transduct. Target. Ther.* **2022**, *7*, 289.
67. Short, S.P.; Pilat, J.M.; Barrett, C.W.; Reddy, V.K.; Haberman, Y.; Hendren, J.R.; Marsh, B.J.; Keating, C.E.; Motley, A.K.; Hill, K.E.; Zemper, A.E.; Washington, M.K.; Shi, C.; Chen, X.; Wilson, K.T.; Hyams, J.S.; Denson, L.A.; Burk, R.F.; Rosen, M.J.; Williams, C.S. Colonic Epithelial-Derived Selenoprotein P Is the Source for Antioxidant-Mediated Protection in Colitis-Associated Cancer. *Gastroenterology* **2021**, *160* (5), 1694–1708.e3. doi: 10.1053/j.gastro.2020.12.059.
68. Ekoue, D.N.; Ansong, E.; Liu, L.; Macias, V.; Deaton, R.; Lacher, C.; Picklo, M.; Nonn, L.; Gann, P.H.; Kajdacsy-Balla, A.; Prins, G.S.; Freeman, V.L.; Diamond, A.M. Correlations of SELENOF and SELENOP Genotypes with Serum Selenium Levels and Prostate Cancer. *Prostate* **2018**, *78*, 279–288. doi: 10.1002/pros.23471.
69. Schweizer, U.; Wirth, E.K.; Klopstock, T.; Hölter, S.M.; Becker, L.; Moskovitz, J.; Grune, T.; Fuchs, H.; Gailus-Durner, V.; de Angelis, M.H.; Köhrle, J.; Schomburg, L. Seizures, Ataxia and Parvalbumin-expressing Interneurons Respond to Selenium Supply in Selenop-deficient Mice. *Redox Biol.* **2022**, *57* , 102490. doi: 10.1016/j.redox.2022.102490.
70. Schomburg, L.; Orho-Melander, M.; Struck, J.; Bergmann, A.; Melander, O. Selenoprotein-P Deficiency Predicts Cardiovascular Disease and Death. *Nutrients* **2019**, *11*, 1852.
71. Santos, L.R.; Durães, C.; Ziros, P.G.; Pestana, A.; Esteves, C.; Neves, C.; Carvalho, D.; Bongiovanni, M.; Renaud, C.O.; Chartoumpekis, D.V.; Habeos, I.G.; Simões, M.S.; Soares, P.; Sykiotis, G.P. Interaction of Genetic Variations in NFE2L2 and SELENOS Modulates the Risk of Hashimoto's Thyroiditis. *Thyroid* **2019**, *29*, 1302–1315. doi: 10.1089/thy.2018.0480.
72. Strauss, E.; Tomczak, J.; Staniszewski, R.; Oszkinis, G. Associations and Interactions Between Variants in Selenoprotein Genes, Selenoprotein Levels and the Development of Abdominal Aortic Aneurysm, Peripheral Arterial Disease, and Heart Failure. *PLoS ONE* **2018**, *13*, e0203350.
73. Varlamova, E.G.; Goltyaev, M.V.; Turovsky, E.A. The Role of Selenoproteins SELENOM and SELENOT in the Regulation of Apoptosis, ER Stress, and Calcium Homeostasis in the A-172 Human Glioblastoma Cell Line. *Biology* **2022**, *11*, 811.
74. Alsharif, I.; Boukhzar, L.; Lefranc, B.; Godefroy, D.; Aury-Landas, J.; Rego, J.D.; Rego, J.D.; Naudet, F.; Arabo, A.; Chagraoui, A.; Maltête, D.; Benazzouz, A.; Baugé, C.; Leprince, J.; Elkahloun, A.G.; Eiden, L.E.; Anouar, Y. Cell-penetrating, Antioxidant SELENOT Mimetic Protects Dopaminergic Neurons and Ameliorates Motor Dysfunction in Parkinson's Disease Animal Models. *Redox Biol.* **2021**, *40*, 101839. doi: 10.1016/j.redox.2020.101839.

75. Rocca, C.; Pasqua, T.; Boukhzar, L.; Anouar, Y.; Angelone, T. Progress in the Emerging Role of Selenoproteins in Cardiovascular Disease: Focus on Endoplasmic Reticulum-resident Selenoproteins. *Cell Mol. Life. Sci.* **2019**, *76*, 3969–3985.
76. Kim, H.; Lee, K.; Kim, J.M.; Kim, M.Y.; Kim, J.-R.; Lee, H.-W.; Chung, Y.W.; Shin, H.-I.; Kim, T.; Park, E.-S.; Rho, J.; Lee, S.H.; Kim, N.; Lee S.Y.; Choi, Y.; Jeong, D. Selenoprotein W Ensures Physiological Bone Remodeling by Preventing Hyperactivity of Osteoclasts. *Nat. Commun.* **2021**, *12*, 2258. https://doi.org/10.1038/s41467-021-22565-7
77. Liao, C.; Hardison, R.C.; Kennett, M.J.; Carlson, B.A.; Paulson, R.F.; Prabhu, K.S. Selenoproteins Regulate Stress Erythroid Progenitors and Spleen Microenvironment During Stress Erythropoiesis. *Blood* **2018**, *131*, 2568–2580.
78. Shreenath, A.P.; Hashmi, M.F.; Dooley, J. Selenium Deficiency. In: StatPearls. Internet. https://www.ncbi.nlm.nih.gov/books/NBK482260/; StatPearls Publishing, **2024** PMID: 29489289. Available from: https://www.ncbi.nlm.nih.gov/books/NBK482260/
79. (a) SenGupta, A.; Lichti, U. F.; Carlson, B. A.; Ryscavage, A. O.; Gladyshev, V. N.; Yuspa, S. H.; Hatfield, D. L. Selenoproteins Are Essential for Proper Keratinocyte Function and Skin Development. *PLoS One* 2010, *5* (8), e12249. DOI: 10.1371/journal.pone.0012249. (b) Kanekura, T.; Yotsumoto, S.; Maeno, N.; Kamenosono, A.; Saruwatari, H.; Uchino, Y.; Mera, Y.; Kanzaki, T. Selenium Deficiency: Report of a Case. *Clin. Exp. Dermatol.* **2005**, *30* (4), 346–348. DOI: 10.1111/j.1365-2230.2005.01746.x
80. a) McKenzie, R. C. Selenium, Ultraviolet Radiation and the Skin. *Clin. Exp. Dermatol.* **2000**, *25* (8), 631–636. DOI: 10.1046/j.1365-2230.2000.00725.x (b) Fleming, J.; Ghose, A.; Harrison, P. R. Molecular Mechanisms of Cancer Prevention by Selenium Compounds. *Nutr. Cancer* **2001**, *40* (1), 42–49. DOI: 10.1207/S15327914NC401_9
81. Rayman, M. P. The Importance of Selenium to Human Health. *Lancet* **2000**, *356* (9225), 233–241. DOI: 10.1016/S0140-6736(00)02490-9
82. a) Drutel, A.; Archambeaud, F.; Caron, P. Selenium and the Thyroid Gland: More Good News for Clinicians. *Clin. Endocrinol. (Oxf.)* 2013, *78* (2), 155–164. DOI: 10.1111/cen.12066. (b) Lacka, K.; Szeliga, A. Significance of Selenium in Thyroid Physiology and Pathology. *Pol. Merkur. Lekarski* **2015**, *38* (228), 348–353.
83. Ventura, M.; Melo, M.; Carrilho, F. Selenium and Thyroid Disease: From Pathophysiology to Treatment. *Int. J. Endocrinol.* **2017**, *2017*, 1297658. DOI: 10.1155/2017/1297658.
84. McGregor, B. The Role of Selenium in Thyroid Autoimmunity: A Review. *J. Restor. Med.* **2015**, *4* (1), 83–92. DOI: 10.14200/jrm.2015.4.0102
85. Guo, X.; Ning, Y.; Wang, X. In *Selenium: Chemistry, Analysis, Function and Effects*, Preedy, V. R., Ed., Royal Society of Chemistry, **2015**, pp. 552–571.
86. Cui, S.; Que, W.; Jiao, Z.; Deng, Q.; Zhang, X.; Cao, Y.; Liu, N.; Li. A.; Sowanou, A.; Li, Z.; Wang, T.; Li, Y.; Yu, J.; Pei, J. Disease and Economic Burden of Kashin-Beck Disease—China, 2021. *China CDC Wkly.* **2024**, *6*(2),40–44. DOI: 10.46234/ccdcw2024.009
87. Schepman, K.; Engelbert, R. H.; Visser, M. M.; Yu, C.; de Vos, R. Kashin Beck Disease: More than Just Osteoarthrosis: A Cross-Sectional Study Regarding the Influence of Body Function-Structures and Activities on Level of Participation. *Int. Orthop.* **2011**, *35* (5), 767–776. DOI: 10.1007/s00264-010-1043-3
88. Mathieu, F.; Begaux, F.; Suetens, C.; De Maertelaer, V.; Hinsenkamp, M. Anthrometry and Clinical Features of Kashin Beck Disease in Central Tibet. *Int. Orthop.* **2001**, *25* (3), 138–141. DOI: 10.1007/s002640000192.

89. Xiong, G. Diagnostic, Clinical and Radiological Characteristics of Kashin Beck Disease in Shaanxi Province, PR China. *Int. Orthop.* **2001**, *25* (3), 147–150. DOI: 10.1007/s002640100248.
90. Guo, X.; Ma, W. J.; Zhang, F.; Ren, F. L.; Qu, C. J.; Lammi, M. J. Recent Advances in the Research of an Endemic Osteochondropathy in China: Kashin-Beck Disease. *Osteoarthr. Cartil.* **2014**, *22* (*11*), 1774–1783. DOI: 10.1016/j.joca.2014.07.023
91. Wen, Y.; Wang, B.; Shi, P.; Chu, X.; Shi, S.; Yao, Y.; Zhang, L.; Zhang, F. A Metabolomics Study of Feces Revealed That a Disturbance of Selenium-Centered Metabolic Bioprocess Was Involved in Kashin-Beck Disease, an Osteoarthropathy Endemic to China. *Nutrients.* **2023,** *15(21),* 4651. doi: 10.3390/nu15214651.
92. (a) Nazıroğlu, M.; Öz, A.; Yıldızhan, K. Selenium and Neurological Diseases: Focus on Peripheral Pain and TRP Channels. *Curr. Neuropharmacol.* 2020, *18* (6), 501–517. DOI: 10.2174/1570159X18666200106152631. (b) Hong, L. K.; Diamond, A. M. Selenium. *Present Knowl. Nutr.* **2020**, 443–456. DOI: 10.1016/b978-0-323-66162-1.00026-3
93. Schweizer, U.; Bohleber, S.; Zhao, W.; Fradejas-Villar, N. The Neurobiology of Selenium: Looking Back and to the Future. *Front Neurosci.* **2021,** *15*, 652099. doi: 10.3389/fnins.2021.652099. PMID: 33732108; PMCID: PMC7959785.
94. Zhang, S.; Rocourt, C.; Cheng, W.H. Selenoproteins and the Aging Brain. *Mech Ageing Dev.* **2010,** *131*(4), 253-60. doi: 10.1016/j.mad.2010.02.006.
95. Cardoso, B. R.; Roberts, B. R.; Malpas, C. B.; Vivash, L.; Genc, S.; Saling, M. M.; Desmond, P.; Steward, C.; Hicks, R. J.; Callahan, J.; Brodtmann, A.; Collins, S.; Macfarlane, S.; Corcoran, N. M.; Hovens, C. M.; Velakoulis, D.; O'Brien, T. J.; Hare, D. J.; Bush, A. I. Supranutritional Sodium Selenate Supplementation Delivers Selenium to the Central Nervous System: Results from a Randomized Controlled Pilot Trial in Alzheimer's Disease. *Neurotherapeutics* **2019**, *16* (1), 192–202. DOI: 10.1007/s13311-018-0662-z
96. Zhang, Z. H.; Peng, J. Y.; Chen, Y. B.; Wang, C.; Chen, C.; Song, G. L. Different Effects and Mechanisms of Selenium Compounds in Improving Pathology in Alzheimer's Disease. *Antioxidants (Basel)* **2023**, *12* (3), 702. DOI: 10.3390/antiox12030702
97. (a) Barrett, C. W.; Reddy, V. K.; Short, S. P.; Motley, A. K.; Lintel, M. K.; Bradley, A. M.; Freeman, T.; Vallance, J.; Ning, W.; Parang, B.; Poindexter, S. V.; Fingleton, B.; Chen, X.; Washington, M. K.; Wilson, K. T.; Shroyer, N. F.; Hill, K. E.; Burk, R. F.; Williams, C. S. Selenoprotein P Influences Colitis-Induced Tumorigenesis by Mediating Stemness and Oxidative Damage. *J. Clin. Investig.* **2015**, *125 (7),* 2646–2660. DOI: 10.1172/JCI76099. (b) Hamid, M.; Abdulrahim, Y.; Liu, D.; Qian, G.; Khan, A.; Huang, K. The Hepatoprotective Effect of Selenium-Enriched Yeast and Gum Arabic Combination on Carbon Tetrachloride-Induced Chronic Liver Injury in Rats. *J. Food Sci.* **2018**, *83(2)*, 525–534. DOI: 10.1111/1750-3841.14030
98. Broome, C. S.; McArdle, F.; Kyle, J. A.; Andrews, F.; Lowe, N. M.; Hart, C. A.; Arthur, J. R.; Jackson, M. J. An Increase in Selenium Intake Improves Immune Function and Poliovirus Handling in Adults with Marginal Selenium Status. *Am. J. Clin. Nutr.* **2004**, *80* (*1*), 154–162. DOI: 10.1093/ajcn/80.1.154
99. Perri, G.; Mendonça, N.; Jagger, C.; Walsh, J.; Eastell, R.; Mathers, J. C.; Hill, T. R. Dietary Selenium Intakes and Musculoskeletal Function in Very Old Adults: Analysis of the Newcastle 85+ Study. *Nutrients* **2020,** *12* (*7*), 2068. DOI: 10.3390/nu12072068
100. Schoenmakers, E.; Carlson, B.; Agostini, M.; Moran, C.; Rajanayagam, O.; Bochukova, E.; Tobe, R.; Peat, R.; Gevers, E.; Muntoni, F.; Guicheney, P.; Schoenmakers, N.; Farooqi, S.; Lyons, G.; Hatfield, D.; Chatterjee, K. Mutation in

Human Selenocysteine Transfer RNA Selectively Disrupts Selenoprotein Synthesis. *J. Clin. Investig.* **2016**, *126* (*3*), 992–996. doi: 10.1172/JCI84747.

101. Mistry, H. D.; Kurlak, L. O., Chapter 24. Selenium in Fertility and Reproduction. In *Handbook of Fertility*, Watson, R. R., Ed.; Academic Press, **2015**, pp. 261–272.

102. Morabbi, A.; Karimian, M. Trace and Essential Elements as Vital Components to Improve the Performance of the Male Reproductive System: Implications in Cell Signaling Pathways. *J. Trace Elem. Med. Biol.* **2024**, *83*, 127403, DOI: 10.1016/j.jtemb.2024.127403

103. (a) Coskun, A.; Arikan, T.; Kilinc, M.; Arikan, D. C.; Ekerbiçer, H. Ç. Plasma Selenium Levels in Turkish Women with Polycystic Ovary Syndrome. *Eur. J. Obstet. Gynecol. Reprod. Biol.* **2013**, *168* (2), 183–186. DOI: 10.1016/j.ejogrb.2013.01.021. (b) Sharma, P.; Khetarpal, P. Genetic Determinants of Selenium Availability, Selenium-Response, and Risk of Polycystic Ovary Syndrome. *Biol. Trace Elem. Res.* **2024**. DOI: 10.1007/s12011-023-04052-w (published online).

104. Lener, M. R.; Gupta, S.; Scott, R. J.; Tootsi, M.; Kulp, M.; Tammesoo, M. L.; Viitak, A.; Metspalu, A.; Serrano-Fernández, P.; Kładny, J.; Jaworska-Bieniek, K.; Durda, K.; Muszyńska, M.; Sukiennicki, G.; Jakubowska, A.; Lubiński, J. Can Selenium Levels Act as a Marker of Colorectal Cancer Risk? *BMC Cancer* **2013**, *13*, 214. DOI: 10.1186/1471-2407-13-214. (b) Vinceti, M.; Crespi, C. M.; Malagoli, C.; Del Giovane, C.; Krogh, V. Friend or Foe? The Current Epidemiologic Evidence on Selenium and Human Cancer Risk. *J. Environ. Sci. Health C Environ. Carcinog. Ecotoxicol. Rev.* **2013**, *31* (4), 305–341. DOI: 10.1080/10590501.2013.844757

105. (a) Rataan, A. O.; Geary, S. M.; Zakharia, Y.; Rustum, Y. M.; Salem, A. K. Potential Role of Selenium in the Treatment of Cancer and Viral Infections. *Int. J. Mol. Sci.* **2022**, *23* (*4*), 2215. DOI: 10.3390/ijms23042215. (b) Hu, X.; Chandler, J.D.; Fernandes, J.; Orr, M.L.; Hao, L.; Uppal, K.; Neujahr, D.C.; Jones, D.P.; Go, Y.M. Selenium Supplementation Prevents Metabolic and Transcriptomic Responses to Cadmium in Mouse Lung. *Biochim Biophys Acta Gen Subj.* **2018**,*1862*(*11*), 2417–2426. doi: 10.1016/j.bbagen.2018.04.009. (c) Zeng, H.; Combs, G. F. Selenium as an Anticancer Nutrient: Roles in Cell Proliferation and Tumor Cell Invasion. *J. Nutr. Biochem.* **2008**, *19* (1), 1–7. DOI: 10.1016/j.jnutbio.2007.02.005

106. (a) Liao, G.; Tang, J.; Wang, D.; Zuo, H.; Zhang, Q.; Liu, Y.; Xiong, H. Selenium Nanoparticles (SeNPs) Have Potent Antitumor Activity Against Prostate Cancer Cells Through the Upregulation of MiR-16. *World J Surg Onc* **2020**, *18*, 81. https://doi.org/10.1186/s12957-020-01850-7 (b) Martínez-Esquivias, F.; Gutiérrez-Angulo, M.; Pérez-Larios, A.; Sánchez-Burgos, J.A.; Becerra-Ruiz, J.S.; Guzmán-Flores, J.M. Anticancer Activity of Selenium Nanoparticles In Vitro Studies. *Anticancer Agents Med Chem.* **2022**, *22*(9), 1658–1673. doi: 10.2174/1871520621666210910084216. c) El-Fakharany, E.M.; Abu-Serie, M.M.; Ibrahim, A. Eltarahony, M. Anticancer Activity of Lactoferrin-coated Biosynthesized Selenium Nanoparticles for Combating Different Human Cancer Cells Via Mediating Apoptotic Effects. *Sci Rep* **2023**, *13*, 9579. https://doi.org/10.1038/s41598-023-36492-8

107. (a) Casanova, P.; Monleon, D. Role of Selenium in Type 2 Diabetes, Insulin Resistance and Insulin Secretion. *World J Diabetes* **2023**; *14(3)*,147–158. (b) Steinbrenner, H.; Duntas, L.H.; Rayman, M.P. The Role of Selenium in Type-2 Diabetes Mellitus and Its Metabolic Comorbidities, *Redox Biol.* **2022**, *50*, 102236.

108. (a) Mita, Y.; Nakayama, K.; Inari, S.; Nishito, Y.; Yoshioka, Y.; Sakai, N.; Sotani, K.; Nagamura, T.; Kuzuhara, Y.; Inagaki, K.; Iwasaki, M.; Misu, H.; Ikegawa, M.; Takamura, T.; Noguchi, N.; Saito, Y. Selenoprotein P-neutralizing Antibodies

Improve Insulin Secretion and Glucose Sensitivity in Type 2 Diabetes Mouse Models. *Nat Commun* **2017**, 8, 1658. https://doi.org/10.1038/s41467-017-01863-z. (b) Zhao, J.; Zou, H.; Huo, Y.; Wei, X.; Li, Y. Emerging Roles of Selenium on Metabolism and Type 2 Diabetes. *Front. Nutr.* 2022, 9, 1027629. https://doi.org/10.3389/fnut.2022.1027629. (c) Campbell, J.E.; Drucker, D.J. Islet A Cells and Glucagon—Critical Regulators of Energy Homeostasis. *Nat Rev Endocrinol.* **2015**, 11(6), 329–38. doi: 10.1038/nrendo.2015.51.

109. (a) Onoguchi, A.; Granata, G.; Haraguchi, D.; Hayashi, H.; Tokoro, C. Kinetics and Mechanism of Selenate and Selenite Removal in Solution by Green Rust-sulfate. *R. Soc. open sci.* **2019**, *6,* 182147. http://dx.doi.org/10.1098/rsos.182147 (b) Schrauzer, G.N. The Nutritional Significance, Metabolism and Toxicology of Selenomethionine. *Adv. Food Nutr. Res.* 2003, *47,* 73–112. doi: 10.1016/s1043-4526(03)47002-2. (c) Eisler, R (2007) *Eisler's Encyclopedia of Environmentally Hazardous Priority Chemicals*. Elsevier. (d) Somogyi, Z. ; Kiss, I.; Kádár, I. ; Bakonyi, G. Toxicity of Selenate and Selenite to the Potworm. *Enchytraeus albidus* (Annelida: Enchytraeidae): A Laboratory Test. *Ecotoxicology* **2007**, *16*, 379–384. https://doi.org/10.1007/s10646-007-0140-6. (e) Mehdi, Y.; Hornick, L.; Istasse, L. ; Dufrasne, I. Selenium in the Environment, Metabolism and Involvement in Body Functions. *Molecules*, 2013, *18*(3), 3292–3311. https://doi.org/10.3390/molecules18033292. (f) Garousi, F. The Toxicity of Different Selenium Forms and Compounds–Review. *Acta Agraria Debreceniensis*, **2015**, *64*, 33–38. https://doi.org/10.34101/actaagrar/64/1859

110. Cai, J.; Su, W.; Chen, X.; Zheng, H. Advances in the Study of Selenium and Human Intestinal Bacteria. *Front Nutr.* **2022,** *9*, 1059358. doi: 10.3389/fnut.2022.1059358.

111. Ferreira, R.L.U.; Sena-Evangelista, K.C.M.; de Azevedo, E.P.; Pinheiro, F.I.; Cobucci, R.N.; Pedrosa, L.F.C. Selenium in Human Health and Gut Microflora: Bioavailability of Selenocompounds and Relationship with Diseases. *Front Nutr.* **2021**, *8*, 685317. doi: 10.3389/fnut.2021.685317.

112. Satarzadeh, N.; Dousari, A.S.; Amirheidari, B.; Shakibaie, M.; Sarbandi, A.R.; Forootanfar, H. An Insight into Biofabrication of Selenium Nanostructures and Their Biomedical Application. *3 Biotech.* **2023**,13(3),79. doi: 10.1007/s13205-023-03476-4.

113. (a) Pouri, S.; Motamedi, H.; Honary, S.; Iraj, K. Biological Synthesis of Selenium Nanoparticles and Evaluation of Their Bioavailability. *Braz. Arch. Biol. Technol.* **2018**, 60: e17160452. doi: 10.1590/1678-4324-2017160452. (b) Wang, H.; Zhang, J.; Yu, H. Elemental Selenium at Nano Size Possesses Lower Toxicity Without Compromising the Fundamental Effect on Selenoenzymes: Comparison with Selenomethionine in Mice. *Free Radic Biol Med.* **2007**, 42:1524–33. doi: 10.1016/j.freeradbiomed.2007.02.013 (c) Li, Q.; Luo, Y.; Sha, A.; Xiao, W.; Xiong, Z.; Chen, X.; He, J.; Peng, L.;Zou, L. Analysis of Synonymous Codon Usage Patterns in Mitochondrial Genomes of Nine Amanita Species. *Front Microbiol.* **2023**, 14, 1134228. doi: 10.3389/fmicb.2023.1134228. (d) Yang, F.; Tang, Q.; Zhong, X.; Bai, Y.; Chen, T.; Zhang, Y.; Li, Y.; Zheng, W. Surface Decoration by Spirulina Polysaccharide Enhances the Cellular Uptake and Anticancer Efficacy of Selenium Nanoparticles. *Int J Nanomedicine.* **2012**, 7, 835–44. doi: 10.2147/IJN.S28278. (e) Tian, J.; Wei, X.; Zhang, W.; Xu, A. Effects of Selenium Nanoparticles Combined

With Radiotherapy on Lung Cancer Cells. *Front Bioeng Biotechnol.* 2020, 8, 598997. doi: 10.3389/fbioe.2020.598997. (f) Gao, F.; Yuan, Q.; Gao, L.; Cai, P.; Zhu, H.; Liu, R.; Wang, Y.; Wei, Y.; Huang, G.; Liang, J.; Gao, X. Cytotoxicity and Therapeutic Effect of Irinotecan Combined with Selenium Nanoparticles. *Biomaterials.* **2014**, 35(31), 8854–8866. doi: 10.1016/j.biomaterials.2014.07.004. g) Nayak, V.; Singh, K.R.B.; Singh, A.K.; Singh, R.P. Potentialities of Selenium Nanoparticles in Biomedical Science. *New J. Chem.*, 2021,45, 2849–2878, doi:10.1039/d0nj05884j

114. Kieliszek, M.; Bano, Iqra. Selenium as an Important Factor in Various Disease States a Review. *EXCLI J.* **2022**, *21*. 948–966. doi: 10.17179/excli2022-5137

8 Role of Vanadium Compounds on Human Health

Sangeeta Bajpai, Monika Kamboj, Garima Pandey and Bimal Krishan Banik

8.1 INTRODUCTION

In trace amounts, vanadium is naturally present in the environment, primarily in rocks, soil, water, and various foods, including mushrooms, shellfish, and black pepper. Its physiological relevance lies in its potential impact on numerous biological processes, such as enzymatic reactions, glucose metabolism, and cellular signalling. This chapter delves into the multifaceted role of vanadium compounds in human health, focusing on both their beneficial and adverse effects.

Vanadium, an element with the atomic number 23, has captured scientific interest for over two centuries. It was first identified in 1801 by the Spanish mineralogist Andrés Manuel del Río in Mexico. Initially mistaken for chromium, it was later rediscovered in 1830 by Swedish chemist Nils Gabriel Sefstrom, who named it after the Norse goddess Vanadis, symbolizing beauty due to its colorful compounds. Vanadium's unique chemical versatility, stemming from its multiple oxidation states, makes it indispensable in both industrial and biological contexts. Vanadium ranks as the 22nd most abundant element in Earth's crust, primarily occurring in minerals such as vanadinite and carnotite. Significant industrial applications of vanadium include the following:

- Steel manufacturing: As an alloy, vanadium enhances the strength, toughness, and heat resistance of steel.
- Catalysis: Vanadium pentoxide (V_2O_5) serves as a catalyst in the production of sulfuric acid and in petrochemical processes.

8.1.1 Biological Relevance of Vanadium

Trace amounts of vanadium are found in various organisms, indicating a possible physiological role. In humans, vanadium is present in bones, teeth, and blood [1]. Although not officially recognized as essential, vanadium influences several biological processes:

 DOI: 10.1201/9781032615110-11

- Modulation of enzymatic activity
- Regulation of phosphate metabolism
- Potential insulin-mimetic effects.

Current research explores vanadium's role in health and disease. Studies investigate its potential in treating metabolic disorders, cancer, and neurodegenerative diseases. Moreover, its environmental impact and sustainable use are emerging as significant areas of study.

8.1.2 Vanadium Chemistry

Vanadium's chemical versatility stems from its ability to exist in multiple oxidation states, ranging from –1 to +5. This flexibility allows vanadium to form a wide variety of compounds, each with unique physical and chemical properties relevant to biological and industrial processes. Vanadium is most commonly found in the +5, + 4, and +3 oxidation states in biological and environmental systems. Each oxidation state exhibits distinct chemical behaviors and colors, making vanadium compounds visually diverse and analytically significant (Table 8.1).

These oxidation states can interconvert under physiological conditions, a property that is critical to vanadium's biological activity. Vanadium forms stable complexes with a variety of ligands, including:

- *Oxygen donors*: for example, vanadate (VO_4^{3-}) and vanadyl (VO^{2+}) species.
- *Nitrogen donors*: for example, amine complexes.
- *Sulfur donors*: for example, thiolate complexes.

These complexes exhibit diverse geometries, such as octahedral and tetrahedral, which influence their reactivity and biological functions (Figure 8.1).

TABLE 8.1
Chemical behavior of select vanadium compounds

Oxidation State	Compound Example	Color	Key Properties
+5	Ammonium metavanadate	Yellow	Strongly oxidizing; prevalent in biological systems
+4	Vanadyl sulfate (VO^{2+})	Blue	Moderately oxidizing; interacts with biomolecules
+3	Vanadium trichloride (VCl_3)	Green	Reducing agent; less common biologically
+2	Vanadium(II) chloride (VCl_2)	Violet	Strong reducing agent; unstable in aqueous environments

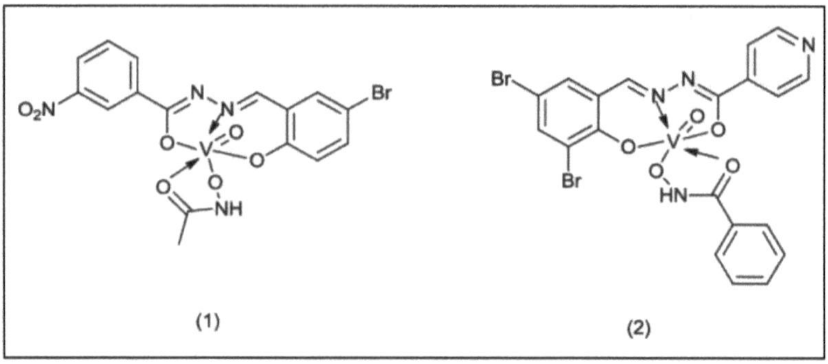

FIGURE 8.1 Structures of vanadium complexes.

8.2 VANADIUM IN BIOLOGICAL SYSTEMS

Vanadium's redox-active nature makes it a key player in biological redox processes. For instance, vanadates (V^{5+}) often mimic phosphate groups, allowing them to inhibit or activate phosphatases and kinases. Vanadyl ions (V^{4+}) are known to interact with proteins and DNA, influencing metabolic and regulatory pathways.

Vanadium plays a significant role in various biological systems, primarily due to its versatile oxidation states, which range from –1 to +5. This ability allows it to participate in diverse biochemical processes, including redox reactions and enzymatic catalysis. One of the most well-documented biological roles of vanadium is its involvement in nitrogen fixation in microorganisms. Certain bacteria, such as *Azotobacter* and *Anabaena*, utilize vanadium-dependent nitrogenases, which help convert atmospheric nitrogen into ammonia [2]. This process is crucial for maintaining the nitrogen cycle and supporting plant growth in ecosystems. Vanadium's role in these microorganisms highlights its importance in global biochemical processes and agricultural productivity.

In addition to its role in microbial systems, vanadium is also found in various marine organisms, particularly in tunicates and algae. For instance, vanadium bromoperoxidase (V-BrPO) is an enzyme present in marine algae like *Ascophyllum nodosum*, where it catalyzes the formation of brominated compounds. These compounds serve as natural defense mechanisms against predators and microbial infections. In higher organisms, vanadium has been shown to mimic insulin's action by regulating glucose metabolism. Research indicates that vanadium compounds, such as vanadyl sulfate, can improve glucose uptake in diabetic models, offering potential therapeutic benefits [3]. Despite these promising findings, the precise mechanisms by which vanadium exerts its biological effects are still under investigation, underscoring the need for further research into its physiological roles and therapeutic potential. Vanadium's biological role is multifaceted, influencing several physiological systems. Although its essentiality for humans remains under debate, its impact on metabolic processes, enzyme function, and cellular health is undeniable.

Vanadium is not officially recognized as an essential trace element, yet it is widely distributed across the human diet. It is most found in plant-based foods, such as

TABLE 8.2
Vanadium content in food

Food Source	Vanadium Content (μg/100g)
Mushrooms	20–40
Shellfish	50–120
Whole grains	10–30
Black pepper	5–10

mushrooms, whole grains, and vegetables, as well as in seafood. The concentration of vanadium in these foods can vary greatly depending on environmental factors like soil composition (Table 8.2).

The total daily intake of vanadium is estimated to range from 10–60 μg, which is far below levels known to cause toxicity. However, due to its potential bioactivity, it is important to understand the effects of both adequate and excessive vanadium consumption.

8.3 ABSORPTION, DISTRIBUTION, METABOLISM, AND EXCRETION (ADME) OF VANADIUM

Vanadium is absorbed through the gastrointestinal tract, primarily in its vanadyl (VO^{2+}) and vanadate (VO_4^{3-}) forms. However, its absorption is inefficient, with bioavailability estimates ranging from 1% to 5% [4]. Once absorbed, vanadium is carried in the bloodstream, bound to plasma proteins like transferrin and albumin, and is distributed to tissues, particularly the liver, kidneys, and bones (Figure 8.2).

Vanadium is metabolized by the liver, where it is converted into more stable forms, such as vanadate and vanadyl complexes. These metabolites can either be stored in tissues or excreted through the urine. The half-life of vanadium in human tissues is generally around 20–40 h.

8.3.1 Interaction with Biomolecules

Vanadium's biological effects are largely mediated through its interactions with various biomolecules. It mimics phosphate in cellular processes, affecting enzymes and proteins involved in cellular signaling and metabolism (Table 8.3).

- *Enzyme inhibition*: Vanadates inhibit protein tyrosine phosphatases (PTPs), which play a crucial role in regulating cell signalling.
- *Insulin-mimetic effects*: Vanadyl compounds have been shown to mimic insulin's action by activating insulin receptors and enhancing glucose uptake in cells.
- *Gene regulation*: Vanadyl complexes can modulate gene expression by interacting with DNA and transcription factors, which influence cellular growth and differentiation.

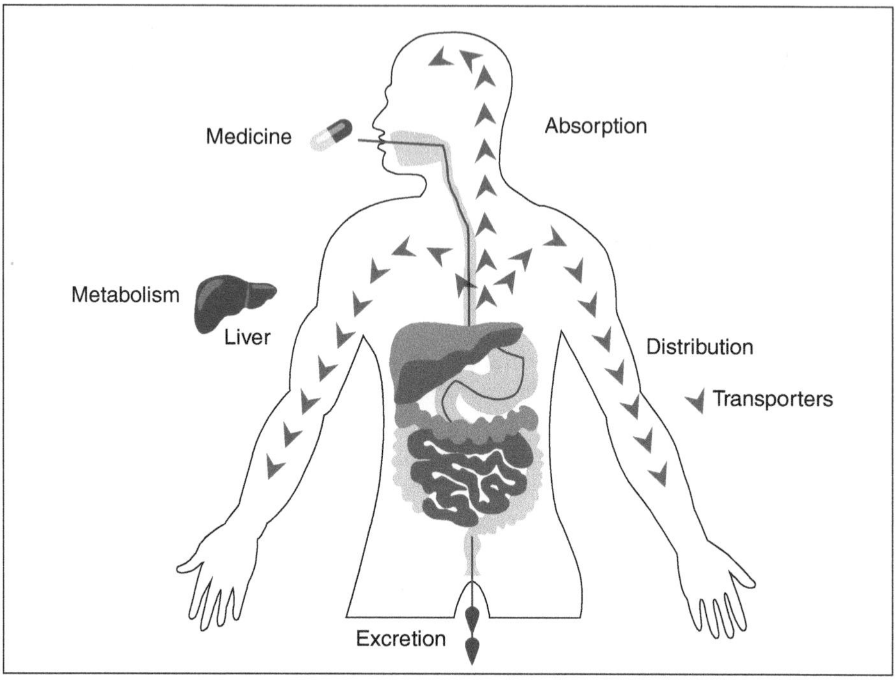

FIGURE 8.2 ADME process in human body.

TABLE 8.3
Vanadium interaction with key biomolecules

Biomolecule	Effect of Vanadium	Mechanism
Protein tyrosine phosphatases	Inhibition	Vanadates mimic phosphate groups, inhibiting phosphatases
Insulin receptors	Activation	Vanadyl complexes activate insulin receptor signalling
DNA and transcription factors	Modulation of gene expression	Vanadyl complexes bind DNA and influence transcription

Vanadium's interactions with biomolecules are diverse and depend on its oxidation state and coordination with ligands. One of the primary modes of interaction is through the formation of complexes with proteins and enzymes, which can modulate their activity. Vanadium ions can bind to the active sites of enzymes, altering their structural conformation and, consequently, their catalytic functions. For example, vanadium is known to interact with phosphate groups in proteins, which is crucial for the regulation of several cellular pathways [6]. This interaction is especially relevant in the context of enzyme regulation in signaling pathways such as those involved in insulin action and glucose metabolism. Vanadium's ability to mimic phosphate groups allows it to interfere with kinase activities, influencing phosphorylation cascades that regulate cellular functions such as cell growth, differentiation, and metabolism.

Additionally, vanadium compounds have been shown to interact with DNA and RNA, affecting gene expression and cellular processes like transcription and translation. Some studies have reported that vanadium can bind to nucleic acids, potentially altering their structure and affecting their ability to participate in cellular processes. This interaction may play a role in vanadium's observed biological effects, both beneficial and toxic. For example, vanadium's binding to nucleotides could influence cellular stress responses or contribute to oxidative damage, depending on the context of exposure and the specific vanadium compound involved [7]. Understanding these interactions in more detail could provide insights into how vanadium may be used therapeutically, as well as its potential risks when misused or overexposed.

Some key chemical reactions related to the interaction of vanadium with biomolecules.

1. Interaction with Proteins (Vanadium Binding to Amino Acids)

Vanadium compounds can interact with amino acid residues in proteins, forming complexes. A typical example involves vanadyl ion (VO^{2+}) binding to histidine or cysteine residues.

Reaction

$$VO^{2+} + 2L \longrightarrow [VO(L)_2]$$

where L represents a ligand such as histidine or cysteine.

- Histidine, with its imidazole group, coordinates with the vanadyl ion through the nitrogen atom, forming stable complexes.
- Similarly, cysteine, with its thiol group (–SH) binds through the sulfur atom.
- These interactions can influence protein function, either activating or inhibiting enzymatic activity.

2. Phosphate Mimicry in Enzymes

Vanadate ($\boldsymbol{VO_4}^{3-}$) mimics phosphate ($\boldsymbol{PO_4}^{3-}$) due to its structural similarity. This mimicry allows vanadate to interact with phosphate-binding sites in enzymes.

Reaction:

$$\boldsymbol{E.\,PO_4^{3-} \longleftrightarrow E.\,VO_4^{3-}}$$

where E is an enzyme such as ATPase or protein phosphatases.

- Vanadate can inhibit ATPase activity by occupying the phosphate site, blocking ATP hydrolysis.
- It also inhibits protein tyrosine phosphatases, affecting signal transduction pathways and cellular regulation.

3. Reduction of Vanadate by Biomolecules

Vanadate (VO^{3-}) is reduced to vanadyl (**VO^{2+}**) in the presence of reducing agents like ascorbic acid, glutathione, or NADH.

Reaction:

$$VO^{3-} + 2H^{+} + 2e^{-} \rightarrow VO^{2+} + H_2O$$

- This reaction typically occurs in biological systems under reducing conditions.
- The resulting vanadyl ion can further interact with other biomolecules, forming complexes that may alter cellular metabolism.

4. Interaction with DNA

Vanadium compounds can bind to DNA, affecting its structure and function. Vanadyl ions often coordinate with phosphate groups in the DNA backbone.

Reaction:

$$VO^{2+} + \text{DNA} \rightarrow [\text{VO–DNA complex}]$$

- Vanadyl ions preferentially bind to the phosphate oxygen atoms in DNA, stabilizing or destabilizing the helical structure.
- This interaction can lead to alterations in DNA replication and transcription.

5. Oxidative Stress Generation

Vanadium compounds catalyze the generation of reactive oxygen species (ROS) through Fenton-like reactions, interacting with cellular biomolecules.

Reaction:

$$VO^{2+} + H_2O_2 \rightarrow VO^{3+} + OH + OH^{-}$$

- The hydroxyl radicals ***(OH·)*** produced in this reaction can damage proteins, lipids, and DNA.
- This oxidative stress is a critical factor in the biological activity of vanadium compounds.

8.4 HEALTH IMPACTS OF VANADIUM COMPOUNDS

Vanadium compounds have garnered significant attention due to their potential therapeutic effects on a variety of health conditions. These range from metabolic disorders, such as diabetes, to cardiovascular and neurodegenerative diseases. However, the potential for toxicity at high doses has also been a concern, emphasizing the need for careful consideration of dosage and long-term safety [8].

TABLE 8.4
Vanadium compounds regulating glucose etabolism

Vanadium Compound	Dosage	Effect on Glucose Levels
Vanadyl sulfate	50–100 mg/day	Significant reduction in blood glucose in diabetics
Vanadium oxide complex	1–3 mg/day	Increased insulin sensitivity

8.4.1 Metabolic Effects and Diabetes

Vanadium compounds have been studied extensively for their potential role in regulating glucose metabolism, particularly for diabetes management. The most well-known vanadium compound in this regard is vanadyl sulfate (VO_2^+), which has been shown to mimic insulin's effects in both animal models and human studies (Table 8.4).

Mechanism of action: Vanadyl sulfate works by activating insulin receptors, enhancing glucose uptake by cells, and improving insulin sensitivity. It also inhibits protein tyrosine phosphatases (PTPs), which play a key role in insulin signaling.

Clinical evidence: Several clinical trials have demonstrated the ability of vanadyl sulfate to lower blood glucose levels in type 2 diabetic patients. In a 2000 study, oral administration of vanadyl sulfate improved insulin sensitivity and decreased blood glucose levels in diabetic individuals. However, the long-term efficacy and safety of vanadium-based compounds for diabetes remain inconclusive.

8.4.2 Chemical Reactions and Structures Related to Metabolic Effects and Diabetes

1. Insulin-Mimetic Action

Vanadium compounds mimic insulin by activating key enzymes involved in glucose uptake and metabolism. This occurs through phosphorylation pathways that mimic insulin receptor activity [9].

Reaction:

$$VO_4^{3-} + \text{ATP} \rightarrow VO_4^{3-} . \text{P}_{\text{tyr}} \text{ (Phosphorylated tyrosine residue)}$$

- Vanadium compounds bind to and activate tyrosine kinases, leading to the phosphorylation of tyrosine residues.
- This phosphorylation cascade mimics the action of insulin, facilitating glucose uptake into cells by translocating glucose transporters (GLUT4) to the cell membrane.

2. Activation of Glycolysis

Vanadium ions can enhance glycolysis by modulating the activity of hexokinase and phosphofructokinase enzymes.

Reaction:

$$\textit{Glucose} + \textit{ATP} \xrightarrow{\textit{Hexokinase}} \textit{Glucose-6-phosphate} + \textit{ADP}$$

- Vanadium compounds upregulate glycolytic enzymes, ensuring efficient glucose utilization.
- By enhancing hexokinase activity, more glucose is phosphorylated and directed into metabolic pathways.

3. Glucose Oxidation Enhancement

Vanadium compounds promote glucose oxidation by activating pyruvate dehydrogenase.

Reaction:

$$\text{Pyruvate} + NAD^+ + \text{CoA} \xrightarrow{\textit{Pyruvate Dehydeogenase}} \text{Acetyl-CoA} + \text{NADH} + CO_2$$

- Acetyl-CoA enters the citric acid cycle, producing ATP and maintaining cellular energy levels.
- Vanadium compounds stimulate this process, helping to normalize blood glucose levels in diabetic conditions.

4. Inhibition of Gluconeogenesis

Vanadium compounds suppress gluconeogenesis, reducing glucose production in the liver.

Reaction:

$$\text{Pyruvate} + \text{ATP} \xrightarrow{\textit{Inhibited}} \text{Phosphoenolpyruvate (PEP)} + \text{ADP}$$

- By inhibiting key gluconeogenic enzymes like phosphoenolpyruvate carboxykinase (PEPCK), vanadium compounds prevent excess glucose production, aiding in blood sugar control.

5. Antioxidant Effects in Diabetes

Vanadium compounds catalyze the scavenging of reactive oxygen species (ROS), which are elevated in diabetes.

Reaction

$$VO^{2+} + 2H_2O_2 \rightarrow VO^{3+} + 2OH. + 2OH$$

ROS, such as hydrogen peroxide (H_2O_2), are reduced or neutralized by vanadium compounds, protecting pancreatic β-cells from oxidative damage.

8.4.3 Cancer Therapy and Anticancer Properties

Vanadium compounds have shown potential as anticancer agents due to their ability to disrupt cellular growth and induce apoptosis in cancer cells. The mechanisms by which vanadium compounds exert anticancer effects include:

Inhibition of tyrosine kinases: Tyrosine kinases are enzymes that regulate cellular signaling pathways associated with cell division and survival. Vanadates (V^{5+}) can inhibit these enzymes, slowing down the proliferation of cancer cells [10].

Oxidative stress induction: Vanadium compounds can generate reactive oxygen species (ROS) within cells, causing oxidative damage that leads to cancer cell death.

Gene regulation: Vanadyl and vanadate complexes can alter the expression of genes involved in apoptosis and cell cycle regulation (Table 8.5).

Studies on vanadium's effects on different types of cancer have produced promising results. For example, vanadate has been shown to inhibit the growth of prostate cancer cells and reduce the viability of breast cancer cell lines.

Vanadium Compound Introduction:

Vanadium compounds, such as vanadate and vanadyl, are administered (or taken up by cells via endocytosis) (Figure 8.3).

Cellular Uptake:

Vanadium enters cancer cells, often through phosphate transporters due to its structural similarity to phosphate.

TABLE 8.5
Vanadium compounds and their anticancer mechanisms

Vanadium Compound	Cancer Type	Mechanism of Action
Vanadyl sulfate	Prostate cancer	Inhibition of cell proliferation
Vanadate complex	Breast cancer	Induction of apoptosis and reactive oxygen species (ROS) generation
Vanadium oxide complex	Leukemia	Activation of pro-apoptotic pathways

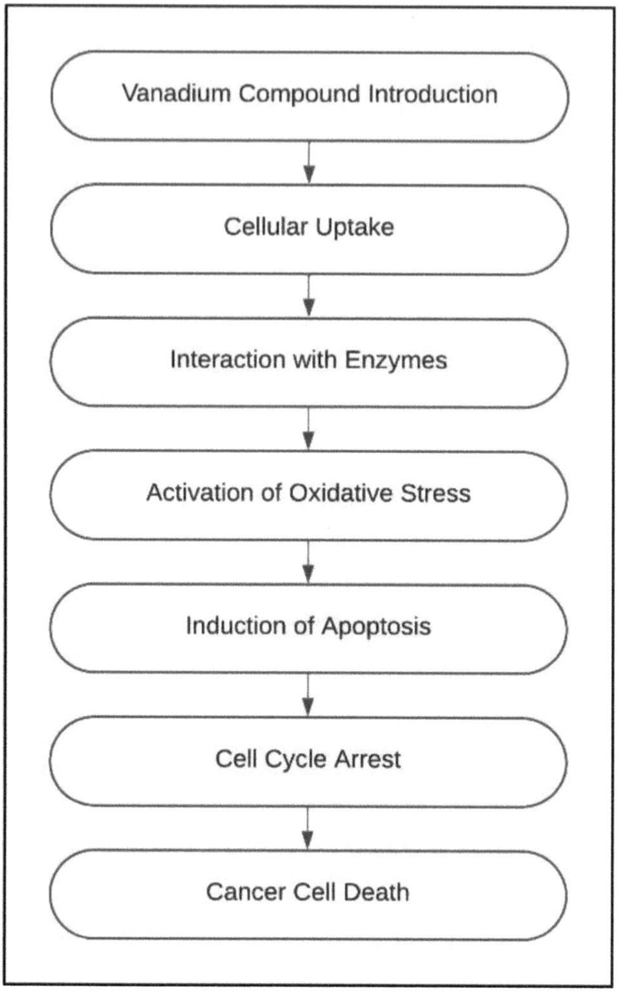

FIGURE 8.3 Vanadium's anticancer effects in biological systems.

Interaction with Enzymes:

Vanadium compounds mimic phosphate and interact with key enzymes such as:

- Protein tyrosine phosphatases (PTP): Inhibits PTPs, leading to activation of tyrosine kinase receptors involved in cancer cell growth.
- Protein kinases: Vanadium compounds activate or inhibit kinases like MAPKs, influencing cell proliferation and apoptosis.

Activation of Oxidative Stress:

Vanadium induces the production of reactive oxygen species (ROS), leading to oxidative stress in cancer cells.

Induction of Apoptosis:

The ROS produced lead to DNA damage, mitochondrial dysfunction, and activation of caspase-mediated apoptosis.

Cell Cycle Arrest:

Vanadium compounds influence cyclin-dependent kinases (CDKs), leading to cell cycle arrest and inhibition of cancer cell division.

Cancer Cell Death:

Ultimately, these effects lead to the death of cancer cells, reducing tumor growth.

8.4.4 Antimicrobial and Antiviral Activities

Vanadium compounds have also been explored for their antimicrobial and antiviral properties, particularly in the context of antibiotic resistance and viral infections (Table 8.6).

Antibacterial properties: Studies show that vanadyl complexes inhibit bacterial growth by interfering with microbial enzyme systems [11]. For example, vanadyl sulfate has demonstrated antimicrobial activity against both gram-positive and gram- negative bacteria, including *Staphylococcus aureus* and *Escherichia coli*.

Antifungal properties: Vanadium compounds have exhibited fungicidal effects, particularly against *Candida albicans*, a common fungal pathogen.

Antiviral activity: Some vanadium compounds have been shown to inhibit viral replication by interfering with viral polymerases. Research has explored their activity against influenza viruses, herpes simplex virus, and even SARS-CoV-2.

8.4.5 Impact on Cardiovascular Health

Vanadium compounds have been investigated for their cardiovascular benefits, particularly in reducing oxidative stress and modulating lipid metabolism [12].

Cholesterol regulation: Some animal studies have demonstrated that vanadate can lower blood cholesterol levels by reducing hepatic cholesterol biosynthesis and increasing bile acid excretion.

TABLE 8.6
Summary of antimicrobial and antiviral activities of vanadium compounds

Vanadium Compound	Microorganism/Virus	Effect
Vanadyl sulfate	*Escherichia coli*	Growth inhibition
Vanadium complex	*Staphylococcus aureus*	Bactericidal effect
Vanadium oxide complex	*Candida albicans*	Fungicidal effect
Vanadyl complex	Influenza virus	Inhibition of replication

Blood pressure: Vanadate has been shown to exert vasodilatory effects by improving endothelial function, potentially contributing to lower blood pressure.

Oxidative stress reduction: Vanadium acts as an antioxidant in the cardiovascular system, scavenging free radicals and reducing inflammation, which are key contributors to cardiovascular diseases such as atherosclerosis.

8.4.6 Neuroprotective Effects

Vanadium's neuroprotective properties have become a topic of increasing interest, particularly for treating neurodegenerative diseases such as Alzheimer's and Parkinson's [13]. Vanadyl sulfate has shown promise in modulating neuronal growth and function (Table 8.7).

Amyloid-beta inhibition: Vanadyl sulfate has been found to inhibit the aggregation of amyloid-beta peptides, a hallmark of Alzheimer's disease, and improve cognitive function in animal models.

Parkinson's disease: In Parkinson's models, vanadium compounds help reduce oxidative damage in dopaminergic neurons, potentially slowing the progression of the disease.

Neurogenesis: Some vanadium compounds have been shown to promote neuronal survival and stimulate neurogenesis in areas of the brain associated with learning and memory.

Vanadium compounds have gained attention in the field of neuroprotection due to their potential to mitigate neurodegenerative diseases, including Alzheimer's disease, Parkinson's disease, and multiple sclerosis. Studies suggest that vanadium can exert protective effects on neurons by acting as an antioxidant [14]. It helps to neutralize ROS, which are often involved in neuronal damage and degeneration. By modulating oxidative stress, vanadium compounds can reduce inflammation and prevent neuronal apoptosis (cell death), which is crucial for the maintenance of healthy brain function. Some vanadium compounds, such as vanadyl sulfate, have shown promise in experimental models by improving cognitive function and reducing brain oxidative damage, potentially offering a therapeutic strategy for neurodegenerative diseases where oxidative stress plays a pivotal role.

TABLE 8.7
Vanadium compounds and their neuroprotective mechanisms

Vanadium Compound	Disease	Effect
Vanadyl sulfate	Alzheimer's disease	Inhibition of amyloid-beta aggregation
Vanadium complex	Parkinson's disease	Reduction of oxidative damage
Vanadate complex	Cognitive decline	Enhanced memory and learning

In addition to their antioxidant properties, vanadium compounds are thought to influence neurotransmission and neuronal signaling pathways. For example, vanadium's ability to mimic insulin has been linked to improved glucose metabolism in the brain, which is essential for maintaining neuronal energy homeostasis [15]. In Alzheimer's disease, insulin resistance is a common issue, and vanadium's insulin-mimetic properties could potentially help restore normal brain function by enhancing glucose uptake in neurons. Moreover, vanadium's interaction with key signaling pathways, such as the MAPK and PI3K/Akt pathways, may support neuronal survival and regeneration. These effects suggest that vanadium compounds could play a role not only in treating neurodegenerative diseases but also in improving recovery from acute neurological injuries, such as stroke or traumatic brain injury. However, while promising, these neuroprotective effects need further exploration in clinical settings to fully understand their therapeutic potential and safety.

8.5 TOXICOLOGY AND SAFETY OF VANADIUM COMPOUNDS

While vanadium compounds hold significant promise in various therapeutic applications, their toxicity remains a critical concern. At high doses, vanadium can be harmful, potentially causing a range of adverse effects [16]. The safety of vanadium compounds in therapeutic settings is thus a subject of ongoing research and regulation.

8.5.1 Toxicity Mechanisms of Vanadium

Vanadium compounds, particularly in their higher oxidation states, can cause cellular damage through several mechanisms, including:

Oxidative stress: Vanadium can induce the generation of ROS within cells, leading to oxidative damage of cellular macromolecules like DNA, proteins, and lipids. This oxidative stress can contribute to various diseases, including cancer, cardiovascular disease, and neurodegenerative disorders.

Disruption of metal homeostasis: Vanadium can interfere with the balance of essential metal ions, such as calcium and magnesium, by competing for binding sites in enzymes and receptors. This can affect cellular signaling, muscle contraction, and neurotransmission.

Mitochondrial dysfunction: Vanadium has been shown to disrupt mitochondrial function, impairing energy production and contributing to cell death. This is particularly relevant in tissues with high energy demands, such as the heart and brain.

Vanadium, while essential in trace amounts for certain biological functions, can become toxic when accumulated in excess. The toxicity mechanisms of vanadium are primarily linked to its ability to generate ROS through redox reactions. These ROS can lead to oxidative stress, damaging cellular components such as lipids, proteins, and DNA. When vanadium is absorbed into the body, it often interacts with cellular membranes, mitochondria, and the endoplasmic reticulum, where it induces oxidative damage and disrupts normal cellular functions [17]. The imbalance between ROS production and antioxidant defenses can result in inflammation, apoptosis (programmed cell death), and in some cases, necrosis. This is particularly concerning tissues such as the liver, kidneys, and nervous system, where oxidative stress plays a significant role in organ damage and dysfunction. As a result, exposure to high levels of vanadium, particularly in industrial settings or through contaminated water sources, can lead to a variety of health issues, ranging from mild irritation to more severe conditions like liver damage and neurological impairment.

In addition to oxidative stress, vanadium toxicity can also involve the disruption of ion homeostasis, particularly the calcium and potassium ions essential for cellular function. Vanadium has been shown to interfere with calcium ion channels and transporters, leading to calcium overload in cells. This disruption can initiate several harmful pathways, including activation of proteases, phospholipases, and endonucleases, all of which contribute to cellular injury. Furthermore, vanadium can disrupt mitochondrial function by impairing the electron transport chain, leading to decreased ATP production and subsequent energy failure. In neurons, this energy deficit exacerbates the damage, contributing to neurodegeneration. The cumulative effect of these mechanisms makes vanadium toxicity a multifaceted issue, requiring careful management in industrial exposure and therapeutic use.

Despite its potential benefits at low concentrations, the narrow therapeutic window and potential toxicity of vanadium demand further research to understand its safe application in medicine and mitigate the risks associated with its misuse [18].

8.5.2 Therapeutic Application of Vanadium Compounds

Below are some key chemical reactions related to the therapeutic applications of vanadium compounds, focusing on their biological interactions:

Insulin-Mimetic Effects

Vanadium compounds mimic the activity of insulin by interacting with enzymes in the insulin-signalling pathway. The reactions generally involve vanadium in its oxidation states (V^{+4} or V+5)

$$[VO^{2+}] + H_2O_2 \rightarrow VO(O_2)^+ + H_2O$$

Vanadyl ion VO^{2+} reacts with hydrogen peroxide to form a peroxovanadyl complex $VO(O_2)^+$, which can inhibit protein tyrosine phosphatases (PTPs) and enhance glucose uptake.

Oxidative Stress Generation in Cancer Therapy

Vanadium compounds induce reactive oxygen species (ROS), causing oxidative damage to cancer cells.

$$V^{5+} + H_2O_2 \rightarrow VO_2^{\ +} + O_2^{\cdot -} + H^+$$

Vanadium in its pentavalent form (V^{5+}) catalyzes the production of superoxide anion ($O_2^{\cdot-}$) and other ROS.

Inhibition of Microbial Enzymes in Antibacterial Therapy
Vanadate (VO_4^{3-}) competes with phosphate (PO_4^{3-}) in binding microbial enzymes, disrupting their function.

$$\text{Enzyme–PO}_4^{3-} + \text{VO}_4^{3-} \rightarrow \text{Enzyme–VO}_4^{3-} + \text{PO}_4^{3-}$$

The phosphate in the enzyme-phosphate complex is replaced by vanadate, leading to enzyme inhibition.

Interaction with DNA in Cancer Therapy
Vanadium compounds can bind to DNA and disrupt replication in cancer cells.

$$[\text{VO(OH}_2)]+ + \text{DNA(phosphate)} \rightarrow [\text{VO-DNA}] + \text{H}_2\text{O}$$

The vanadyl ion binds to the phosphate backbone of DNA, inhibiting DNA replication and transcription.

Reduction of Vanadium Compounds in Biological Systems
In biological environments, vanadium compounds undergo redox reactions, switching between V^{+4} (vanadyl) and V^{+5} (vanadate) forms.

$$V^{5+} + e^- \rightarrow V^{4+}$$

This redox cycling contributes to ROS generation and therapeutic effects in various diseases.

Vanadium in Cancer Treatment
Vanadium compounds have also gained attention for their potential in cancer therapy due to their ability to disrupt cellular growth and induce apoptosis in cancer cells. Vanadium's ability to mimic phosphate in cellular processes makes it a promising candidate for affecting critical pathways involved in cancer progression.

TABLE 8.8
Vanadium compounds and their anticancer activities

Vanadium Compound	Cancer Type	Therapeutic Effect
Vanadyl sulfate	Prostate cancer	Reduced tumor size, apoptosis induction
Vanadium oxide complex	Breast cancer	Inhibition of metastasis, cell cycle arrest
Vanadate	Lung cancer	Reduced cell migration, apoptosis induction

- *Cell cycle arrest and apoptosis*: Vanadium compounds, particularly vanadate (V^{5+}), can induce cell cycle arrest and trigger apoptotic pathways in cancer cells. They interact with key enzymes and proteins involved in cell division, preventing uncontrolled cell proliferation.
- *Inhibition of cancer cell migration*: Studies have shown that vanadate inhibits the migration of cancer cells, a key step in metastasis. By interfering with the actin cytoskeleton and integrin signalling, vanadium compounds reduce the ability of cancer cells to invade surrounding tissues [19].
- *Clinical trials*: Some early-phase clinical trials have explored the use of vanadium compounds as adjuncts to chemotherapy, particularly in breast, prostate, and lung cancers.

Vanadium compounds have shown significant potential in cancer treatment due to their ability to target multiple cellular pathways involved in tumor growth and progression. One of the primary mechanisms by which vanadium exerts its anticancer effects is through the induction of oxidative stress in cancer cells. Unlike normal cells, cancer cells have a higher baseline level of oxidative stress, making them more vulnerable to additional ROS generated by vanadium compounds. This oxidative stress can lead to mitochondrial dysfunction, DNA damage, and activation of apoptosis pathways, ultimately resulting in cancer cell death. Furthermore, vanadium compounds have been found to inhibit key signaling pathways such as PI3K/Akt and MAPK, which are often dysregulated in cancer and promote cell proliferation, survival, and metastasis [20]. By disrupting these pathways, vanadium compounds can effectively suppress tumor growth and prevent the spread of cancer cells to other parts of the body.

In addition to their direct anticancer effects, vanadium compounds have also demonstrated the ability to enhance the efficacy of conventional cancer treatments, such as chemotherapy and radiotherapy. Some studies have reported that vanadium compounds sensitize cancer cells to these treatments by impairing their DNA repair mechanisms, thereby increasing their susceptibility to treatment-induced damage [21]. Moreover, vanadium compounds exhibit selective toxicity, meaning they tend to target cancer cells more effectively than normal cells, reducing the risk of adverse effects on healthy tissues. Despite these promising results, the clinical application of vanadium in cancer therapy remains in the early stages. Future research should focus on optimizing the formulation and delivery of vanadium compounds to maximize their therapeutic benefits while minimizing potential side effects.

Vanadium in Antimicrobial and Antiviral Applications

In addition to its potential for treating chronic diseases, vanadium compounds have been investigated for their antimicrobial and antiviral activities. Their ability to disrupt microbial enzyme systems and inhibit viral replication makes them promising candidates for novel therapeutic agents, particularly in the face of rising antibiotic resistance (Table 8.9).

- *Antimicrobial*: Vanadium compounds, such as vanadyl sulfate, have shown antibacterial effects against both gram-positive and gram-negative bacteria.

TABLE 8.9
Summary of vanadium compounds' antimicrobial and antiviral properties

Vanadium Compound	Microorganism/Virus	Effect
Vanadyl sulfate	Staphylococcus aureus	Antibacterial growth inhibition
Vanadium oxide complex	Influenza virus	Inhibition of viral replication
Vanadate	Herpes simplex virus (HSV)	Reduced viral load, inhibited replication

This has implications for treating infections caused by resistant bacteria, such as *S. aureus* and *E. coli* [22].

- *Antiviral*: Vanadium complexes have demonstrated antiviral activity against several viruses, including influenza, herpes simplex virus (HSV), and SARS-CoV-2. These compounds work by inhibiting viral replication and reducing viral load in infected cells.

Vanadium compounds have demonstrated significant potential in combating microbial and viral infections due to their ability to interfere with critical biological processes in pathogens. Research has shown that vanadium exhibits broad-spectrum antimicrobial activity against various bacterial strains, including antibiotic-resistant species like *Staphylococcus aureus* (MRSA) and *Pseudomonas aeruginosa* [23]. These effects are attributed to vanadium's capacity to disrupt bacterial cell membranes, inhibit key enzymatic pathways, and generate ROS, leading to oxidative stress and cell death in microbes. In addition to its antibacterial properties, vanadium compounds have shown promise in antiviral applications by targeting viral replication processes. For instance, vanadium complexes have been found to inhibit the replication of viruses such as HIV, herpes simplex virus (HSV), and influenza by interfering with viral DNA or RNA synthesis and modulating host immune responses. This dual antimicrobial and antiviral action makes vanadium an attractive candidate for developing novel therapies, particularly in addressing the growing threat of drug-resistant infections and emerging viral diseases [24].

8.6 ENVIRONMENTAL AND INDUSTRIAL APPLICATIONS OF VANADIUM

8.6.1 Industrial Catalysis and Material Science

Vanadium is widely used as a catalyst in numerous industrial processes, particularly in the production of sulfuric acid. Vanadium pentoxide (V_2O_5) is a highly effective catalyst in the contact process, which is used to oxidize sulfur dioxide (SO_2) to sulfur trioxide (SO_3), a critical step in the production of sulfuric acid. This industrial process is crucial to the manufacture of fertilizers, chemicals, and petrochemicals. Vanadium's catalytic properties are attributed to its ability to change oxidation states, enabling efficient electron transfer and facilitating the oxidation reactions [25]. The

TABLE 8.10
Industrial applications of vanadium

Application	Vanadium Compound Used	Industry	Key Benefits
Sulfuric acid production	Vanadium pentoxide (V_2O_5)	Chemical manufacturing	Catalyzes oxidation of SO_2 to SO_3
Steel production	Ferrovanadium	Metallurgical industry	Improves strength, hardness, and corrosion resistance
Alloys for aerospace	Vanadium-alloyed steel	Aerospace, automotive	High tensile strength, heat resistance
Superconducting materials	Vanadium Oxide (V_2O_3)	Electronics, medical technology	Potential use in MRI, quantum computing
Energy storage	Vanadium Redox Flow Batteries (VRFBs)	Renewable energy, grid storage	High scalability, long cycle life, low environment impact

effectiveness and longevity of vanadium pentoxide catalysts have made them indispensable in large-scale chemical production.

In material science, vanadium is used to produce high-strength, corrosion-resistant alloys. These alloys are used in applications that require materials to withstand extreme conditions, such as in the aerospace and automotive industries, where lightweight yet durable materials are critical. Vanadium is also a key component in steel production, as it enhances the hardness and wear resistance of steel, making it ideal for use in tools, engines, and heavy machinery. Additionally, vanadium-based superconducting materials are under investigation for use in high-performance magnets and electronic devices, promising advancements in magnetic resonance imaging (MRI) and quantum computing [26] (Table 8.10).

In addition to its role in sulfuric acid production and steel manufacturing, vanadium is also crucial in the production of specialty chemicals and polymerization processes. Vanadium catalysts are used in the production of certain high-value chemicals, such as acrylic acid and maleic anhydride, which are essential in the production of plastics, coatings, and adhesives. These catalysts provide efficient reactions that increase yield while reducing energy consumption [27]. In polymer science, vanadium-based catalysts are used to polymerize olefins, contributing to the development of specialized plastics with unique properties like increased strength and thermal stability. Furthermore, vanadium alloys have found significant applications in high-temperature environments, such as in the nuclear and aerospace industries, where their combination of heat resistance, durability, and low weight makes them ideal for manufacturing advanced turbine blades, reactor components, and aerospace parts.

8.7 FUTURE PERSPECTIVES AND RESEARCH DIRECTIONS

Vanadium-based compounds have demonstrated significant potential in various biomedical and industrial applications, but many aspects remain underexplored. Future research will likely focus on overcoming current limitations, including toxicity, bioavailability, and specificity. This section outlines the promising avenues and emerging trends in vanadium research, with an emphasis on improving therapeutic efficacy and expanding industrial applications [28] (Table 8.11)

8.7.1 Innovations in Drug Design and Delivery

Future drug development involving vanadium will aim at creating more selective and less toxic formulations. Nanotechnology and bioengineering will play a pivotal role in this transformation. For instance, nanoparticle-mediated delivery systems and targeted prodrug approaches are expected to revolutionize how vanadium compounds are administered.

One promising avenue in vanadium drug development involves the integration of stimuli-responsive delivery systems, which release the therapeutic agent in response to specific environmental triggers such as pH, temperature, or enzymatic activity. For example, vanadium- loaded nanoparticles that respond to the acidic microenvironment of tumors can release their payload selectively at the cancer site, enhancing therapeutic efficacy while minimizing systemic toxicity [29]. Similarly, smart hydrogels that release vanadium compounds in response to inflammation could offer targeted treatment for inflammatory diseases. These systems not only improve the precision of drug delivery but also allow for real-time adjustments in dosing, paving

TABLE 8.11
Emerging drug delivery technologies for vanadium compounds

Technology	Description	Potential Benefits	Current Challenges
Nanoparticles	Engineered particles for targeted drug delivery.	Increased bioavailability, reduced side effects.	Costly manufacturing, scalability issues.
Hydrogel systems	Gel-based carriers for controlled vanadium release.	Prolonged release, localized delivery.	Stability in biological environments.
Microsphere delivery	Micron-sized particles for slow, controlled drug release.	Enhanced patient compliance, reduced dosing frequency.	Complex formulation processes.
Prodrug approaches	Inactive compounds activated at target sites.	Reduced systemic toxicity, enhanced selectivity.	Precise targeting mechanisms required.

the way for more personalized and dynamic treatment regimens. The incorporation of such cutting-edge technologies highlights the potential for vanadium-based therapies to evolve into highly sophisticated and adaptive medical solutions.

Furthermore, researchers are exploring the use of biodegradable carriers for vanadium delivery, which break down naturally after releasing the drug, leaving no toxic residues. These carriers, such as biodegradable polymers and lipid-based systems, not only ensure sustained and controlled drug release but also reduce the risk of long-term side effects associated with residual materials [30]. This approach aligns with the growing emphasis on environmentally friendly and patient-safe drug formulations, offering a promising direction for future vanadium-based therapies.

8.7.2 Exploring New Therapeutic Areas

Emerging evidence suggests that vanadium could play a role in treating diseases beyond its current focus areas. Potential applications include immunomodulation, anti-inflammatory therapies, and rare metabolic disorders. Research in these areas could significantly expand the therapeutic scope of vanadium (Table 8.12).

Vanadium's potential in immunomodulation is gaining increasing attention, particularly in the context of autoimmune diseases and inflammatory disorders. Research suggests that vanadium compounds may have the ability to regulate immune cell activation and cytokine production, which could help control the overactive immune responses observed in conditions like rheumatoid arthritis, lupus, and multiple sclerosis [31]. Through modulating immune pathways, vanadium could provide a novel therapeutic approach to rebalancing immune functions without the side effects associated with traditional immunosuppressive drugs. Early preclinical studies have shown promising results, with vanadium complexes demonstrating the ability to

TABLE 8.12
Potential new therapeutic applications of vanadium

Therapeutic Area	Potential Mechanism of Action	Current Research Status
Immunomodulation	Modulating immune cell activity to enhance or suppress responses.	Preliminary studies in autoimmune disease models.
Anti-inflammatory therapy	Inhibiting inflammatory pathways to treat chronic inflammation.	Ongoing trials for arthritis and inflammatory diseases.
Rare metabolic disorders	Correcting enzyme deficiencies or metabolic imbalances.	Early preclinical studies in metabolic syndromes.
Wound healing	Promoting tissue regeneration and angiogenesis.	Experimental phase with promising in vitro results.

reduce inflammation and tissue damage in animal models, warranting further investigation in clinical settings.

Additionally, vanadium's role in treating rare metabolic disorders holds significant promise. Some metabolic diseases, particularly those involving enzyme deficiencies or dysfunctions in metabolic pathways, could benefit from vanadium's ability to regulate key enzymes or influence metabolic processes. For instance, vanadium compounds have been explored for their potential to restore normal glucose metabolism in individuals with disorders like glycogen storage diseases or maturity-onset diabetes of the young (MODY). These conditions, often characterized by abnormalities in glucose production and insulin sensitivity, could see improvements through vanadium's impact on insulin signaling and cellular energy regulation [32]. Ongoing research is aiming to identify specific metabolic pathways influenced by vanadium, which could lead to novel treatments for these challenging and often underserved diseases.

8.7.3 Advanced Biomolecular Research

The interaction of vanadium with key biomolecules such as DNA, proteins, and enzymes will remain a critical research focus. Advances in structural biology and molecular dynamics simulations are expected to provide deeper insights into these interactions, aiding in the design of more effective and safer drugs (Table 8.13).

One of the most exciting frontiers in vanadium research lies in understanding its interactions with biomolecules, particularly proteins and nucleic acids. The precise mechanisms by which vanadium binds to proteins, alters their structure, or modulates their function remain a critical area of study. Recent studies have suggested that vanadium can bind to metalloproteins, influencing enzyme activity and potentially altering metabolic pathways. For example, vanadium's interaction with protein tyrosine phosphatases (PTPs) is an area of growing interest, as PTPs are critical regulators of cellular signaling and have been implicated in diseases like cancer and diabetes [33]. Understanding how vanadium compounds can influence these biomolecules

TABLE 8.13
Future research directions in biomolecular studies

Research Focus	Key Questions	Impact on Therapeutics
Vanadium–DNA Interactions	How does vanadium bind to DNA and influence gene expression?	could lead to new cancer or gene therapy treatments.
Protein targeting	Which proteins can vanadium selectively target?	Improved drug design for metabolic and neurological diseases.
Enzyme modulation	How does vanadium affect enzymatic pathways?	Potential for novel enzyme-based treatments.
Membrane transport	How is vanadium transported across cellular membranes?	Insights into improving bioavailability and uptake.

could open new therapeutic avenues, especially for conditions that involve dysfunctional signaling pathways.

Additionally, vanadium's potential to interact with DNA could provide novel insights into its role in gene expression regulation. Studies have shown that vanadium compounds may induce changes in DNA structure, either by intercalating into the helix or by binding to specific sites, potentially influencing gene transcription. This property could be harnessed in cancer therapy, where vanadium's ability to modulate gene expression may enhance the effectiveness of other chemotherapeutic agents. Molecular dynamics simulations and advanced spectroscopic techniques, such as X-ray crystallography and nuclear magnetic resonance (NMR) spectroscopy, are crucial in revealing the atomic-level details of vanadium's interactions with these biological macromolecules. These studies could lead to the design of vanadium-based drugs that specifically target gene expression pathways, making them a powerful tool in precision medicine.

Moreover, ongoing research into vanadium's role in membrane transport mechanisms could offer insights into how vanadium compounds cross biological membranes to reach their targets. By understanding the transport processes at the cellular level, researchers can design more effective drug delivery systems and improve the bioavailability of vanadium-based therapies [34]. This aspect of vanadium research also ties into the broader field of drug resistance, as vanadium compounds may provide a means of overcoming cellular barriers that limit the effectiveness of conventional drugs.

8.8 CONCLUSION

The role of vanadium compounds in human health presents a fascinating and multifaceted research frontier with vast potential for therapeutic applications, yet many challenges remain to be fully understood and addressed. Throughout this review, we have explored the diverse biological effects of vanadium, highlighting its potential in treating various diseases, including diabetes, cancer, neurodegenerative conditions, and cardiovascular diseases. Additionally, vanadium's applications in industrial and environmental processes, such as catalysis and energy storage, underline its broader significance beyond human health.

Vanadium's therapeutic potential stems from its ability to interact with key biomolecules, modulating critical pathways involved in cellular processes such as signal transduction, metabolism, and oxidative stress management. However, despite its promise, the clinical translation of vanadium-based therapies has been hindered by concerns over toxicity, bioavailability, and limited understanding of its interactions at the molecular level. Ongoing research efforts are focused on overcoming these challenges through innovative drug delivery systems, the development of less toxic vanadium derivatives, and targeted therapies that maximize the compound's therapeutic effects while minimizing adverse outcomes. The incorporation of nanotechnology, biodegradable carriers, and stimuli-responsive systems into vanadium drug formulations will be pivotal in advancing its clinical applications.

In addition to therapeutic applications, the exploration of vanadium's role in industrial catalysis, green chemistry, and sustainable energy production also presents exciting opportunities. Vanadium-based catalysts, particularly in areas such as hydrogen production and CO_2 reduction, hold the potential to contribute to environmental sustainability and renewable energy solutions. As the demand for clean energy and efficient industrial processes continues to grow, vanadium's role in these sectors will become increasingly significant.

Looking forward, there are ample opportunities to further explore the biomolecular interactions of vanadium compounds, specifically their impact on proteins, enzymes, and DNA. This detailed understanding could enable the design of more precise and effective therapies for a range of diseases, especially those that are currently underserved by traditional treatments. Moreover, expanding research into new therapeutic areas, such as immunomodulation and rare metabolic disorders, will likely unlock new avenues for vanadium-based treatments.

As with any novel therapeutic agent, future research must prioritize the long-term safety and efficacy of vanadium-based treatments. A comprehensive understanding of its toxicity mechanisms, including organ-specific damage and dose-dependent effects, will be crucial in determining the appropriate clinical applications and establishing clear regulatory guidelines. Risk assessments will be essential in ensuring that vanadium compounds can be used safely in human populations, particularly for long-term therapies.

While vanadium compounds show great promise in diverse scientific and medical fields, their successful application will depend on continued research efforts aimed at improving their safety, efficacy, and delivery mechanisms. As advancements in biomolecular research, drug design, and industrial applications unfold, vanadium could emerge as a cornerstone in both medicine and industry. However, careful consideration of its toxicological profiles, environmental impact, and regulatory compliance will be necessary to fully realize its potential, ensuring that vanadium-based solutions benefit human health and the environment in a sustainable and ethical manner.

Vanadium compounds hold significant promise in advancing human health and contributing to a range of scientific fields, yet their clinical and industrial application is still in its developmental stages. Over the course of this review, we have delved into the various facets of vanadium's biological effects, therapeutic potentials, and challenges. Vanadium, once considered an industrial metal, is now emerging as a powerful bioactive compound with the potential to address a broad spectrum of diseases, including diabetes, cancer, cardiovascular disorders, and neurodegenerative diseases. This shift in perspective opens up new possibilities for its use as a therapeutic agent, leveraging its ability to interact with critical biological targets such as proteins, enzymes, and DNA, which are integral to human health.

A central theme throughout this review is the complex interplay between vanadium compounds and the body's cellular machinery. Vanadium's ability to modulate key metabolic pathways, alter protein functions, and affect gene expression positions it as a potential treatment for diseases that are driven by abnormal cellular signaling. For instance, in diabetes, vanadium's insulin-like effects provide a potential therapeutic alternative to traditional treatments. Similarly, its anticancer and neuroprotective properties are being actively explored for their potential to treat conditions like Alzheimer's disease and various forms of cancer. Despite these promising results, the translation of vanadium-based therapies to the clinic remains a challenge due to concerns about toxicity, bioavailability, and the mechanisms of action at the molecular level.

One of the most promising directions for advancing vanadium therapeutics is through innovative drug delivery systems. Research into nanotechnology, biodegradable carriers, and stimuli-responsive materials offers the potential to deliver vanadium more effectively and with reduced toxicity. By enhancing vanadium's ability to target specific tissues or cells, researchers are working to maximize its therapeutic effects while minimizing side effects. Furthermore, the development of low-toxicity vanadium compounds and advanced drug formulations will be crucial in ensuring that vanadium-based treatments can be safely administered to patients over extended periods. This shift toward safer and more targeted delivery systems is essential to overcoming the hurdles of toxicity and improving the clinical applicability of vanadium.

Beyond its medicinal applications, vanadium's potential in industrial catalysis and green chemistry is also of great significance. Vanadium-based catalysts are used in a variety of processes, from energy production to environmental protection. As global demands for sustainable energy and cleaner industrial processes increase, vanadium's role in hydrogen production, carbon capture, and environmentally friendly chemical synthesis is becoming more relevant. By reducing the environmental impact of industrial activities, vanadium could play a key role in promoting a sustainable future. Moreover, the use of vanadium in energy storage systems, such as in vanadium redox flow batteries, holds great promise for improving renewable energy storage and distribution, contributing to more efficient use of green energy sources.

However, despite the exciting potential for vanadium, several challenges remain that need to be addressed to fully unlock its therapeutic and industrial applications. A critical area of research must focus on understanding the toxicity mechanisms of vanadium, including its organ-specific effects and dose-dependent toxicity. Ensuring that vanadium compounds are both safe and effective will require comprehensive studies on their long-term safety profiles and potential for cumulative toxicity. Moreover, regulatory guidelines will need to evolve to accommodate vanadium-based treatments, ensuring that they meet the rigorous standards for safety and efficacy required for clinical approval. Ongoing risk assessments and toxicological studies will be crucial in mitigating potential risks associated with vanadium's use in human health applications.

In the context of future research directions, vanadium's role in biomolecular interactions, gene expression modulation, and immune system regulation could reveal new therapeutic avenues for a range of diseases, particularly those involving immune dysregulation or metabolic disturbances. Further studies are needed to identify novel biomarkers that could indicate the most effective therapeutic use of vanadium, as well as to explore its interactions with other drugs and therapies in combination treatments. Vanadium's neuroprotective and immunomodulatory effects, for example, could be further investigated to develop novel treatments for diseases such as Parkinson's disease or multiple sclerosis. Additionally, vanadium's potential in rare metabolic diseases could become a key area of focus, offering new treatments for conditions that currently lack effective therapeutic options.

Looking forward, vanadium's future in both clinical medicine and industrial applications seems highly promising, but its journey will be guided by continued interdisciplinary research. The convergence of biochemistry, nanotechnology, environmental science, and materials engineering will play a pivotal role in advancing vanadium-based therapies and applications. By further refining vanadium's therapeutic formulations, enhancing its delivery mechanisms, and addressing its toxicological concerns, the scientific community can unlock the full potential of this remarkable element. As our understanding of vanadium's complex biological and chemical properties deepens, it may very well become a cornerstone in the treatment of chronic diseases and a key player in the development of sustainable technologies, driving both medical and industrial innovation forward.

In conclusion, the transformative potential of vanadium compounds lies in their ability to bridge the gap between medicine and industry, contributing to both human health and environmental sustainability. As research advances, the integration of vanadium into clinical therapies and industrial processes will likely expand its impact, offering solutions to some of the most pressing health and environmental challenges of the modern world.

REFERENCES

1. Treviño, S., & Durán, C. (2018). Vanadium in biological action: Chemical, pharmacological aspects, and metabolic implications in diabetes mellitus. *Biological Trace Element Research, 177*(2), 1–12. https://doi.org/10.1007/s12011-018-1540-6
2. Rehder, D. (2015). The role of vanadium in biology. *Metallomics*, 7(5), 730–742. https://doi.org/10.1039/C4MT00304G
3. Chasteen, N. D. (Ed.). (1990). *Vanadium in Biological Systems: Physiology and Biochemistry*. Springer. https://doi.org/10.1007/978-94-009-2023-1
4. Yadav, A., Yadav, R., Patel, D., & Patle, D. (2023). Synthesis, characterization, predicted ADME value and molecular docking of vanadium containing metal complexes as potential antimicrobial activity. *International Journal of Current Research and Technology*, 11(5), 123–135. https://ijcrt.org/papers/IJCRT2408676.pdf
5. Pessoa, J. C., Garribba, E., Santos, M. F. A., & Santos-Silva, T. (2015). Vanadium and proteins: Uptake, transport, structure, activity and function. *Coordination Chemistry Reviews, 301–302*, 49–86. https://doi.org/10.1016/j.ccr.2015.03.016
6. Ścibior, A., Adamczyk, A., Gołębiowska, D., & Kurus, J. (2018). Evaluation of lipid peroxidation and the level of some elements in rat erythrocytes during separate and combined vanadium and magnesium administration. *Chemico-Biological Interactions, 293*, 1–10. https://doi.org/10.1016/j.cbi.2018.07.014
7. Tracey, A.S., Willsky, G.R., & Takeuchi, E.S. (2007). *Vanadium: Chemistry, Biochemistry, Pharmacology and Practical Applications*, 1st ed. CRC Press. https://doi.org/10.1201/9781420046144
8. Zhang, S., & Kim, S. M. (2019). Synthesis, characterization, antioxidant and antidiabetic activities of a novel protein–vanadium complex. *Applied Organometallic Chemistry, 10*(2), 123–135. https://doi.org/10.1002/aoc.5102

9. Anitha, C., Sheela, C. D., Tharmaraj, P., & Sumathi, S. (2012). Spectroscopic studies and biological evaluation of some transition metal complexes of azo Schiff-base ligand derived from (1-phenyl-2,3-dimethyl-4-aminopyrazol-5-one) and 5-((4-chlorophenyl)diazenyl)-2- hydroxybenzaldehyde. *Spectrochimica Acta Part A: Molecular and Biomolecular Spectroscopy, 96*, 493–500. https://doi.org/10.1016/j.saa.2012.05.053
10. R. Renirie, A. Dewilde, C. Pierlot, R. Wever, D. Hober, J.-M. Aubry (2008). Bactericidal and virucidal activity of the alkalophilic P395D/L241V/T343A mutant of vanadium chloroperoxidase, *Journal of Applied Microbiology*, *105*(1), 264–270, https://doi.org/10.1111/j.1365-2672.2008.03742.x
11. Sheikhshoaie, I., Ebrahimipour, S. Y., Lotfi, N., Mague, J. T., & Khaleghi, M. (2016). Synthesis, spectral characterization, X-ray crystal structure and antimicrobial activities of two cis dioxido-vanadium(V) complexes incorporating unsymmetrical dimalonitrile-based (NNO) Schiff base ligands. *Inorganica Chimica Acta, 442*, 151–157. https://doi.org/10.1016/j.ica.2015.11.026
12. Elberg, G, He, Z., Li, J., Sekar, N., & Shechter Y. (1997). Vanadate activates membranous nonreceptor protein tyrosine kinase in rat adipocytes. *Diabetes 46*(11), 1684–1690. https://doi.org/10.2337/diab.46.11.1684
13. McLauchlan, C. C., Hooker, J. D., Jones, M. A., Dymon, Z., Backhus, E. A., Greiner, B. A., Dorner, N. A., Youkhana, M. A., & Manus, L. M. (2010). Inhibition of acid, alkaline, and tyrosine (PTP1B) phosphatases by novel vanadium complexes. *Journal of Inorganic Biochemistry, 104*(3), 274–281. https://doi.org/10.1016/j.jinorgbio.2009.12.001
14. Vilter, H. (1984). Peroxidases from phaeophyceae: A vanadium(V)-dependent peroxidase from Ascophyllum nodosum. *Phytochemistry, 23*(7), 1387–1390. https://doi.org/10.1016/S0031-9422(00)80471-9
15. Vlasiou, M. C., & Pafiti, K. S. (2021). Cell arrest and apoptosis induced by the next generation of vanadium based drugs: Action mechanism to structure relation and future perspectives. *Anti-Cancer Agents in Medicinal Chemistry, 21*(16), 2111–2116. https://doi.org/10.2174/1871520621666201222143839
16. Scior, T., Guevara-Garcia, J. A., Do, Q.-T., Bernard, P., & Laufer, S. (2016). Why antidiabetic vanadium complexes are not in the pipeline of "Big Pharma" drug research? A critical review. *Current Medicinal Chemistry, 23*(25), 2874–2891. https://doi.org/10.2174/0929867323666160321121138
17. Pessoa, J. C., & Correia, I. (2021). Misinterpretations in evaluating interactions of vanadium complexes with proteins and other biological targets. *Inorganics*, *9*(2), 17. https://doi.org/10.3390/inorganics9020017
18. Tolman, E. L., Barris, E., Burns, M., Pansini, A., & Partridge, R. (1979). Effects of vanadiumon glucose metabolismin vitro. *Life Sciences, 25*(13), 1159–1164. https://doi.org/10.1016/0024-3205(79)90138-3
19. Shechter, Y., & Karlish, S. (1980). Insulin-like stimulation of glucose oxidation in rat adipocytes by vanadyl (IV) ions. *Nature, 284*(556–558). https://doi.org/10.1038/284556a0
20. Damena, T., Alem, M. B., Zeleke, D., Demissie, T. B., & Desalegn, T. (2023). Synthesis and computational studies of novel cobalt(II) and oxovanadium(IV) complexes of quinoline carbaldehyde derivative ligand for antibacterial and antioxidant applications. *Journal of Molecular Structure, 1280*, 134994. https://doi.org/10.1016/j.molstruc.2023.134994

21. Pessoa, J. C., Etcheverry, S., & Gambino, D. (2015). Vanadium compounds in medicine. *Coordination Chemistry Reviews, 301–302*, 24–48. https://doi.org/10.1016/j.ccr.2014.12.002
22. Tolman, E. L., Barris, E., Burns, M., Pansini, A., & Partridge, R. (1979). Effects of vanadiumon glucose metabolism in vitro. *Life Sciences, 25*(13), 1159–1164. https://doi.org/10.1016/0024-3205(79)90138-3
23. Domingo, J. L. (2002). Vanadium and tungsten derivatives as antidiabetic agents. *Biological Trace Element Research, 88*, 97–112. https://doi.org/10.1385/BTER:88:2:097
24. Rehder, D. (2012). The potentiality of vanadium in medicinal applications. *Future Medicinal Chemistry*, *4*(14), 1823–1837. https://doi.org/10.4155/fmc.12.103
25. Roy, S., Banerjee, S. & Chakraborty, T. Vanadium quercetin complex attenuates mammary cancer by regulating the P53, Akt/mTOR pathway and downregulates cellular proliferation correlated with increased apoptotic events. *Biometals* **31**, 647–671 (2018). https://doi.org/10.1007/s10534-018-0117-3
26. Ścibior, A., Pietrzyk, Ł., Plewa, Z., & Skiba, A. (2020). Vanadium: Risks and possible benefits in the light of a comprehensive overview of its pharmacotoxicological mechanisms and multi-applications with a summary of further research trends. *Journal of Trace Elements in Medicine and Biology, 61*, 126508. https://doi.org/10.1016/j.jtemb.2020.126508
27. Younes, M., & Strubelt, O. (1991). Vanadate-induced toxicity towards isolated perfused rat livers: the role of lipid peroxidation. *Toxicology, 66*(1), 63–74. https://doi.org/10.1016/0300-483X(91)90178-4
28. Fatola, O. I., Olaolorun, F. A., Olopade, F. E., & Olopade, J. O. (2019). Trends in vanadium neurotoxicity. *Brain Research Bulletin, 145*, 75–80. https://doi.org/10.1016/j.brainresbull.2018.03.010
29. Ghosh, S. K., Saha, R., & Saha, B. (2015). Toxicity of inorganic vanadium compounds. *Research on Chemical Intermediates, 41*(6), 4873–4887. https://doi.org/10.1007/s11164-014-1573-1
30. Rojas-Lemus, M., López-Valdez, N., Bizarro-Nevares, P., González-Villalva, A., Ustarroz- Cano, M., Zepeda-Rodríguez, A., Pasos-Nájera, F., Garcí a-Peláez, I., & Rivera-Fernández, N. (2021). Toxic effects of inhaled vanadium attached to particulate matter: A literature review. *International Journal of Environmental Research and Public Health, 18*(16), 8457. https://doi.org/10.3390/ijerph18168457
31. Somerville, J., & Davies, B. (1962). Effect of vanadium on serum cholesterol. *American Heart Journal, 64*(1), 54–56. https://doi.org/10.1016/0002-8703(62)90091-1
32. Zhao, Y., Ye, L., Liu, H., Xia, Q., Zhang, Y., Yang, X., & Wang, K. (2010). Vanadium compounds induced mitochondria permeability transition pore (PTP) opening related to oxidative stress. *Journal of Inorganic Biochemistry, 104*(4), 371–378. https://doi.org/10.1016/j.jinorgbio.2009.11.007
33. Jandhyala, B. S., & Hom, G. J. (1983). Physiological and pharmacological properties of vanadium. *Life Sciences, 33*(14), 1325–1340. https://doi.org/10.1016/0024-3205(83)90816-0
34. Domingo, J. L., Gomez, M., Llobet, J. M., Corbella, J., & Keen, C. L. (1991). Oral vanadium administration to streptozotocin-diabetic rats has marked negative side-effects which are independent of the form of vanadium used. *Toxicology, 66*(3), 279–287.https://doi.org/10.1016/0300-483X(91)90199-B

9 Chromium Compounds and Their Interaction with Humankind

Garima Pandey, Monika Kamboj, Sangeeta Bajpai and Bimal Krishan Banik

9.1 INTRODUCTION

Chromium compounds, particularly in their trivalent (Cr(III)) and hexavalent (Cr(VI)) forms, significantly impact human health and the environment. Cr(III) is an essential trace element necessary for glucose and lipid metabolism, while Cr(VI) is a toxic pollutant linked to severe health issues, including respiratory problems, skin ulcers, and increased cancer risk. Industrial activities, especially in electroplating, leather tanning, and pigment production, release substantial amounts of Cr(VI), leading to contamination of soil and water. The transport of chromium in soils is influenced by various factors, including weathering of parent materials and anthropogenic activities, such as improper disposal of chromium-bearing wastes. This contamination can lead to elevated chromium concentrations in freshwater and groundwater, which can severely impact plant growth and animal health, indirectly harming terrestrial and aquatic ecosystems. Long-term exposure to Cr(VI) has mutagenic, teratogenic, and carcinogenic effects. In India, chromite deposits account for approximately 2 percent of global resources, with Odisha hosting the majority. However, mining activities, illegal waste disposal, and leaching from chromite mines have resulted in significant soil and water pollution. The tanning industry is particularly problematic, with high load of Cr(VI) concentrations in effluents. States like Tamil Nadu, Uttar Pradesh, and Gujarat are heavily impacted, with untreated wastewater from tanneries contaminating rivers and groundwater. In Kanpur, where over 350 tanneries operate, chromium-laden sludge is routinely dumped into the Ganges River, compounding the toxicity of municipal waste with heavy metals. The lack of adequate treatment and regulatory oversight in the tanning industry contributes to the release of Cr(VI) as either dichromate or chromate, depending on pH levels. This unchecked disposal of hazardous effluents poses serious risks to human health, leading to illnesses such as lung cancer, kidney disease, and liver malfunction. Moreover, the presence of chromium compounds in sludge can release toxic substances into the air, exacerbating health risks in nearby communities. Addressing these challenges requires stringent environmental regulations, effective waste management practices, and remediation efforts to protect public health and restore affected ecosystems.

DOI: 10.1201/9781032615110-12

9.2 IMPACT ON BODY COMPOSITION

Research on chromium's effects on body composition has produced mixed results. Researchers conducted a randomized controlled trial showing that chromium picolinate supplementation increased lean body mass and reduced body fat in overweight individuals [1]. However, inconsistencies were observed in the evidence, with some studies reporting no significant changes in body composition. More recently, a study conducted a meta-analysis that found modest but consistent benefits of chromium supplementation on body composition, particularly in reducing body fat [2]. Chromium exists in several forms, most notably trivalent chromium (Cr(III)) and hexavalent chromium (Cr(VI)), each with distinct health implications depending on exposure levels.

Trivalent Chromium (Cr(III))

Cr(III) is a vital nutrient necessary in small quantities for human health, aiding in the metabolism of carbohydrates, fats, and proteins by enhancing insulin function [3]. However, excessive amounts can pose health risks.

Health Effects of Cr(III)
Beneficial in small amounts: Essential for normal metabolism and maintaining blood sugar levels.

Potential toxicity at high levels: Excessive intake can cause skin irritation, allergic reactions, and possible kidney or liver damage, though such cases are uncommon [4].

Hexavalent Chromium (Cr(VI))
Cr(VI) is highly toxic and recognized as a carcinogen. It is used in various industrial processes, including chrome plating, stainless steel production, and as a pigment. Exposure can occur via inhalation, ingestion, or skin contact.

Health Effects of Cr(VI)
Chromium(VI) compounds are highly toxic and present serious health risks due to their oxidative and carcinogenic effects. The respiratory system is one of the primary areas affected, with chronic exposure leading to irritation of the respiratory tract, causing symptoms like coughing, wheezing, and difficulty breathing. Long-term inhalation of Cr(VI) compounds is strongly linked to an elevated risk of lung cancer, as Cr(VI) is a potent carcinogen. Furthermore, direct contact with Cr(VI) compounds can cause severe skin irritation, leading to painful ulcers, often referred to as "chrome ulcers," as well as allergic dermatitis. Table 9.1 summarizes the common health effects of chromium.

9.3 GENERAL HEALTH EFFECTS OF CHROMIUM EXPOSURE

Occupational Risks: Workers in industries such as welding, chrome plating, and textile manufacturing face higher exposure risks. Using proper safety equipment and procedures is essential to reduce these risks.

TABLE 9.1
Health effects of Cr(VI)

Respiratory problems	Nasal irritation, nosebleeds, ulcers, and perforation of the nasal septum, chronic bronchitis, asthma, and lung cancer
Skin contact	Skin irritation, ulcers, allergic reactions, and "chrome ulcers," which are small sores on the skin
Gastrointestinal issues	Stomach ulcers, convulsions, kidney and liver damage, and potentially death in extreme cases
Cancer risk	Risk of lung cancer

Environmental Exposure: Contaminated water and soil can lead to chromium exposure. Regulatory limits aim to minimize this risk, though pollution incidents still occur.

9.3.1 Safety Measures and Regulations

Workplace Safety: OSHA (Occupational Safety and Health Administration) sets exposure limits for chromium in the workplace.

Environmental Protection: The Environmental Protection Agency (EPA) regulates chromium levels in drinking water and industrial emissions to protect public health.

In summary, Cr(III) is a necessary nutrient beneficial at low levels, while Cr(VI) poses severe health risks and is a known carcinogen. Adhering to safety measures and regulatory standards is crucial to minimize the adverse health effects associated with chromium exposure. The health effects of Cr (VI) compounds will be quite different from that of Cr (III) compounds.

Minimal quantities of chromium (III) compounds are essential for your health and well-being. Chromium VI compounds are usually highly toxic. Breathing in chromium (VI) compounds can damage and irritate your nose, throat, lungs, stomach, and intestines. It may lead to asthma and other allergic reactions. Exposure to chromium (VI) can cause stomach upsets, ulcers, convulsions, kidney and liver damage. Long-term exposure to airborne chromium (VI) can have adverse effects on the respiratory and immune systems and can cause cancer. Skin contact with liquids and solids may lead to skin ulcers, redness, and swelling. A US expert panel investigated chromium-contaminated soils in a residential area and concluded that soil concentrations of 2,800 mg/kg (i.e. 2.8 g/kg) chromium(III) and 180 mg/kg (i.e. 0.18 g/kg) chromium(VI) do not pose a health hazard following acute or chronic exposure for residents living on or near the contaminated sites. These findings and recent reports of health-based soil action levels and case studies involving human exposure to Cr(VI) in soil and groundwater put the relative toxicity of Cr(III) and Cr(VI) into perspective. Cr(VI) is more toxic than Cr(III) but low levels can be tolerated. Human studies showed that the gastrointestinal tract can reduce ingested Cr(VI) to Cr(III) at concentrations up to 10 mg Cr(VI)/L and soil concentrations of 1,240 ppm Cr (VI) do not result in allergic contact dermatitis in 99.9% of the general population.

9.3.2 Entering the Body

Airborne particles can be breathed in. Skin exposure to Cr(VI) will be especially noticed by those individuals with skin allergies. Dangerous exposure to Cr(VI) happens mostly from breathing workplace air, from welding, chrome plating and handling some chromate chemicals.

9.3.3 Entering the Environment

Chromium(VI) released into the atmosphere is carried in the air as particles or dust for a period of time. As it settles, it will contaminate soil and water with only small amounts dissolving. Rain will remove Cr(VI) particles from the atmosphere, depositing them in the ground, or they may be transported over long distances by wind.

9.3.4 Where It Ends Up

Chromium(VI) is released into the air, soil, and water during the manufacture and disposal of products and chemicals containing Cr(VI) or the burning of fossil fuels (coal, oil, or gas). Airborne Cr(VI) particles will settle in less than 10 days, depending on particle size, and will stick strongly to soil particles. Small amounts may travel from the soil to groundwater. Chromium(VI) in water will stick to dirt particles that fall to the bottom with only a small amount of dissolving.

Chromium is ubiquitous in the environment as Cr(III) and Cr(VI) oxidation states, which interconvert under environmentally and biologically relevant conditions (although Cr(III) usually predominates). While Cr(VI) is an established human carcinogen and a major occupational and environmental hazard, Cr(III) has long been regarded as an essential human micronutrient, although recent literature has cast serious doubts on the validity of this postulate. Despite five decades of research, no functional Cr-containing enzymes or cofactors have been characterized conclusively, and several hypotheses on their possible structures have been refuted. Gastrointestinal absorption pathways for both Cr(III) and Cr(VI) are apparent and whole-blood speciation can involve Cr(VI) uptake and reduction by red blood cells, as well as Cr(III) binding to both proteins and low-molecular-mass ligands in the plasma. DNA-damaging effects of Cr(VI) and anti-diabetic activities of Cr(III) are likely to arise from common mechanistic pathways that involve reactive Cr(VI/V/IV) intermediates and kinetically inert Cr(III)–protein and Cr(III)–DNA adducts. Both Cr(III) and Cr(VI) are toxic to plants and microorganisms, particularly Cr(VI) due to its higher bioavailability and redox chemistry. Some bacteria reduce Cr(VI) to Cr(III) without the formation of toxic Cr(V) intermediates and these bacteria are being considered for use in the bioremediation of Cr(VI)-polluted environments.

9.4 BIOLOGICAL ACTIVITY OF CHROMIUM

9.4.1 Chromium (Cr) as an Important Trace Element

Metal ions have been categorized into distinct classes based on their biological function, specifically as essential, helpful, abiological, or deleterious. Interestingly, other authors have classified Cr(III) into each of these categories. Cr(III) is well recognized by nutritionists as a crucial micronutrient for human health. Animal tests conducted by Mertz and colleagues in the 1950s provided the initial data indicating the participation of Cr(III) in glucose metabolism [5,6]. Chromium(III) is regarded as an important trace element for humans. Prior work has revealed that Cr(III) serves as a second messenger in modulating the activity of enzymes, the effects of insulin, glucose tolerance factors (GTF), as well as peptide and nucleic acid stability. It also improves aberrant arterial smooth muscle cell activities linked to hyperglycemia-related vascular problems [7]. Typical blood chromium levels in babies are around 9.0 nM, dropping to 5.0–6.0 nM in grownups and 3.8 nM in type II diabetes individuals [8]. Cr(III) consumption should range from 25 and 200 g per day. However, today's diets high in refined and processed meals frequently fall below of necessary chromium levels. Cardiovascular illness and diabetes can be caused by a lack of chromium, which manifests as impaired glucose tolerance, higher circulating insulin levels, fewer insulin receptors, and high cholesterol as well as triglyceride levels [9]. To remedy the shortage, supplements containing chromium such as chromium picolinate, nicotinate chromium, amino acid chromium, and chromium yeast can be found in the marketplace [10]. It can help improve the serotonergic characteristics of the brain in diabetics[11].

Cr(VI) appears mainly as dichromate ($Cr_2O_7^{2-}$) in acidic aqueous solutions or chromate (CrO_4^{2-}) in alkali aqueous solutions at its maximum oxidation state [12]. Because of its capacity to easily enter the membranes of cells, Cr(VI) causes a significantly greater toxicity threat to individuals than Cr(III). It is classed as a carcinogenic and mutagenic ion due to its intensive reduction activity in cells that generate free radicals, causing considerable DNA damage. As a consequence, even slight disturbances in Cr homeostasis can cause acute and chronic illnesses such as ulcers, airway irritations, headaches, ocular damage, vomiting, gastrointestinal troubles, hematological, hepatic, and kidney issues, and possibly cancer[13].

9.4.2 Absorption and Metabolism

Cr(III) is frequently ingested in the form of carboxylate complexes, such as oxalato or citrato, or amino acid complexes, through a plant or animal-based food. However, there is limited information available regarding the specific forms of Cr(III) in various dietary sources [14].The gastrointestinal absorption of Cr(III) in humans and experimental animals is minimal, ranging from 0.4 to 2 percent. It appears that the absorption levels of Cr(III) are influenced by organic acid ligands, such as nicotinate or ascorbate. However, certain investigations have not shown a significant correlation between the type of Cr(III) complexation and the absorption levels [15]. Although numerous models of Cr(III) pharmacokinetics in mammals have been established, there is limited understanding regarding the speciation of Cr(III) in organisms.

Transferrin, a prominent protein responsible for transporting various metal ions, is very probable to serve as the primary carrier of Cr(III) in the bloodstream[16][17].

9.4.3 Biological Properties

The influence of acute and chronic Cr exposure upon heat shock protein genes as well as the antioxidant enzyme activity of *Orthetrum albistylum* larvae were investigated. In general, the findings suggest that HSP gene expression along with antioxidant enzyme activity in *O. albistylum* larvae might be employed as biomarkers to assess the condition of ecosystems in freshwater [18]. Dong et al. synthesized the Cr complex with drugs and performed acute oral toxicity, which demonstrated that the complex had zero lesions on biopsies of the kidneys, liver, and pancreas. The MTT test also demonstrates that the complex has reduced cytotoxicity and increased biocompatibility [19]. In another study, Cr complexes with picolinate were used to improve their biological activities. The functional groups were discovered to affect the physicochemical characteristics of these complexes, including OH- and Cr(VI)-generation [20]. On streptozocin-induced diabetic mutant mice, the biological functions

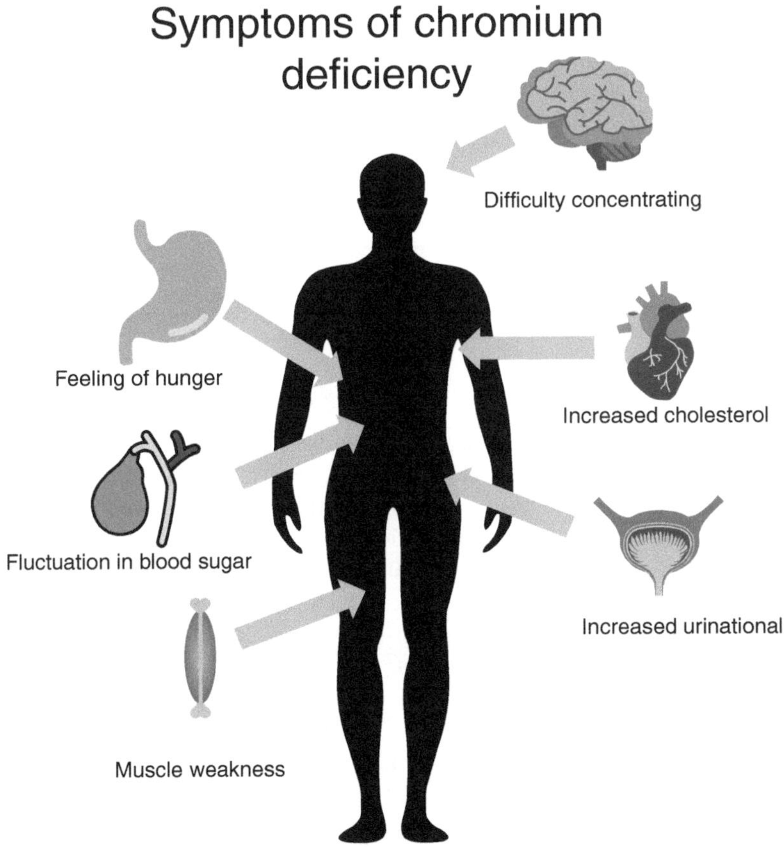

FIGURE 9.1 Symptoms of chromium deficiency.

of Cr and metformin complex (CrMet) have been investigated in depth in terms of fasting blood glucose levels, fasting serum insulin, triglyceride, total cholesterol, low-density lipoprotein cholesterol, as well as high-density lipoprotein cholesterol. These findings suggested that CrMet showed a favorable effect on the level of blood glucose alongside lipid metabolism in diabetics [21]. The interaction using chromium(III) chloride hexahydrate with metformin HCl (Mfn.HCl) results in a diabetes mimetic model. The chromium(III)–metformin combination was successful in lowering blood glucose levels in rats with diabetes and demonstrating its anti-diabetic effectiveness, demonstrating the efficacy of metformin and the Cr(III) complex in increasing antioxidant capabilities [22]. More findings suggested that Cr complexes may have a great blood-glucose-lowering impact at 40 mg/kg dosage with a significant reduction of 65.74 and 66.76 percent for two complexes. Diabetic mice treated with Cr complexes survived at 83.33–100 percent, demonstrating a sufficient level of safety [23]. Figure 9.1 highlights the common chromium deficiency.

9.4.4 Damage to DNA

Currently, the main cause of worry about Cr(III)-induced harm to DNA is the consumption of $[Cr(pic)_3]$ as a dietary supplement. Several researchers have proposed that the negative effects of $[Cr(pic)_3]$ are solely due to the presence of picolinato ligand and therefore other Cr(III) compounds should be used instead [24–26]. Because of their octahedral geometries and kinetic inertness, Cr(III) complexes usually permeate cell membranes extremely slowly via diffusion that is passive. This has been postulated that transferrin (Tf) interacts with Cr(III) in the bloodstream at Fe(III) binding locations or on protein side chains to enhance Cr(III) intracellular absorption [27,28]. Several investigations have been published about the in vitro oxidative cleavage of DNA with H_2O_2 under the presence of Cr(III) complexes such as aquated $CrCl_3$, trinuclear Cr(III) oxo-carboxylates, and Cr(III)-ascorbate and Cr(III)-glucose complexes [29,30].

9.4.5 Diabetes Control

The rapidly increasing prevalence of the condition is becoming a global health problem because of the aging of the population, the urbanization of the population, and lifestyle changes. There has been a significant rise in the number of people who have diabetes mellitus over the course of the last three decades, making it one of the most significant threats to public health in every country. It is anticipated that the number of individuals who have diabetes mellitus will reach 439 million by the year 2030. Nevertheless, it is important to point out that the question of whether or not chromium is essential is still subject to debate. This is because there are neither known metabolic pathways nor known proteins that require chromium [31,32].

However, supplemental trivalent chromium intake appears to be a useful instrument in the world's fight against epidemic-like presentations of various aspects of metabolic syndrome, including obesity and diabetes. Certain forms of trivalent chromium have been shown to have a variety of characteristics, including the ability to overcome insulin resistance, alleviate diabetes, suppress the generation of free radicals, and lower blood pressure. These capabilities have been demonstrated in both laboratory and clinical investigations. Although the first and most commonly

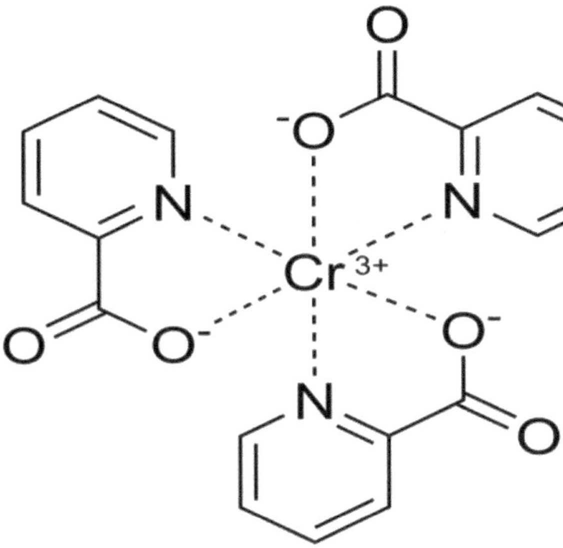

FIGURE 9.2 Structure of $Cr(pic)_3$.

employed chromium complex, CrPic has been successfully advertised for many years, novel complexes of trivalent chromium with distinctive ligands have been continuously prepared and studied, suggesting that the ligand plays an essential role on enhancing the efficiency of this kind of chromium complexes [33].

The anti-diabetic and anti-obese chromium complex known as CrPic, which is composed of chromium and picolinic acid, has been the subject of the significant amount of research. Pic is a metabolite that plays a significant role in the field of neurology and brain tumors. It has been demonstrated that Pic is capable of inhibiting the neurotoxic, making it a neuroprotective agent [34]. For several decades, researchers have been looking into the complex of chromium niacin (CrNIA), which is employed as a chromium micronutrient. According to the findings of an early investigation, CrNIA has the potential to boost the immune response and contribute to an increase in disease resistance in calves. On a later occasion, Jain and colleagues discovered that CrNIA has the capability to reduce the levels of TNF-α, IL-6, and plasma lipid peroxidation in diabetic rats that were administered streptozotocin medication. An interesting finding is that CrNIA led to a considerable reduction in the levels of the inflammation protein CRP, GHb, plasma cholesterol, and triglycerides, as well as an improvement in the ratio of cholesterol to HDL [35,36]. Several investigations conducted in recent years have demonstrated that the compounds of phenylalanine (Phe) are powerful DPP-4 inhibitors that exhibit a unique biological function that is anti-diabetic. On the other hand, CrPhe has been directly improved to have a significant anti-diabetic activity, as demonstrated by several different lines of evidence. This complex is capable of increasing insulin sensitivity by amplifying the phosphorylation of insulin-induced Akt and increasing insulin-stimulated glucose uptake in adipocytes, according to a study that was conducted in vivo [37–39].

9.4.6 Sensing

Different Cr compounds and complexes can be used to modify electrodes for sensing of various compounds. A nanocomposite-modified chromium vanadate (CrV) and graphene oxide (rGO) electrode could serve as a successful electrochemical electrode for a Nifedipine (NF) monitoring technique [40].

9.5 TOXICITY OF CHROMIUM IN PLANTS

Soil chromium (Cr) contamination is a severe and growing environmental concern. Excess Cr in the food chain causes a variety of human health problems, including respiratory ailments, cardiovascular disease, renal failure, and cancer. Cr contamination can be controlled using a variety of physical, chemical, and biological remediation techniques. Chromium is considered a significant source of soil contamination due to its widespread use in a variety of industrial processes, including plating, alloying, stainless steel welding, leather tanning, and the production of pigments, cement, steel, and nuclear weapons. One of the principal causes of Cr contamination in soil is the excessive discharge of untreated wastewater (Cr-bearing wastewater and solid sludge) from these businesses [41,42]. Figure 9.4 shows the uptake of chromium by plants in the environment.

Chromium has a negative influence on seed germination and, consequently, seedling growth. Rediske et al. studied the effect of different chromium doses on the germination of barley seeds. They found that chromium (Cr^{6+}) at a concentration of 500 mg kg^{-1} completely inhibited seed germination, while a dose of 100 mg kg^{-1} inhibited the enzyme diatase responsible for mobilizing starch stores necessary for initial growth [44]. It has frequently been shown to produce wilting inside the roots after root cell plasmolysis. Exposure to Cr^{6+} was found to be beneficial in reducing

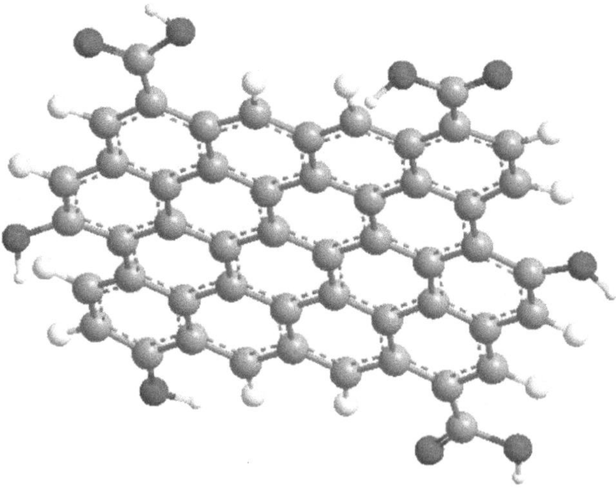

FIGURE 9.3 Structure of rGO.

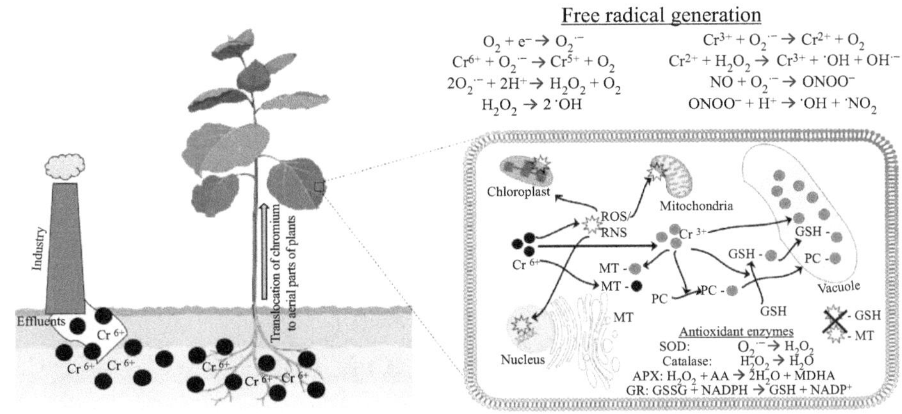

FIGURE 9.4 Chromium in the environment and its uptake by plants[43].

the number of root cells, which are responsible for altering root characteristics such as surface area, root diameters, and the number of root hairs [45]. When compared to the effect that was detected on the germination of seeds, the influence on the growth of seedlings is far more pronounced. Consequently, it causes a severe impact on the uptake of water and nutrient acquisition at the rhizospheric plane, which is accompanied by a decline in the general growth parameters of the plant. This is because it interferes with the process of cell division, which results in the shortening of the root length [46]. When compared to the controls, an increase in chromium concentration is accompanied by a considerable change in the morphological characteristics, including a significant reduction in the root length. A noteworthy alteration in the morphological characteristics of maize was found when it was subjected to 300 mg L^{-1} of $K_2Cr_2O_7$. This included the epinasty of the leaves after a 6-h exposure, as well as the withering of the leaves after a prolonged exposure of more than 12 h, which was induced by water stress[47]. In a similar vein, Bishehkolaei et al. showed that changes in chloroplast structure (lamellar structure and granum number) occurring as a result of chromium exposure frequently resulted in a reduction in the rate of photosynthesis through the deterioration of the pigment system. Although the plasma membrane is the first structure to be exposed to chromium, it is presumed to play a significant part in determining the degree of tolerance to this metal [48,49]. Table 9.2 defines the effect of chromium on different plant species.

Plants play a vital role in the overall ecological cycle, and Cr(VI) contamination in soil can inhibit plant growth and metabolism and even lead to plant death. In nature, Cr(III) exists as $FeOCr_2O_3$, and Cr(VI) combines with oxygen to form CrO_4^{2-} or $Cr_2O_7^{2-}$. Cr (III) is the most stable while Cr(VI) is the most phytotoxic form of Cr [63]. When Cr slag accumulates, Cr redox consumes a large number of anions, causing the pH to rise and the soil to become alkaline [64], which can inhibit soil alkaline phosphatase activity and soil dehydrogenase activity. Cr in the soil is eventually absorbed by green plants growing in the soil and accumulates in plant tissues.

TABLE 9.2
Effects of chromium on different plant species

Plant	Effect of Chromium	References
Vignaradiata L	Reduced germination of seeds and growth of seedlings; increased generation of reactive oxygen species (ROS) and subsequent elevation of malondialdehyde (MDA) levels	[50]
Lactucasativa	a decrease in the production of pigments that are produced by photosynthesis, primarily chlorophyll "b" and carotenoids	[51]
Solanumlycopersicum	There is a correlation between an increase in the levels of oxidative stress and a subsequent decrease in the production of photosynthetic pigments, carotenoids, and lycopene	[52]
Triticumaestivum	The levels of chlorophyll pigment, the generation of biomass, and the grain yield all experienced a significant decrease	[51][53]
Medicagosativa L.	An increase in the formation of reactive oxygen species and reactive nitrogen species that occurs in conjunction with an increase in stress	[54]
Citrus aurantium	The output of biomass has been significantly reduced	[55]
Pistia stratiotes	Significant suppression of the growth of the roots	[56]
Hordeumvulgare L.	A decrease in the production of pigments that are produced by photosynthesis, as well as overall growth and productivity	[45]
Spartinaargentinensis	A decrease in the consumption of many vital nutrients	[57]
O. sativa	There is a decrease in the growth of roots, which is followed by a decrease in the intake of critical nutrients, grain weight, and overall yield	[58]
Prosopislaevigatar	A reduction in the number of leaves, which leads to a drop in the amount of photosynthetic activity	[59]
Zea mays	A slowdown in the general rate of business expansion and productivity	[60]
Loliumperenne	The overall growth and productivity have both slowed down	[61]
Avinasativa	a slowdown in the overall rate of growth and performance	[62]

The heavy metals accumulated in plants can affect nutrient uptake and physiological balance, thereby disrupting development and growth. Metal uptake in plants is concentrated in the roots, but some species also redistribute a greater proportion of metals to aboveground tissues, especially leaves. Specifically, Cr(VI) disrupts plant metabolic functions and enzyme activities, leading to oxidative stress, photosynthesis and respiration inhibition, delayed germination, premature defoliation, and reduced yields. It may also induce genotoxicity and even cause plant death [65]. Cr(VI) toxicity affects plant growth and impedes its essential metabolic processes. It reduces plant growth and productivity, photosynthetic pigments, and antioxidant enzyme activity [66]. The reduced growth rate and smaller leaves after treatment with potassium dichromate may be due to oxidative damage to photosynthetic pigments caused by Cr(VI), thereby decreasing photosynthesis. In peas, excessive Cr(VI) exposure disrupts the structure and function of the pea root plasma membrane, leading to reduced photosynthesis and poor plant growth [67]. In cotton, Cr(VI) toxicity significantly reduces plant height, root length, leaf, stem, and root fresh and dry weights [68]. In maize, Cr(VI) stress reduces leaf area, rachis formation, 100-grain weight, and shoot biomass [69]. These heavy metals that accumulate in plants can eventually harm animals and humans through the food chain.

9.5.1 Toxicity to Microorganisms

Microorganisms play an important role in soil material cycle, and soil Cr(VI) contamination inhibits the growth and metabolism of microorganisms. When assessing the toxicity of Cr(VI), microorganisms are more sensitive indicators of ecotoxicity than soil invertebrates and plants [70]. The alpha diversity (species richness and evenness in samples) of microbial communities is usually associated with ecosystem stability and function, and Cr-contaminated soils exhibit reduced bacterial alpha diversity [71]. The significant differences in bacterial abundance and diversity between the 0 and >20.00 cm depth layers suggest that the heavy metal content in the soil has a certain depth gradient that affects the bacterial groups at different depths [64]. High concentrations of Cr(VI) can be lethal to microorganisms that are not tolerant to Cr(VI) growth. However, microorganisms that are Cr tolerant can resist Cr toxicity and convert the more toxic Cr(VI) to the less toxic Cr(III). Thus, people have thought of using Cr-tolerant strains to clean Cr-contaminated soil and water.

9.6 CHROMIUM TOXICITY IN HUMANS

Chromite (Cr) is a heavy metal that occurs naturally and is the seventeenth most common element in the core of the earth [72]. Cr is both a vital micronutrient and a potential health hazard, associated with several diseases. Exposure to chromium can lead to various health conditions, including skin allergies, respiratory problems, kidney damage, and even cancer [73]. Dermal exposure to chromium can activate inflammasomes, triggering allergic reactions and skin irritations. Moreover, chromium exposure has been linked to respiratory, hepatic, renal, reproductive, and neurological disorders [74].

The toxic effects of chromium extend across multiple physiological systems in the body. Soluble Cr(VI) compounds, in particular, can cause a range of adverse effects. In the skin, chromium exposure can lead to allergic reactions, rashes, and swelling. Respiratory effects may include nasal and throat irritation, bronchitis, and even lung cancer. Gastrointestinal effects renal damage, cardiovascular diseases, liver damage, and genetic abnormalities are also observed. The severity of these effects depends on the extent and duration of exposure [75]. Exposure to chromium in the general population primarily occurs through air emissions and water contamination. Industries like cement manufacturing, construction, electroplating, metal finishing, and leather tanning are major sources of chromium pollution [76]. Workers in these industries are at high risk of exposure, as are those living near contaminated sites. Oral exposure to contaminated water poses a significant risk to the entire population [77].

These risks associated with Cr toxicity and the methods by which it occurs are gradually being undiscovered. An example of this is the inflammatory response that can be triggered by Cr(VI), which in turn can lead to oxidative stress, autophagy, apoptosis, pyroptosis, and stress in the endoplasmic reticulum (ER). This cascade of events leads to cellular damage due to the excessive generation of reactive oxygen species (ROS) [78]. The cellular mechanisms underlying chromium toxicity involve the interaction of different chromium species with biological molecules. Cr(III) enters cells through passive diffusion or phagocytosis, leading to intracellular accumulation and DNA damage [79]. Meanwhile, Cr(VI) can interact with histones and penetrate the nucleus, potentially causing epigenetic changes and cancer development [80].

The formation of Cr(III) from intracellular reduction of Cr(VI) plays a critical role in its mutagenicity and carcinogenicity. Reactive species generated during Cr(VI) reduction can induce DNA damage, cell cycle checkpoints, and apoptosis. Chromosome abnormalities and DNA adducts are among the genetic lesions induced by chromium exposure, increasing the risk of cancer development [81,82]. In **Table 9.3**, the potential toxicity of Cr, their exposure routes, and biological processes have been defined.

TABLE 9.3
Potential toxicity of Cr

Exposure Route	Potential Toxicity Mechanism	Biological Process (BP)	Hub Genes
Dermal contact	DNA damage, metastasis, immune system disorder, allergic reaction	RNA modification, sterol biosynthesis process, DNA replication, intrinsic apoptotic signaling pathway	CXCL8, PTGS2, FOS, HMOX1, ATF3, IRS1, E2F1
Inhalation	DNA damage, metastasis, cell cycle alteration, immune system disorder	Ribosome biogenesis, rRNA metabolic process, cell–substrate adhesion, macrophage cytokine production	TLR4, TGM2, TNFRSF11B, BDKRB1, KIT, ZEB1
Ingestion	DNA damage, metastasis	Nucleotide biosynthesis process, nucleotide phosphate biosynthesis process, alcohol metabolic process, steroid metabolic process	VEGFA, EGFR, APP, SGK1, JUN, TLR2

9.6.1 Accumulative Poisoning

In general, Cr poisoning can be divided into acute and chronic poisoning. Cr(VI) and Cr(III) can cause poisoning, but Cr(VI) is more toxic than Cr(III). China stipulates that the maximum allowable concentration of chromium in groundwater is Cr(III) 0.50 mg L^{-1} and Cr(VI) 0.05 mg L^{-1} (Ministry of Ecology and Environment of the People's Republic of China, GB5749-85), and the maximum allowable discharge standard of Cr(VI) and its compounds in industrial wastewater is 0.50 mg L^{-1} (The People's Republic of China Ministry of Ecology and Environment Integrated Wastewater Discharge Standards GB 8978-1996).

The role of Cr(III), which was once considered essential for mammals, is controversial. Some believe that Cr deficiency is associated with impaired glucose tolerance, fasting hyperglycemia, increased body fat, dyslipidemia, and impaired fertility [83]. Cr(III) may regulate the metabolism of carbohydrates, lipids, and even proteins by enhancing the function of insulin. However, this claim is not supported by sufficient evidence. The mainstream view is that excessive intake of Cr(III) can cause some harm to the body. For example, excessive intake of Cr(III) may disrupt the absorption of trace elements in the serum and brain [84]. Cr accumulates in large quantities in the heart, liver, and kidneys over a long period of time and leads to impaired metal absorption and metabolism in chickens [85]. Oral administration of $CrCl_3$ can reduce the growth performance of chickens, cause pathological lesions such as renal tubular atrophy and structural degeneration of the glomerulus, and affect the antioxidant capacity of the kidney [86]. The addition of different concentrations of $CrCl_3$ to drinking water induces oxidative stress in the chicken brain, and Cr(III) at doses up to 50.00 percent can significantly increase the levels of malondialdehyde and hydrogen peroxide in brain tissue [87]. High doses and prolonged exposure to Cr(III) induce oxidative stress and hepatocyte injury [88].

The acute oral toxicity of Cr(VI) (1,900–3,300 mg kg^{-1}) is significantly higher than that of Cr(III) and can damage several tissues throughout the body, and post-mortem examination of rats receiving lethal doses of Cr(VI) results in diffuse hemorrhage, gastritis, and enteritis [89]. The effects of occupational and non-occupational Cr(VI) ingestion include gastrointestinal symptoms, hypotension, and liver and kidney failure [90]. Cr(VI) poisoning can damage multiple tissues throughout the body, including liver, kidney, reproductive organs. Cr(VI) can cause oxidative damage to the liver in chickens [91], inflammatory heart tissue damage, mitochondrial impairment, autophagy [92], and oxidative damage to the kidney in roosters [93]. Cr(VI) enters the body through drinking water and accumulates in the brain and serum over time, disrupting the absorption of trace elements in the brain and serum of animals [94]. Excessive intake of Cr(VI) can cause inflammatory brain damage in chickens [95]. Damage to the male reproductive system, including sperm, has been observed in experimental animals exposed to Cr(VI) [83]. Cr(VI) can damage dog hearts by causing oxidative damage and altering A-TPase content [96]. Cr(VI) is also known to have toxic effects on the immune system, causing a series of changes such as a dramatic decrease in the levels of leukocytes, erythrocytes, and neutrophils and an increase in lymphocytes in mice [97]. Cr accumulation in fish may enter the body through food and cause serious health hazards [98]. Cr can also cause hair loss, headaches, diarrhea, nausea, and vomiting in humans. In addition, Cr(VI) increases the expression of tumor necrosis factoralpha, interleukin 6, and heat shock proteins in chicken liver, triggering inflammatory damage [99]. Recently, the neurotoxicity caused by Cr(VI) exposure has also received attention. Rats exposed to Cr(VI) exhibit long aggressive behavior toward "intruder" rats and take long to recognize previously encountered rats, thus confirming the social memory deficit observed in humans [100]. In addition, Cr(VI) is neurotoxic, and Cr(VI) exposure can cause various behavioral abnormalities in fish, such as irritability and rapid surfacing. The abnormal behavior may be due to the neurotoxic effects and stimulation of the body's sensory system after Cr(VI) exposure [101].

9.6.2 Carcinogenicity

The carcinogenicity of Cr(VI) has also been studied experimentally, and chronic or high exposure to Cr in animals and humans can increase the risk of cancer. In humans, long-term exposure to Cr-containing environments can lead to an increased probability of developing stomach, lung, bladder, and pancreatic cancers. In 1987, investigators in Liaoning Province, China, studied the association between Cr exposure and cancer. Data showed that between 1970 and 1978, the mortality rate from stomach cancer was higher in Liaoning Province than in nearby uncontaminated areas, and the mortality rate from lung cancer also increased [102]. Similarly, exposure to Cr(VI) in drinking water increases the likelihood of stomach tumors in humans and animals [83]. Cancer mortality or incidence rates in industries associated with Cr(VI) exposure have been analyzed, and results showed that workers exposed to Cr(VI) have an approximately 7 percent higher risk of cancer and an increased risk of respiratory, oral, throat, prostate, and gastric cancers compared with the general population of the same age and sex [103]. In animals, Cr toxicity increases the risk of developing cancer. Animal experiments showed that exposure to Cr(VI) alone does not cause skin tumors but increases the sensitivity of mice to UV light, whereas the presence of Cr(VI) in drinking water increases the frequency of skin tumors in UV-irradiated hairless mice [104]. The National Toxicology Program designed a two-year rodent study to determine the potential hazards of human exposure to Cr(VI) in drinking water. The study found a significant increase in the incidence of oral squamous cell carcinoma in F344/N rats and in the combined incidence of adenoma and small intestine cancer in B6C3F1 mice in animals exposed to Cr-containing compounds [105].

9.6.3 Genotoxicity of Cr(III)

Cr(III) cannot enter the cell through the cell membrane, but Cr(VI) can enter the cell through non-specific ion channels and is eventually reduced to Cr(III). The reduction of Cr(VI) generates unstable radicals, such as thiol groups, hydroxide radicals, hydrogen peroxide, and superoxide anions, which have a wide range of DNA-damaging effects. Cr–DNA formation adducts can lead to chromosome fragmentation and mutagenesis [104]. Earlier studies have shown that Cr(VI) compounds can produce different types of DNA damage, such as single-strand breaks, base instability sites, and DNA–protein crosslinks, while selectively inhibiting the activity of enzymes, such as glutathione reductase [106]. An increase in Cr(VI) level leads to oxidative damage to tissues, and oxidative damage generates large amounts of ROS; overproduction of ROS an lead to deleterious biological effects and irreversible damage by promoting the oxidation of proteins, lipids, carbohydrates, and DNA [107]. Cr(VI) is genotoxic, and nearly 400 out of 450 samples from the experiment showed that soluble Cr(VI) compounds are mutagenic in bacteria. In addition, soluble Cr(VI) compounds can induce a wide range of genetic effects in yeast and insects.

In summary, chromium toxicity presents significant health risks to humans, affecting multiple physiological systems and increasing the likelihood of various diseases, including cancer. Understanding the sources, mechanisms, and health effects of chromium exposure is essential for implementing effective preventive measures and protecting public health.

9.7 SOURCES AND TRANSPORT OF CHROMIUM AND ITS FATE INTO THE ENVIRONMENT

9.7.1 Sources and Transport of Chromium and Its Fate into the Environment

Sources of chromium in the environment Cr is one of the heavy metals whose concentration is continuously rising because of industrial expansion and combustion processes, particularly the rise of the metal, chemical, and tanning sectors [63]. Industrial processes like leather tanning, Cr plating, pigment production, wood preservation, and the use of Cr as a corrosion inhibitor in cooling towers are examples of anthropogenic sources [46]. Natural sources include the leaching of Cr during the weathering of ultramafic rocks is another [108]. Other environmental sources of Cr include power plants using liquid fuels, brown, and hard coal, industrial and municipal trash, and rocks eroded by water and air [109]. Cr pollution is not a concern on a global basis, but it may cause excessive concentrations of this pollutant to circulate in the biogeochemical cycle locally due to metal permeability into soil, water, or the atmosphere [110].

Chromium is a natural component of the Earth's crust. The Earth's crust has Cr concentration >200 mg kg^{-1} in ultramafic (ultrabasic) rocks and ophiolite serpentinites complexes, which constitute approximately 1 percent of the landscape of the terrestrial ecosystem, mainly within the high inhabited areas of Mediterranean regions and Circum-Pacifi. Cr exists in several oxidation states in the environment, of them the most stable and common forms are Cr(0); for example, elemental metallic chromium that does not arise in nature but exists in ores essentially chromite; Cr (III), for example, ferrochrome ($FeCr_2O_4$) or chromate; and Cr(VI) species, for example, ammonium dichromate [$(NH_4)_2Cr_2O_7$]. All chromite deposits are supposed to have been shaped by initial crystal settling or through slow gravitative fluid buildup from magma. Cr(III) in resultant soils and sediments are intemperate by the weathering of rocks. Diagenetic chemical reactions, where total Cr values range from 1000 to 60,000 mg kg^{-1}, are common as a value of silicate mineral alteration relative to

TABLE 9.4
Type of hexavalent chromium (Cr(VI form used in various industries

Cr Used in Industries	Chemical Form
Anti-corrosion agent (chrome, spray coatings)	Barium chromate ($BaCrO_4$), calcium chromate ($CaCrO_4$), zinc chromate ($ZnCrO_4$), strontium chromate ($SrCrO_4$), sodium chromate ($CrNaO_4$),
Tanning of leather products	Ammonium dichromate[$(NH_4)_2Cr_2O_7$]
Wood preservatives	Chromium trioxide (CrO_3)
Stainless steel	Potassium chromate ($KCrO_4$), sodium chromate ($CrNa_2O_4$), ammonium dichromate [$(NH_4)_2Cr_2O_7$], potassium dichromate ($K_2Cr_2O_7$)
Paints, inks, and plastics, pigments	Barium chromate ($BaCrO_4$), calcium chromate (CaCrO4), lead chromate ($PbCrO_4$), zinc chromate ($ZnCrO_4$), potassium dichromate ($K_2Cr_2O_7$), sodium chromate ($CrNa_2O_4$)

chromite [111]. Cr(III) released from weathering is readily adsorbed by clay minerals and precipitates with iron and aluminum hydroxides. Chromium is transported in the environment primarily through weathering of rock, atmosphere fallout, and wet and dry precipitation. It is also washed out from the terrestrial systems. A study found higher Cr in soil and rock samples, indicating that the loess and mudstone are important sources of Cr(VI) pollution in the middle Chinese Loess Plateau area. The concentration of Cr in the environment is also known to increase due to human-made activities. Chromium is extensively used in various industrial applications such as the production processing of metals, chrome, chromate, leathers, textiles, chrome plating, stainless steel welding, steel slags, and ferrochrome pigment. Several industries use Cr compounds and various forms, which are shown in Table 9.4.

9.7.2 Transport of Chromium in Water

Chromium can frequently enter the surface and groundwater bodies by natural and anthropogenic sources. The major potential sources are the weathering of rocks, mudstone, microbial interaction with mafic and ultramafic rocks, and the geogenic process, which naturally releases Cr(VI) in the environment by chromite oxidation. Moreover, Cr(VI) pollution is widely related to wastewater and sludge from industrial operations, such as chromate manufacturing, metal finishing, textile dyeing and processing, and leather tanning industries. These industries contribute a significant amount of Cr(VI) toxic residues in freshwater streams and polluting soil and creating a severe negative impact on plant growth and animal health, indirectly damaging terrestrial and aquatic ecosystems. Its long-term exposure creates many harmful consequences and acts as mutagenic, teratogenic, and carcinogenic substances. Cr concentrations ranged from 0.1 to 117 $\mu g\ L^{-1}$ in freshwater and 0.2–50 $\mu g L^{-1}$ in seawater. Groundwater pollution could arise due to Cr(IV) seepage from chromite mines or inappropriate waste mining products, and industrial equipment finishing. In the aquatic environment, Cr can go through a transformation in various forms through sorption, desorption, reduction, oxidation, dissolution, and precipitation processes. Most Cr(III) compounds are soluble in water at the pH range's normal function. Under impartial to basic pH, Cr(III) precipitate, and an acidic pH, it's solubilized. Cr(VI), mainly chromate and dichromate, is soluble at all pH levels. Nevertheless, it could be precipitated with divalent cations. The prescribed limits for Cr level in drinking water are 8.0 μg L^{-1} for Cr(III) and 1.0 μg L^{-1} for hexavalent chromium [Cr(VI)]. In the wastewater effluents in the closeness of Cr industries, the limits of Cr range from 2 to 5 g L^{-1}. The concentrations of total Cr, Cr(III), and Cr(VI) ranged from 1.240 to 823.87 mg L^{-1}, 1.207–821.86 mg L^{-1}, and 0.033–2.014 mg L^{-1}, respectively in tannery wastewater collected from Kombolcha, South Wollo, Ethiopia.

9.7.2 Transport of Chromium in Soil

Transport of chromium in soils can vary significantly because of the weathering of parental material, rocks, and sediments that make up the particular soil. The concentration of Cr in the soil may be enhanced by anthropogenic activities, especially the dumping of Cr-bearing liquids and solid wastes and mining by-products, or even atmospheric deposition. Inappropriate handling of Cr-bearing solid waste from mining and unlawful waste disposal from these specific industries causes severe transport of chromium in water. Chromium can frequently enter the surface and groundwater bodies by natural and anthropogenic sources. The major potential sources are the weathering of rocks, mudstone, microbial interaction with mafic and ultramafic rocks, and the geogenic process, which naturally releases Cr(VI) in the environment by chromite oxidation. Moreover, Cr(VI) pollution is widely related to wastewater and sludge from industrial operations, such as chromate manufacturing, metal finishing, textile dyeing and processing, leather tanning industries. These industries contribute a significant amount of Cr(VI) toxic residues in freshwater streams and polluting soil and creating a severe negative impact on plant growth and animal health, indirectly damaging terrestrial and aquatic ecosystems. Its long-term exposure creates many harmful consequences and acts as mutagenic, teratogenic, carcinogenic substances. Cr concentrations ranged from 0.1 to 117 μg L^{-1} in freshwater and 0.2–50 μg L^{-1} in seawater. Groundwater pollution could arise due to Cr(IV) seepage from chromite mines or inappropriate waste mining products, and industrial equipment finishing. In the aquatic environment, Cr can go through a transformation in various forms through sorption, desorption, reduction, oxidation, dissolution, and precipitation processes. Most Cr(III) compounds are soluble in water at the pH range's normal function. Under impartial to basic pH, Cr(III) precipitate, and an acidic pH, it's solubilized. Cr(VI), mainly chromate and dichromate, is soluble at all pH levels. Nevertheless, it could be precipitated with divalent cations. The prescribed limits for Cr level in drinking water are 8.0 μg L^{-1} for Cr(III) and 1.0 μg L^{-1} for hexavalent chromium [Cr(VI)]. In the wastewater effluents in the closeness of Cr industries, the limits of Cr range from 2 to 5 g L^{-1}. The concentrations of total Cr, Cr(III), and Cr(VI) ranged from 1.240 to 823.87 mg L^{-1}, 1.207–821.86 mg L^{-1}, and 0.033–2.014 mg L^{-1}, respectively in tannery wastewater collected from Kombolcha, South Wollo, Ethiopia.

9.7.3 Indian Scenario of Chromium Contamination

Indian chromite deposits contribute around 2 percent of total world resources. Odisha state holds about 98 percent of the total proved chromite (Cr ore) reserves in India, with about 97 percent found in the Sukinda Valley. The main chromite deposits have been found in the part of states: (i) Sukinda-Nausahi Belt, Orissa, (ii) Nuggihalli, Sindhuvalli, TalurBelt, Karnataka, (iii) Bhandara, Nagpur-Chandrapur belt, Maharashtra, (iv) Sindhudurg belt, Maharashtra, (v) Jojohau, Roroburu belt, Jharkhand, (vi) Jannaram, Konnyapalem, and Lingannapeta, Andhra Pradesh, and (vii) Namakkal and Thiruchengodu, Tamil Nadu. The occurrence of Cr(VI) pollution from the mining of chromites, dumping of chromium-bearing waste, leaching and mine-tailing infiltration, tannery effluent, and other expanding industrial activities are the significant source of water pollution and soil contamination in India.

9.8 CHROMIUM CONTAMINATION DUE TO TANNERY EFFLUENT

The tanning industry in India is one of the prime sources of water pollution due to high Cr(IV). These industries are centered in Tamil Nadu, Uttar Pradesh, Gujarat, and West Bengal. Nearly 80 percent of the Indian tanneries are involved in the chrome tanning processes. They alone contribute around 2000–3000 tons of Cr in the environment, thereby polluting water bodies. In wastes, effluent Cr concentration ranges from 2000 to 5000 mg L^{-1}. Most of the leather industries release untreated wastewater into rivers and streams, lakes, and oceans.

In such waste effluents, Cr(VI) is found as either dichromate ($Cr_2O_7^{2-}$) in acidic conditions or as chromate (CrO_4^{2-}) in alkaline conditions. Unchecked disposal of the Cr-laden effluent has led to soil and groundwater pollution. Kanpur is among India's most polluted cities; its eastern part has around 350 tanneries and uses a massive amount of basic chrome sulfate (BCS). Every year, almost 1500 metric tons of chromium sulfate produced by tanneries as waste, which is directly dumped in open lands and rivers. These tanneries release a massive amount of untreated sludge and effluents into the Ganges River flowing nearby. These pollutants mix with other municipal solid waste containing mercury, arsenic, and other metalloids, making those waste effluents very toxic. Chromium is vital to the tanning industry, as it makes leather products stronger and more durable. Waste generated from tanneries contains a considerable amount of Cr(VI), which is proven to raise many health issues like lung cancer, kidney illness, liver malfunction, and early dementia. The Cr-containing sludge leaks into the river, sub-soil, and groundwater is potentially dangerous and likely to cause human health problems.

The sludge releases deadly toxins and combustible methane into the air during the hot summer months and sometimes catches fire. In those areas, winds diffuse more Cr(VI), carrying dust particles from the sludge that is hazardous when inhaled. A study was conducted by the Central Pollution Control Board (CPCB) in Kanpur to assess the water quality and summarized Cr(VI)concentration of 6.2 mg L^{-1} in groundwater. However, for water quality, polluting water bodies. In wastes, effluent Cr concentration ranges from 2000 to 5000 mg L^{-1}. Most of the leather industries release untreated wastewater into rivers and streams, lakes, and oceans.

REFERENCES

[1] W. W. Campbell, L. J. O. Joseph, S. L. Davey, D. Cyr-Campbell, R. A. Anderson, and W. J. Evans, "Effects of resistance training and chromium picolinate on body composition and skeletal muscle in older men," *J. Appl. Physiol.*, vol. 86, no. 1, pp. 29–39, Jan. 1999, doi: 10.1152/jappl.1999.86.1.29.

[2] J. B. Vincent, "The potential value and toxicity of chromium picolinate as a nutritional supplement, weight loss agent and muscle development agent," *Sport. Med.*, vol. 33, no. 3, pp. 213–230, 2003, doi: 10.2165/00007256-200333030-00004.

[3] A. Swaroop, M. Bagchi, H. G. Preuss, S. Zafra-Stone, T. Ahmad, and D. Bagchi, "Chapter 8 – Benefits of chromium(III) complexes in animal and human health," Ed. J. B. Vincent, *The Nutritional Biochemistry of Chromium (III)* (Second Ed.. Elsevier, 2019, pp. 251–278. doi: https://doi.org/10.1016/B978-0-444-64121-2.00008-8.

[4] F. Baruthio, "Toxic effects of chromium and its compounds," *Biol. Trace Elem. Res.*, vol. 32, no. 1, pp. 145–153, 1992, doi: 10.1007/BF02784599.

[5] K. Schwarz and W. Mertz, "A glucose tolerance factor and its differentiation from factor 3.," *Arch. Biochem.*, vol. 72, pp. 515–518, 1957.

[6] A. Levina, R. Codd, C. T. Dillon, and P. A. Lay, *Chromium in biology: Toxicology and nutritional aspects*, vol. 51. 2003. doi: 10.1016/s0162-0134(03)80417-8.

[7] S. Liu, L. Zhang, H. Kim, J. Sun, and J. Yoon, "Recent advances and challenges in monitoring chromium ions using fluorescent probes," *Coord. Chem. Rev.*, vol. 501, no. November 2023, p. 215575, 2024, doi: 10.1016/j.ccr.2023.215575.

[8] E. Wysocka, M. Cymerys, G. Mielcarz, W. Bryl, S. Dzięgielewska, and L. Torliński, "The way of serum chromium utilization may contribute to cardiovascular risk factors in centrally obese persons," *Arch. Med. Sci.*, vol. 7, no. 2, pp. 257–263, 2011, doi: 10.5114/aoms.2011.22076.

[9] M. Valko, H. Morris, and T. D. M. Cronin, "Metals, toxicity and oxidative stress," *Current Medicinal Chemistry*, vol. 12, no. 10. pp. 1161–1208, 2005. doi: http://dx.doi.org/10.2174/0929867053764635.

[10] W. T. Cefalu, Z. Q. Wang, X. H. Zhang, L. C. Baldor, and J. C. Russell, "Oral chromium picolinate improves carbohydrate and lipid metabolism and enhances skeletal muscle glut-4 translocation in obese, hyperinsulinemic (JCR-LA corpulent) rats," *J. Nutr.*, vol. 132, no. 6, pp. 1107–1114, 2002, doi: https://doi.org/10.1093/jn/132.6.1107.

[11] J. R. Komorowski et al., "Chromium picolinate modulates serotonergic properties and carbohydrate metabolism in a rat model of diabetes," *Biol. Trace Elem. Res.*, vol. 149, no. 1, pp. 50–56, 2012, doi: 10.1007/s12011-012-9393-x.

[12] M. Szabó et al., "Equilibria and kinetics of chromium(VI) speciation in aqueous solution – A comprehensive study from pH 2 to 11," *Inorganica Chim. Acta*, vol. 472, pp. 295–301, 2018, doi: https://doi.org/10.1016/j.ica.2017.05.038.

[13] J. Briffa, E. Sinagra, and R. Blundell, "Heavy metal pollution in the environment and their toxicological effects on humans," *Heliyon*, vol. 6, no. 9, p. e04691, 2020, doi: 10.1016/j.heliyon.2020.e04691.

[14] D. E. Kimbrough, Y. Cohen, A. M. Winer, L. Creelman, and C. Mabuni, "A critical assessment of chromium in the environment," *Crit. Rev. Environ. Sci. Technol.*, vol. 29, no. 1, pp. 1–46, Jan. 1999, doi: 10.1080/10643389991259164.

[15] B. Gammelgaard, K. Jensen, and B. Steffansen, "In vitro metabolism and permeation studies in rat jejunum: organic chromium compared to inorganic chromium," *J. Trace Elem. Med. Biol.*, vol. 13, no. 1, pp. 82–88, 1999, doi: https://doi.org/10.1016/S0946-672X(99)80028-5.

[16] T. H. Lim, T. Sargent, and N. Kusubov, "Kinetics of trace element chromium(III) in the human body," *Am. J. Physiol. Integr. Comp. Physiol.*, vol. 244, no. 4, pp. R445–R454, Apr. 1983, doi: 10.1152/ajpregu.1983.244.4.R445.

[17] H. Sun, H. Li, and P. J. Sadler, "Transferrin as a metal ion mediator," *Chem. Rev.*, vol. 99, no. 9, pp. 2817–2842, Sep. 1999, doi: 10.1021/cr980430w.

[18] T. Ma et al., "Effects of acute and chronic chromium stress on the expression of heat shock protein genes and activities of antioxidant enzymes in larvae of Orthetrum albistylum," *Environ. Pollut.*, vol. 340, no. P2, p. 122712, 2024, doi: 10.1016/j.envpol.2023.122712.

[19] J. long Dong et al., "Potential antidiabetic molecule involving a new chromium(III) complex of dipicolinic and metformin as a counter ion: Synthesis, structure, spectroscopy, and bioactivity in mice," *Arab. J. Chem.*, vol. 14, no. 7, p. 103236, 2021, doi: 10.1016/j.arabjc.2021.103236.

[20] J. Chai, Y. Liu, J. Dong, B. Liu, and B. Yang, "Synthesis, structure, chemical and bioactivity behavior of eight chromium(III) picolinate derivatives Cr(R-pic)3," *Inorganica Chim. Acta*, vol. 466, pp. 151–159, 2017, doi: 10.1016/j.ica.2017.05.041.

[21] J. Dong, B. Liu, G. Liang, and B. Yang, "Synthesis, biological activity and toxicity of chromium(III) metformin complex as potential insulin-mimetic agent in C57BL/6 mice," *J. Coord. Chem.*, vol. 71, no. 10, pp. 1526–1541, 2018, doi: 10.1080/00958972.2018.1459580.

[22] S. M. El Megharbel, "Synthesis, characterization and antidiabetic activity of chromium (III) metformin complex," *J. Microb. Biochem. Technol.*, vol. 07, no. 02, pp. 65–75, 2015, doi: 10.4172/1948-5948.1000184.

[23] M. A. Mahmoud, S. A. Zaitone, A. M. Ammar, and S. A. Sallam, "Synthesis, structure and antidiabetic activity of chromium(III) complexes of metformin Schiff-bases," *J. Mol. Struct.*, vol. 1108, pp. 60–70, 2016, doi: 10.1016/j.molstruc.2015.11.055.

[24] A. Levina and P. A. Lay, Redox chemistry and biological activities of chromium(III) complexes, Ed., J. B. Vincent, *The Nutritional Biochemistry of Chromium (III)* (Second Edition), Elsevier B.V., 2019. doi: 10.1016/b978-0-444-64121-2.00009-x.

[25] M. S. Mozaffari et al., "Renal and glycemic effects of high-dose chromium picolinate in db/db mice: assessment of DNA damage," *J. Nutr. Biochem.*, vol. 23, no. 8, pp. 977–985, 2012, doi: https://doi.org/10.1016/j.jnutbio.2011.05.004.

[26] L. Jiang, J. B. Vincent, and M. M. Bailey, "$[Cr_3O(O_2CCH_2CH_3)6(H_2O)3]NO_3{\cdot}H_2O$ (Cr3) toxicity potential in bacterial and mammalian cells," *Biol. Trace Elem. Res.*, vol. 183, no. 2, pp. 342–350, 2018, doi: 10.1007/s12011-017-1132-x.

[27] G. Deng, K. Wu, A. A. Cruce, M. K. Bowman, and J. B. Vincent, "Binding of trivalent chromium to serum transferrin is sufficiently rapid to be physiologically relevant," *J. Inorg. Biochem.*, vol. 143, pp. 48–55, 2015, doi: https://doi.org/10.1016/j.jinorgbio.2014.12.004.

[28] G. Deng, S. L. Dyroff, M. Lockart, M. K. Bowman, and J. B. Vincent, "The effects of the glycation of transferrin on chromium binding and the transport and distribution of chromium in vivo," *J. Inorg. Biochem.*, vol. 164, pp. 26–33, 2016, doi: https://doi.org/10.1016/j.jinorgbio.2016.08.008.

[29] A. Parand et al., "DNA nicking by a trinuclear chromium complex," *Inorganica Chim. Acta*, vol. 268, no. 2, pp. 211–219, 1998, doi: https://doi.org/10.1016/S0020-1693(97)05739-3.

[30] T.- C. Tsou and J.-L. Yang, "Formation of reactive oxygen species and DNA strand breakage during interaction of chromium(III) and hydrogen peroxide in vitro: evidence for a chromium(III)-mediated Fenton-like reaction," *Chem. Biol. Interact.*, vol. 102, no. 3, pp. 133–153, 1996, doi: https://doi.org/10.1016/S0009-2797(96)03740-4.

[31] J. E. Shaw, R. A. Sicree, and P. Z. Zimmet, "Global estimates of the prevalence of diabetes for 2010 and 2030," *Diabetes Res. Clin. Pract.*, vol. 87, no. 1, pp. 4–14, 2010, doi: https://doi.org/10.1016/j.diabres.2009.10.007.

[32] G. Danaei et al., "National, regional, and global trends in fasting plasma glucose and diabetes prevalence since 1980: Systematic analysis of health examination surveys and epidemiological studies with 370 country-years and 2·7 million participants," *Lancet*, vol. 378, no. 9785, pp. 31–40, 2011, doi: https://doi.org/10.1016/S0140-6736(11)60679-X.

[33] M. Peng and X. Yang, "Controlling diabetes by chromium complexes: The role of the ligands," *J. Inorg. Biochem.*, vol. 146, pp. 97–103, 2015, doi: 10.1016/j.jinorgbio.2015.01.002.

[34] R. J. Beninger, A. M. Colton, J. L. Ingles, K. Jhamandas, and R. J. Boegman, "Picolinic acid blocks the neurotoxic but not the neuroexcitant properties of quinolinic acid in the rat brain: Evidence from turning behaviour and tyrosine hydroxylase immunohistochemistry," *Neuroscience*, vol. 61, no. 3, pp. 603–612, 1994, doi: https://doi.org/10.1016/0306-4522(94)90438-3.

[35] E. B. Kegley, J. W. Spears, and T. T. Brown, "Immune response and disease resistance of calves fed chromium nicotinic acid complex or chromium chloride," *J. Dairy Sci.*, vol. 79, no. 7, pp. 1278–1283, 1996, doi: https://doi.org/10.3168/jds.S0022-0302(96)76482-2.

[36] S. K. Jain, J. L. Rains, and J. L. Croad, "Effect of chromium niacinate and chromium picolinate supplementation on lipid peroxidation, TNF-α, IL-6, CRP, glycated hemoglobin, triglycerides, and cholesterol levels in blood of streptozotocin-treated diabetic rats," *Free Radic. Biol. Med.*, vol. 43, no. 8, pp. 1124–1131, 2007, doi: https://doi.org/10.1016/j.freeradbiomed.2007.05.019.

[37] Y. Liu et al., "Synthesis and biological evaluation of novel benzyl-substituted (S)-phenylalanine derivatives as potent dipeptidyl peptidase 4 inhibitors," *Bioorg. Med. Chem.*, vol. 21, no. 18, pp. 5679–5687, 2013, doi: https://doi.org/10.1016/j.bmc.2013.07.034.

[38] X. Yang et al., "A newly synthetic chromium complex – chromium(phenylalanine)3 improves insulin responsiveness and reduces whole body glucose tolerance," *FEBS Lett.*, vol. 579, no. 6, pp. 1458–1464, 2005, doi: https://doi.org/10.1016/j.febslet.2005.01.049.

[39] Y. Liu et al., "Design, synthesis, biological evaluation, and docking studies of (S)-phenylalanine derivatives with a 2-cyanopyrrolidine moiety as potent dipeptidyl peptidase 4 inhibitors," *Chem. Biol. Drug Des.*, vol. 82, no. 2, pp. 140–146, Aug. 2013, doi: https://doi.org/10.1111/cbdd.12139.

[40] S. Alagarsamy et al., "Chromium vanadate nanoparticles entrapped reduced graphene oxide nanosheets: A nanocomposite with high catalytic activity for the detection of hypertension-drug nifedipine in biological samples," *Colloids Surfaces A Physicochem. Eng. Asp.*, vol. 674, no. April, p. 131903, 2023, doi: 10.1016/j.colsurfa.2023.131903.

[41] S. Prasad et al., "Chromium contamination and effect on environmental health and its remediation: A sustainable approaches," *J. Environ. Manage.*, vol. 285, p. 112174, 2021.

[42] Z. Bao, H. Feng, W. Tu, L. Li, and Q. Li, "Method and mechanism of chromium removal from soil: a systematic review," *Environ. Sci. Pollut. Res.*, vol. 29, no. 24, pp. 35501–35517, 2022.

[43] M. A. Bhat et al., "Chromium and plant health: A prospective study of toxicity to remediation and restoration of the environment," *South African J. Bot.*, vol. 162, pp. 505–518, 2023, doi: https://doi.org/10.1016/j.sajb.2023.09.043.

[44] P. D. Parr, F. G. Taylor Jr, and J. J. Beauchampt, "Sensitivity of tobacco to chromium from mechanical draft cooling tower drift," *Atmos. Environ.*, vol. 10, no. 6, pp. 421–423, 1976.

[45] S. Ali, F. Zeng, L. Qiu, and G. Zhang, "The effect of chromium and aluminum on growth, root morphology, photosynthetic parameters and transpiration of the two barley cultivars," *Biol. Plant.*, vol. 55, pp. 291–296, 2011.

[46] A. K. Shanker, C. Cervantes, H. Loza-Tavera, and S. Avudainayagam, "Chromium toxicity in plants," *Environ. Int.*, vol. 31, no. 5, pp. 739–753, 2005.

[47] R. Wang, F. Gao, B.-Q. Guo, J.-C. Huang, L. Wang, and Y.-J. Zhou, "Short-term chromium-stress-induced alterations in the maize leaf proteome," *Int. J. Mol. Sci.*, vol. 14, no. 6, pp. 11125–11144, 2013.

[48] R. Bishehkolaei, H. Fahimi, S. Saadatmand, T. Nejadsattari, M. Lahouti, and F. T. Yazdi, "Ultrastructural localisation of chromium in Ocimum basilicum," *Turk. J. Botany*, vol. 35, no. 3, pp. 261–268, 2011.

[49] S. Hayat, G. Khalique, M. Irfan, A. S. Wani, B. N. Tripathi, and A. Ahmad, "Physiological changes induced by chromium stress in plants: an overview," *Protoplasma*, vol. 249, pp. 599–611, 2012.

[50] D. Singh et al., "Chromium (VI)-induced alterations in physio-chemical parameters, yield, and yield characteristics in two cultivars of mungbean (Vigna radiata L.)," *Front. Plant Sci.*, vol. 12, p. 735129, 2021.

[51] A. Christou et al., "Uptake of hexavalent chromium by Lactuca sativa and Triticum aestivum plants and mediated effects on their performance, linked with associated public health risks," *Chemosphere*, vol. 267, p. 128912, 2021.

[52] A. Christou et al., "Uptake of hexavalent chromium by tomato (*Solanum lycopersicum* L.) plants and mediated effects on their physiology and productivity, along with fruit quality and safety," *Environ. Exp. Bot.*, vol. 189, p. 104564, 2021.

[53] D. Subrahmanyam, "Effects of chromium toxicity on leaf photosynthetic characteristics and oxidative changes in wheat (*Triticum aestivum* L.)," *Photosynthetica*, vol. 46, no. 3, pp. 339–345, 2008.

[54] A. Christou et al., "Hexavalent chromium leads to differential hormetic or damaging effects in alfalfa (*Medicago sativa* L.) plants in a concentration-dependent manner by regulating nitro-oxidative and proline metabolism," *Environ. Pollut.*, vol. 267, p. 115379, 2020.

[55] S. Shiyab, "Morphophysiological effects of chromium in sour orange (*Citrus aurantium* L.)," *HortScience*, vol. 54, no. 5, pp. 829–834, 2019.

[56] S. B. Kakkalameli, A. Daphedar, N. Hulakoti, B. N. Patil, and T. C. Taranath, "Azolla filiculoides lam as a phyto tool for remediation of heavy metals from sewage," *Int. J. Pharm*, vol. 8, pp. 282–287, 2018.

[57] S. Redondo-Gómez, E. Mateos-Naranjo, I. Vecino-Bueno, and S. R. Feldman, "Accumulation and tolerance characteristics of chromium in a cordgrass Cr-hyperaccumulator, Spartina argentinensis," *J. Hazard. Mater.*, vol. 185, no. 2–3, pp. 862–869, 2011.

[58] M. Ahmad, A. Wahid, S. S. Ahmad, Z. A. Butt, and M. Tariq, "Ecophysiological responses of rice (*Oryza sativa* L.) to hexavalent chromium," *Pak J Bot*, vol. 43, no. 6, pp. 2853–2859, 2011.

[59] L. Buendía-González, J. Orozco-Villafuerte, F. Cruz-Sosa, C. E. Barrera-Díaz, and E. J. Vernon-Carter, "*Prosopis laevigata* a potential chromium (VI) and cadmium

(II) hyperaccumulator desert plant," *Bioresour. Technol.*, vol. 101, no. 15, pp. 5862–5867, 2010.

[60] S. E. Golovatyj and E. N. Bogatyreva, "Effect of levels of chromium content in a soil on its distribution in organs of corn plants.," *Soil Res. use Fertil.*, pp. 197–204, 1999.

[61] P. Vernay, C. Gauthier-Moussard, and A. Hitmi, "Interaction of bioaccumulation of heavy metal chromium with water relation, mineral nutrition and photosynthesis in developed leaves of *Lolium perenne* L.," *Chemosphere*, vol. 68, no. 8, pp. 1563–1575, 2007.

[62] J. López-Luna, M. C. González-Chávez, F. J. Esparza-García, and R. Rodríguez-Vázquez, "Toxicity assessment of soil amended with tannery sludge, trivalent chromium and hexavalent chromium, using wheat, oat and sorghum plants," *J. Hazard. Mater.*, vol. 163, no. 2–3, pp. 829–834, 2009.

[63] A. Sharma et al., "Chromium bioaccumulation and its impacts on plants: An overview," *Plants*, vol. 9, no. 1. 2020. doi: 10.3390/plants9010100.

[64] B. Liu et al., "Vertical distribution of microbial communities in chromium-contaminated soil and isolation of Cr(VI)-Reducing strains," *Ecotoxicol. Environ. Saf.*, vol. 180, pp. 242–251, 2019, doi: https://doi.org/10.1016/j.ecoenv.2019.05.023.

[65] P. Wang and D. Chao, "Phytoremediation of heavy metal contamination and related molecular mechanisms in plants," *Sheng Wu Gong Cheng Xue Bao*, vol. 36, no. 3, pp. 426–435, 2020, doi: 10.13345/j.cjb.190332.

[66] S. Ali et al., "Fulvic acid mediates chromium (Cr) tolerance in wheat (*Triticum aestivum* L.) through lowering of Cr uptake and improved antioxidant defense system," *Environ. Sci. Pollut. Res.*, vol. 22, no. 14, pp. 10601–10609, 2015, doi: 10.1007/s11356-015-4271-7.

[67] B. K. Kushwaha and V. P. Singh, "Glutathione and hydrogen sulfide are required for sulfur-mediated mitigation of Cr(VI) toxicity in tomato, pea and brinjal seedlings," *Physiol. Plant.*, vol. 168, no. 2, pp. 406–421, Feb. 2020, doi: https://doi.org/10.1111/ppl.13024.

[68] M. A. Farooq et al., "Cadmium stress in cotton seedlings: Physiological, photosynthesis and oxidative damages alleviated by glycinebetaine," *South African J. Bot.*, vol. 104, pp. 61–68, 2016, doi: https://doi.org/10.1016/j.sajb.2015.11.006.

[69] S. A. Anjum et al., "Phyto-toxicity of chromium in maize: oxidative damage, osmolyte accumulation, anti-oxidative defense and chromium uptake," *Pedosphere*, vol. 27, no. 2, pp. 262–273, 2017, doi: https://doi.org/10.1016/S1002-0160(17)60315-1.

[70] X. Zhang et al., "The toxicity of hexavalent chromium to soil microbial processes concerning soil properties and aging time," *Environ. Res.*, vol. 204, p. 111941, 2022, doi: https://doi.org/10.1016/j.envres.2021.111941.

[71] C. S. Sheik et al., "Exposure of soil microbial communities to chromium and arsenic alters their diversity and structure," *PLoS One*, vol. 7, no. 6, p. e40059, Jun. 2012, [Online]. Available: https://doi.org/10.1371/journal.pone.0040059

[72] D. Srivastava et al., "Chromium stress in plants: toxicity, tolerance and phytoremediation," *Sustainability*, vol. 13, no. 9. 2021. doi: 10.3390/su13094629.

[73] P. Sharma, S. P. Singh, S. K. Parakh, and Y. W. Tong, "Health hazards of hexavalent chromium (Cr (VI)) and its microbial reduction," *Bioengineered*, vol. 13, no. 3, pp. 4923–4938, Mar. 2022, doi: 10.1080/21655979.2022.2037273.

[74] D. Y. Shin et al., "Adverse human health effects of chromium by exposure route: A comprehensive review based on toxicogenomic approach," *International Journal of Molecular Sciences*, vol. 24, no. 4. 2023. doi: 10.3390/ijms24043410.

[75] H. Hossini et al., "A comprehensive review on human health effects of chromium: insights on induced toxicity," *Environ. Sci. Pollut. Res.*, vol. 29, no. 47, pp. 70686–70705, 2022, doi: 10.1007/s11356-022-22705-6.

[76] E. Vaiopoulou and P. Gikas, "Regulations for chromium emissions to the aquatic environment in Europe and elsewhere," *Chemosphere*, vol. 254, p. 126876, 2020, doi: https://doi.org/10.1016/j.chemosphere.2020.126876.

[77] M.-N. Georgaki and M. Charalambous, "Toxic chromium in water and the effects on the human body: a systematic review," *J. Water Health*, vol. 21, no. 2, pp. 205–223, Dec. 2022, doi: 10.2166/wh.2022.214.

[78] G. Yan et al., "Toxicity mechanisms and remediation strategies for chromium exposure in the environment," *Front. Environ. Sci.*, vol. 11, no. February, pp. 1–10, 2023, doi: 10.3389/fenvs.2023.1131204.

[79] T. L. DesMarias and M. Costa, "Mechanisms of chromium-induced toxicity," *Curr. Opin. Toxicol.*, vol. 14, pp. 1–7, 2019, doi: https://doi.org/10.1016/j.cotox.2019.05.003.

[80] Q. Y. Chen, A. Murphy, H. Sun, and M. Costa, "Molecular and epigenetic mechanisms of Cr(VI)-induced carcinogenesis," *Toxicol. Appl. Pharmacol.*, vol. 377, p. 114636, 2019, doi: https://doi.org/10.1016/j.taap.2019.114636.

[81] R. Mezencev and C. Gibbons, "Interactions between chromium species and DNA in vitro and their potential role in the toxicity of hexavalent chromium," *Metallomics*, vol. 15, no. 8, p. mfad045, Aug. 2023, doi: 10.1093/mtomcs/mfad045.

[82] L. Zhang, N. Li, X. Zhang, H. Wu, and S. Yu, "Hexavalent chromium caused DNA damage repair and apoptosis via the PI3K/AKT/FOXO1 pathway triggered by oxidative stress in the lung of rat," *Ecotoxicol. Environ. Saf.*, vol. 267, p. 115622, 2023, doi: https://doi.org/10.1016/j.ecoenv.2023.115622.

[83] K. Jomova and M. Valko, "Advances in metal-induced oxidative stress and human disease," *Toxicology*, vol. 283, no. 2, pp. 65–87, 2011, doi: https://doi.org/10.1016/j.tox.2011.03.001.

[84] Y. Liu, P. Hao, X. Zhang, X. Zhao, Y. Liu, and J. Liu, "Effects of excess cr3+ on trace element contents in the brain and serum in chicken," *Biol. Trace Elem. Res.*, vol. 177, no. 1, pp. 180–186, 2017, doi: 10.1007/s12011-016-0875-0.

[85] Y. Liu, X. Zhao, X. Zhang, X. Zhao, Y. Liu, and J. Liu, "Effects of oral administration of $CrCl_3$ on the contents of Ca, Mg, Mn, Fe, Cu, and Zn in the liver, kidney, and heart of chicken," *Biol. Trace Elem. Res.*, vol. 171, no. 2, pp. 459–467, 2016, doi: 10.1007/s12011-015-0559-1.

[86] Y. Liu, C. Liu, J. Cheng, W. Fan, X. Zhang, and J. Liu, "Growth performance and oxidative damage in kidney induced by oral administration of Cr(III) in chicken," *Chemosphere*, vol. 139, pp. 365–371, 2015, doi: https://doi.org/10.1016/j.chemosphere.2015.07.032.

[87] J. Cheng et al., "Oxidative stress and histological alterations of chicken brain induced by oral administration of chromium(III)," *Biol. Trace Elem. Res.*, vol. 173, no. 1, pp. 185–193, 2016, doi: 10.1007/s12011-016-0640-4.

[88] W.-T. Fan, X.-N. Zhao, J. Cheng, Y.-H. Liu, and J.-Z. Liu, "Oxidative stress and hepatocellular injury induced by oral administration of Cr^{3+} in chicken," *J. Biochem. Mol. Toxicol.*, vol. 29, no. 6, pp. 280–287, Jun. 2015, doi: https://doi.org/10.1002/jbt.21697.

[89] S. A. Katz and H. Salem, "The toxicology of chromium with respect to its chemical speciation: A review," *J. Appl. Toxicol.*, vol. 13, no. 3, pp. 217–224, May 1993, doi: https://doi.org/10.1002/jat.2550130314.

[90] C. C. Alvarez, M. E. Bravo Gómez, and A. Hernández Zavala, "Hexavalent chromium: Regulation and health effects," *J. Trace Elem. Med. Biol.*, vol. 65, p. 126729, 2021, doi: https://doi.org/10.1016/j.jtemb.2021.126729.
[91] P. Chen et al., "Effects of the oral administration of K2Cr2O7 and Na_2SeO_3 on Ca, Mg, Mn, Fe, Cu, and Zn contents in the heart, liver, spleen, and kidney of chickens," *Biol. Trace Elem. Res.*, vol. 180, no. 2, pp. 285–296, 2017, doi: 10.1007/s12011-017-0999-x.
[92] Y. Wang et al., "Inflammatory injury and mitophagy in the cock heart induced by the oral administration of hexavalent chromium," *Biol. Trace Elem. Res.*, vol. 200, no. 3, pp. 1312–1320, 2022, doi: 10.1007/s12011-021-02715-0.
[93] H. Wan et al., "Effect of various selenium doses on chromium(IV)-induced nephrotoxicity in a male chicken model," *Chemosphere*, vol. 174, pp. 306–314, 2017, doi: https://doi.org/10.1016/j.chemosphere.2017.01.143.
[94] Y. Zhu et al., "Selenium–chromium(VI) interaction regulates the contents and correlations of trace elements in chicken brain and serum," *Biol. Trace Elem. Res.*, vol. 181, no. 1, pp. 154–163, 2018, doi: 10.1007/s12011-017-1038-7.
[95] S. Guo et al., "Inflammatory injury and mitophagy of the brain in chicken exposed to Cr(VI)," *Environ. Sci. Pollut. Res.*, vol. 28, no. 31, pp. 42353–42361, 2021, doi: 10.1007/s11356-021-13675-2.
[96] J. Lu et al., "Effects of Cr(VI) exposure on electrocardiogram, myocardial enzyme parameters, inflammatory factors, oxidative kinase, and ATPase of the heart in Chinese rural dogs," *Environ. Sci. Pollut. Res.*, vol. 26, no. 29, pp. 30444–30451, 2019, doi: 10.1007/s11356-019-06253-0.
[97] A. Monga, A. B. Fulke, and D. Dasgupta, "Recent developments in essentiality of trivalent chromium and toxicity of hexavalent chromium: Implications on human health and remediation strategies," *J. Hazard. Mater. Adv.*, vol. 7, p. 100113, 2022, doi: https://doi.org/10.1016/j.hazadv.2022.100113.
[98] M. S. Rahman, A. H. Molla, N. Saha, and A. Rahman, "Study on heavy metals levels and its risk assessment in some edible fishes from Bangshi River, Savar, Dhaka, Bangladesh," *Food Chem.*, vol. 134, no. 4, pp. 1847–1854, 2012, doi: https://doi.org/10.1016/j.foodchem.2012.03.099.
[99] Y. Wang et al., "Inflammatory injury and mitophagy induced by Cr(VI) in chicken liver," *Environ. Sci. Pollut. Res.*, vol. 27, no. 18, pp. 22980–22988, 2020, doi: 10.1007/s11356-020-08544-3.
[100] J. P. Wise, J. L. Young, J. Cai, and L. Cai, "Current understanding of hexavalent chromium [Cr(VI)] neurotoxicity and new perspectives," *Environ. Int.*, vol. 158, p. 106877, 2022, doi: https://doi.org/10.1016/j.envint.2021.106877.
[101] M. K. Ahmed, G. K. Kundu, M. H. Al-Mamun, S. K. Sarkar, M. S. Akter, and M. S. Khan, "Chromium (VI) induced acute toxicity and genotoxicity in freshwater stinging catfish, *Heteropneustes fossilis*," *Ecotoxicol. Environ. Saf.*, vol. 92, pp. 64–70, 2013, doi: https://doi.org/10.1016/j.ecoenv.2013.02.008.
[102] J. J. Beaumont et al., "Cancer mortality in a chinese population exposed to hexavalent chromium in drinking water," *Epidemiology*, vol. 19, no. 1, 2008, [Online]. Available: https://journals.lww.com/epidem/fulltext/2008/01000/cancer_mortality_in_a_chinese_population_exposed.4.aspx
[103] Y. Deng et al., "The effect of hexavalent chromium on the incidence and mortality of human cancers: A meta-analysis based on published epidemiological cohort studies," *Front. Oncol.*, vol. 9, no. FEB, 2019, doi: 10.3389/fonc.2019.00024.

[104] A. Zhitkovich, "Chromium in drinking water: sources, metabolism, and cancer risks," *Chem. Res. Toxicol.*, vol. 24, no. 10, pp. 1617–1629, Oct. 2011, doi: 10.1021/tx200251t.

[105] R. Mezencev and S. S. Auerbach, "Inferred inactivation of the Cftr gene in the duodena of mice exposed to hexavalent chromium (Cr(VI)) in drinking water supports its tumor-suppressor status and implies its potential role in Cr(VI)-induced carcinogenesis of the small intestines," *Toxicol. Appl. Pharmacol.*, vol. 433, p. 115773, 2021, doi: https://doi.org/10.1016/j.taap.2021.115773.

[106] A. VonHandorf, H. A. Zablon, and A. Puga, "Hexavalent chromium disrupts chromatin architecture," *Semin. Cancer Biol.*, vol. 76, pp. 54–60, 2021, doi: https://doi.org/10.1016/j.semcancer.2021.07.009.

[107] B. Scharf et al., "Molecular analysis of chromium and cobalt-related toxicity," *Sci. Rep.*, vol. 4, no. 1, p. 5729, 2014, doi: 10.1038/srep05729.

[108] M. Ao et al., "Chromium biogeochemical behaviour in soil-plant systems and remediation strategies: A critical review," *J. Hazard. Mater.*, vol. 424, p. 127233, 2022, doi: https://doi.org/10.1016/j.jhazmat.2021.127233.

[109] M. Shahid et al., "Chromium speciation, bioavailability, uptake, toxicity and detoxification in soil-plant system: A review," *Chemosphere*, vol. 178, pp. 513–533, 2017, doi: https://doi.org/10.1016/j.chemosphere.2017.03.074.

[110] Y. Yang et al., "microRNAs: Key players in plant response to metal toxicity," *International Journal of Molecular Sciences*, vol. 23, no. 15. 2022. doi: 10.3390/ijms23158642.

[111] C. Oze, D. K. Bird, and S. Fendorf, "Genesis of hexavalent chromium from natural sources in soil and groundwater," *Proc. Natl. Acad. Sci.*, vol. 104, no. 16, pp. 6544–6549, Apr. 2007, doi: 10.1073/pnas.0701085104.

Index

D

E

F

G

H

I

K

L

M

N

S

T

FSC
www.fsc.org
MIX
Papier | Fördert
gute Waldnutzung
FSC® C083411

Kolibri 360 GmbH
Ferdinand-Jühlke-Straße 7
99095 Erfurt, Deutschland
produktsicherheit@kolibri360.de